D0056259

The Rising Curve

The Rising Curve

Long-Term Gains in IQ and Related Measures

Edited by Ulric Neisser

American Psychological Association • Washington, DC

BF
431
.R477
1998

Copyright © 1998 by the American Psychological Association. All rights reserved. Except as permitted under the United States Copyright Act of 1976, no part of this publication may be reproduced or distributed in any form or by any means, or stored in a database or retrieval system, without the prior written permission of the publisher. An exception to this copyright is chapter 3. This chapter was prepared by a government employee and is, therefore, in the public domain.

Published by
American Psychological Association
750 First Street, NE
Washington, DC 20002

Copies may be ordered from
APA Order Department
P.O. Box 92984
Washington, DC 20090-2984

In the United Kingdom, Europe, Africa, and the Middle East, copies may be ordered from American Psychological Association
3 Henrietta Street, Covent Garden
London WC2E 8LU, England

Typeset in Minion by EPS Group Inc., Easton, MD
Printer: Edward Brothers, Inc., Ann Arbor, MI
Jacket designer: Kachergis Book Design, Inc., Pittsboro, NC
Technical/production editor: Amy J. Clarke

Library of Congress Cataloging-in-Publication Data

The rising curve : long-term gains in IQ and related measures / edited by Ulric Neisser
p. cm.
Based on paper presented at a conference held in the spring of 1996 at Emory University.
Includes bibliographical references and indexes.
ISBN 1-55798-503-0 (hardcover : alk. paper)
1. Intelligence levels—History—20th century. 2. Afro-Americans—Intelligence levels. 3. Intellect—Genetic aspects. I. Neisser, Ulric. II. American Psychological Association.
BF431.R477 1998
153.9′3—DC21 97-52183
for Library of Congress CIP

British Library Cataloging-in-Publication Data

A CIP record is available from the British Library.

Printed in the United States of America
First edition

APA Science Volumes

Attribution and Social Interaction: The Legacy of Edward E. Jones
Best Methods for the Analysis of Change: Recent Advances, Unanswered Questions, Future Directions
Cardiovascular Reactivity to Psychological Stress and Disease
The Challenge in Mathematics and Science Education: Psychology's Response
Changing Employment Relations: Behavioral and Social Perspectives
Children Exposed to Marital Violence: Theory, Research, and Applied Issues
Cognition: Conceptual and Methodological Issues
Cognitive Bases of Musical Communication
Conceptualization and Measurement of Organism–Environment Interaction
Converging Operations in the Study of Visual Selective Attention
Creative Thought: An Investigation of Conceptual Structures and Processes
Developmental Psychoacoustics
Diversity in Work Teams: Research Paradigms for a Changing Workplace
Emotion and Culture: Empirical Studies of Mutual Influence
Emotion, Disclosure, and Health
Evolving Explanations of Development: Ecological Approaches to Organism–Environment Systems
Examining Lives in Context: Perspectives on the Ecology of Human Development
Global Prospects for Education: Development, Culture, and Schooling
Hostility, Coping, and Health
Measuring Patient Changes in Mood, Anxiety, and Personality Disorders: Toward a Core Battery

Occasion Setting: Associative Learning and Cognition in Animals
Organ Donation and Transplantation: Psychological and Behavioral Factors
The Perception of Structure
Perspectives on Socially Shared Cognition
Psychological Testing of Hispanics
Psychology of Women's Health: Progress and Challenges in Research and Application
Researching Community Psychology: Issues of Theory and Methods
The Rising Curve: Long-Term Gains in IQ and Related Measures
Sleep and Cognition
Sleep Onset: Normal and Abnormal Processes
Stereotype Accuracy: Toward Appreciating Group Differences
Stereotyped Movements: Brain and Behavior Relationships
Studying Lives Through Time: Personality and Development
The Suggestibility of Children's Recollections: Implications for Eyewitness Testimony
Taste, Experience, and Feeding: Development and Learning
Temperament: Individual Differences at the Interface of Biology and Behavior
Through the Looking Glass: Issues of Psychological Well-Being in Captive Nonhuman Primates
Uniting Psychology and Biology: Integrative Perspectives on Human Development
Viewing Psychology as a Whole: The Integrative Science of William N. Dember

APA expects to publish volumes on the following conference topics:

Computational Modeling of Behavior Processes in Organizations
Dissonance Theory 40 Years Later: A Revival With Revisions and Controversies
Experimental Psychopathology and Pathogenesis of Schizophrenia
Marital and Family Therapy Outcome and Process Research

Models of Gender and Gender Differences: Then and Now
Psychosocial Interventions for Cancer

As part of its continuing and expanding commitment to enhance the dissemination of scientific psychological knowledge, the Science Directorate of the APA established a Scientific Conferences Program. A series of volumes resulting from these conferences is produced jointly by the Science Directorate and the Office of Communications. A call for proposals is issued twice annually by the Scientific Directorate, which, collaboratively with the APA Board of Scientific Affairs, evaluates the proposals and selects several conferences for funding. This important effort has resulted in an exceptional series of meetings and scholarly volumes, each of which has contributed to the dissemination of research and dialogue in these topical areas.

The APA Science Directorate's conferences funding program has supported 47 conferences since its inception in 1988. To date, 39 volumes resulting from conferences have been published.

WILLIAM C. HOWELL, PHD
Executive Director

VIRGINIA E. HOLT
Assistant Executive Director

Contents

CONCORDIA UNIVERSITY LIBRARY
2811 NE HOLMAN ST.
PORTLAND, OR 97211-6099

Contributors

Mark Berends, RAND Institute for Education and Training, Washington, DC

Stephen J. Ceci, Department of Human Development, Cornell University

James R. Flynn, Political Studies, University of Otago, Dunedin, New Zealand

Patricia M. Greenfield, Department of Psychology, University of California–Los Angeles

David W. Grissmer, RAND Institute for Education and Training, Washington, DC

Robert M. Hauser, Center for Demography, University of Wisconsin

Min-Hsiung Huang, Institute of European and American Studies, Academia Sinica, Taipei, Taiwan

Sheila Nataraj Kirby, RAND Institute for Education and Training, Washington, DC

Matthew Kumpf, Department of Human Development, Cornell University

John C. Loehlin, Department of Psychology, University of Texas

Richard Lynn, University of Ulster at Coleraine, Northern Ireland

Reynaldo Martorell, Rollins School of Public Health, Emory University

Ulric Neisser, Department of Psychology, Cornell University

Samuel H. Preston, Population Studies Center, University of Pennsylvania

Tina B. Rosenblum, Department of Human Development, Cornell University

Carmi Schooler, Laboratory of Environmental Studies, National Institute of Mental Health, Rockville, MD

Marian Sigman, Department of Psychology, University of California–Los Angeles
Irwin D. Waldman, Department of Psychology, Emory University
Shannon E. Whaley, Department of Psychology, University of California–Los Angeles
Wendy M. Williams, Department of Human Development, Cornell University
Stephanie Williamson, RAND Institute for Education and Training, Washington, DC

Preface

Psychometric research rarely makes the headlines, but when it does the news is mostly bad. Herrnstein and Murray's *The Bell Curve*, a national bestseller, is only the most recent and best publicized summary of the depressing "facts" about IQ and group differences. Intelligence is hereditary, so nothing much can be done to raise it; the mean difference between the IQs of Black and Whites is apparently incorrigible; and the people with the lowest IQs have the most children, so the gene pool is bound to get worse. But can we take these facts at face value? It turns out that there is another side to the story, one that leads to quite different conclusions:

- Scores on intelligence tests are rising, not falling.
- The Black–White gap in school achievement has closed dramatically in recent years.
- There is no convincing evidence that any dysgenic trend exists.

If these encouraging findings were more widely understood, the intelligence debate might take on a different tone. The purpose of this book is to contribute to such a transformation.

I myself learned about all this only recently. In late 1994 it fell to my lot to head an American Psychological Association task force charged with assessing the present status of the scientific study of intelligence. At that point, I was not fully familiar with the literature of the field, having worked mostly in other areas of cognitive psychology. (This turned out to be an advantage in preparing the report because I had no established position to defend.) My experience with the task force taught me many things, and two of the most remarkable have become the focus of this volume: the steady worldwide rise in test scores and the diminishing Black–White difference in school achievement.

As soon as the task force report was completed, I took steps to convene a conference to help clarify these phenomena. That conference was held in the spring of 1996 at Emory University, where I was then teaching. At the suggestion of Richard Lynn, its scope was broadened to include a third topic: the threat of negative trends in the population pool of genotypes for intelligence. Because this issue has lurked in the background of the intelligence debate for more than a century but has rarely been the subject of open debate, I welcomed Lynn's suggestion. Such a debate seemed especially timely because recent work by Preston and Campbell (see chapter 15, this volume) has undercut the most basic assumption of the dysgenic hypothesis. It turns out, counterintuitively, that differential birth rates (for groups scoring high and low on a trait) do *not* necessarily produce changes in the population mean.

These three issues were addressed at Emory by an impressive group of scholars. James Flynn opened the discussion by presenting the basic facts about the rise in test scores that now bears his name. He also offered a skeptical analysis of various explanations of the effect, explanations that were then discussed and defended by other speakers. In subsequent sessions, Robert Hauser and David Grissmer described the recent dramatic gains in Black children's school achievement and reviewed their implications; this topic, too, was extensively discussed. Finally, Lynn presented his version of the dysgenic hypothesis, which was vigorously debated by Preston and several other commentators.

Most of the participants in that conference have contributed to this book. In almost every case, however, their chapters go well beyond the presentations that were made at Emory. Many of them have been written and rewritten several times, in response to detailed critiques. The result is a coherent and scholarly volume of which I am extremely proud, one that may make an important contribution to the ongoing debate about tests and their implications.

I am grateful to the American Psychological Association for its consistent support of this enterprise, including both the conference and the publication of the present volume. I also very much appreciate the contributions of Emory University and the Emory Cognition Project, which made the conference possible in the first place. Many of my friends and colleagues at Emory, who were so kind to me in so many ways for so many years, provided help and encouragement at critical moments of this project as well. Special thanks go to Earl Hunt and Nathan Brody, whose thoughtful critiques have made this volume much better than it would otherwise have been. Most of all, I am grateful to the contributors themselves; they gave their talks, wrote and rewrote their chapters, put up with my many delays and sometimes awkward requests, and finally brought a fine book into existence. Thanks, folks; here it is.

Ulric Neisser
Ithaca, New York

The Rising Curve

1

Introduction: Rising Test Scores and What They Mean

Ulric Neisser

For the better part of a century, Americans have been giving intelligence and achievement tests and viewing the results with alarm. As early as the 1920s, for example, the country was dismayed to learn that the average American man had the mental age of a 14-year-old (Lippmann, 1976). In the 1970s, Scholastic Aptitude Test (SAT) scores seemed to be in free fall; in the 1980s, children were found to lag behind their counterparts in much of the world in their knowledge of science and math. Throughout this time, the average IQ and school achievement scores of Black Americans remained substantially below those of Whites—a gap so persistent that some theorists came to regard it as inevitable. In such a context, the findings reported in this book come as a welcome surprise. Scores on intelligence tests are *rising*, not falling; indeed, they have been going up steeply for years. This rapid rise is not confined to the United States; comparable gains have occurred all over the industrialized world. A second major finding—that the gap in school achievement between Black and White children has diminished sharply in recent years—is at least equally important. These remarkable, environmentally driven trends deserve our most thoughtful attention.

Psychometrics, the study of mental tests, is a rather esoteric business. From time to time, however, new findings or new claims about test scores seem to erupt into public consciousness. The most recent of those eruptions occurred in 1994, when Richard Herrnstein and Charles Murray published a best-selling book called *The Bell Curve.* Their analysis was profoundly pessimistic. Like many psychometricians, they began by assuming that scores on intelligence tests chiefly reflect a single underlying ability, called *g.* Some people have more *g* and others less, depending in large part on their genetic endowment. According to Herrnstein and Murray, an individual's *g* largely determines what he or she can achieve in today's complex society. They also thought it likely that genetic factors contribute to the difference between the average IQs of Whites and Blacks, a gap which shows few signs of narrowing. As if all that were not discouraging enough, *The Bell Curve* also devoted a chapter to so-called *dysgenic trends*: If *g* is highly heritable and if low-*g* people consistently have more children than high-*g* people do, is not the overall intelligence of the population bound to decline in the long run?

This book has a very different take on all three facets of that argument: on what intelligence tests measure, on the difference between the test scores of Black and White Americans, and on the likelihood that there are long-term dysgenic trends. Our starting point is the surprising and continuing rise in test scores: Performance on broad-spectrum tests of intelligence has been going up about 3 IQ points per decade ever since testing began. (The practice of restandardizing the major tests from time to time has kept the mean IQ at about 100 despite these gains, making these changes harder to see.) Even more surprising is that scores on specialized tests of abstract reasoning like the Raven Progressive Matrices—often described as the very best measure of *g*—are rising still faster. Herrnstein and Murray were aware of these gains but gave them short shrift—an understandable decision, considering how profoundly they undermine many of the claims of *The Bell Curve.*

IMPLICATIONS OF THE GAINS

Psychometricians have long known that test performance tends to rise from one generation to the next. The average scores of American draftees in the second world war, for example, were far higher than in the first world war (Tuddenham, 1948). Students of adult intellectual development have also noted the existence of generational differences (Schaie, 1983, 1997). Nevertheless, the size and significance of the gains were not widely appreciated until they were systematically documented by James Flynn, a political scientist at the University of Otago in New Zealand (Flynn, 1984, 1987). Herrnstein and Murray christened them the "Flynn effect," and the name has stuck. No matter how the Flynn effect is eventually explained, it presents grave difficulties for all three facets of the pessimistic argument in *The Bell Curve.* Let us consider them one at a time:

1. Is there a single underlying *g*, largely determined by genetic factors? If this were true, there would be only two possibilities: The Flynn rise either does or does not reflect real increases in *g*. If it does reflect real increases, *g* clearly is affected by environmental factors because no genetic process could produce such large changes so quickly. Whatever those environmental factors may be, we can at least reject the hypothesis that intelligence is genetically fixed. But if it does not reflect real increases—if, as Flynn himself believes, the gains only reflect some trivial artifact—then the tests are evidently flawed, and all arguments based on test scores become suspect. Either way, things look bad for *g* and the arguments of *The Bell Curve.*
2. Does the mean Black–White difference on IQ tests reflect a genetic—or at least very firmly entrenched—limitation on the mental abilities of average Black Americans? In the 1930s, the average Black–White IQ difference was about 15 points. Half a century later in the 1980s, it was still about 15 points. Do these findings imply that nothing much had changed? On the contrary, given the Flynn rise, *both* groups gained some 15 points during those 50 years! The gains made by Blacks are especially impressive because they closed what was once the entire gap: Blacks in the 1980s performed at the level of Whites in the 1930s. Perhaps it was once hard to believe that environmental factors could sustain (or elimi-

nate) a difference of this magnitude, but in the light of these gains it is not hard to believe today. In addition there are new findings, based on school achievement tests rather than intelligence tests, which show substantial reductions in an analogous Black–White gap. Part 2 of this volume reviews those findings and their significance.

3. Is the overall genetic potential of the population bound to decline because people of low IQ have more children than people of high IQ? This is an old bogey. First conceived by Francis Galton in the 19th century, it continued to worry his successors in the "eugenics" movement for decades. Flynn's findings suggest that the concern was misplaced: Intelligence has in fact been rising, not falling. Nevertheless, the issue is not quite closed. Because people of low IQ *do* tend to have more children, scholars who worry about eugenic issues can argue that the decline *must* be occurring. Although that (hypothetical) downward trend is currently masked by an environmentally driven rise, it might still be something to worry about. In Part 3 of this book, Richard Lynn insists that the danger is real, whereas a number of critics argue to the contrary. One of those critics, Samuel Preston, shows mathematically that Galton's basic assumption was simply wrong: Differential birth rates do *not* necessarily produce changes in a population mean.

One way or another, all these issues revolve around test scores. For that reason, it is important to say a few words about the tests themselves and the concepts of intelligence that they help to support. There have been two major strands in the history of psychometrics, both now about a century old. One strand, which can be traced back to the British psychologist Charles Spearman (1904), has focused on the theoretical analysis of relations among tests; the concept of *g* arose in this tradition. The other strand began with Alfred Binet in France but is now primarily an American enterprise; it has been more closely driven by practical considerations. The latter tradition had given us the concept of IQ as well as the distinction between tests of intelligence and of achievement.

FACTOR ANALYSIS AND *g*

Theoretical analysis in the first of these traditions begins with correlations. Suppose that two different tests have been administered to the same group of individuals; did people who got high scores on A also tend to get high scores on B? To get a quantitative answer to this question, one uses the paired test scores to calculate the correlation coefficient *r*. Any substantial positive value of *r* indicates that people who scored high on A tend to score high on B as well. If *r* reaches its maximum value of 1.00, A and B are very closely related indeed: Scores on either one of the tests can be perfectly predicted by scores on the other. A negative value of *r* (not likely with tests of mental ability) implies that high scores on A go with *low* scores on B and vice versa. An *r* near zero means that the tests are essentially independent.

When more than two tests are involved, the analysis gets more complicated. Three tests generate three different values of *r* (A with B, A with C, B with C), and the pattern of those correlations may be interesting. If r_{AB} is very high and r_{AC} and r_{BC} are both near zero, for example, we might conclude that Tests A and B measure roughly the same thing, whereas Test C measures something different. This would amount to saying that there are only two underlying *factors* even though three tests were given. A larger battery of tests would generate too many *r*s for this kind of intuitive reasoning, but one can still arrange them in a correlation matrix like that in Table 1. The data illustrated there are typical: The *r*s are all positive but well below 1.00. It is an empirical fact that almost every battery of mental tests produces just such a *positive manifold* (at least if the sample of test takers is representative of the general population). This means that people who do well on one test tend also to do well on others, in varying degrees that depend on the tests in question. Why should this be true?

Spearman's (1904) approach to this problem was to invent a new statistical method, one that is still widely used today. The purpose of *factor analysis* is to determine how well the data in any given correlation matrix can be fit by a model based on a smaller number of underlying factors. Spearman's own analyses convinced him that only a single factor—he called it *g* for "general ability"—was needed to account for

Table 1

Intercorrelations of Six U.S. Navy Classification Tests

Test	Reading	Arithmetic Reasoning	Mechanical Aptitude	Electrical Knowledge	Mechanical Knowledge
General Classification	.81	.69	.60	.53	.49
Reading		.69	.56	.51	.46
Arithmetic Reasoning			.61	.47	.41
Mechanical Aptitude				.53	.55
Electrical Knowledge					.78
Mechanical Knowledge					

Note. An illustration of the positive manifold. From *Essentials of Psychological Testing* (p. 312, Table 10.3), by L. J. Cronbach, 1970, New York: Harper & Row. Copyright 1970 by HarperCollins. Reprinted with permission.

the positive correlations in the matrix. Individuals differed in how much *g* they had, and *g* contributed in varying degrees to performance on different tests (each of which also measured a more specific ability). Individual differences in *g* itself were most important for tests involving some form of abstract reasoning, so Spearman concluded that *g* represents a core ability to extract (or "educe") logical relations. He and his successors regarded *g* as the central essence of intelligence.

The hypothesis that there is a single basic form of intelligence—and that some of us have more of it than others—seems to fit a lot of people's intuitions. (As we have seen, it is central to the argument of *The Bell Curve.*) Nevertheless, it is by no means a necessary conclusion. Many different forms of factor analysis have been devised over the years, and they do not all find the same structure; in particular, they do not all find a single *g*. Some analysts have identified a half-dozen partly independent forms of intelligence, while others have argued that all such forms still incorporate something like *g*. There are theorists who regard *g* and all other products of factor analysis as little more than statistical artifacts; on the other side are many who believe, even more strongly than Spearman, that it reflects an inherited and very basic property of the brain.

THE RAVEN

Whatever *g* may be, we at least know how to measure it. The accepted best measure, which has played a central role in analyses of the worldwide rise in test scores, is the Raven Progressive Matrices. This test, devised by Spearman's student John C. Raven, was first published in 1938 and is now available at several levels of difficulty. Arthur Jensen has said that Raven's test "apparently measures *g* and little else" (1980, p. 646) and that it "is probably the surest instrument we now possess for discovering intellectually gifted children from disadvantaged backgrounds" (p. 648).

Raven's test consists of a series of items whose difficulty varies systematically. Each item consists of a 3 × 3 matrix with one empty cell; the test taker must decide which of eight candidate entries (shown below the matrix) would best complete it. Figure 1 shows a relatively difficult Raven-type item. One way to solve this particular item is to discover that the following principle applies to the first two rows (as well as to the first two columns): The entry in the third cell results from superimposing those in the first two cells, while deleting all line segments that they have in common. Only one of the candidate entries (No. 5) would complete the matrix in a way that fits this principle.

The Raven is of particular interest because it shows such large IQ gains over time. In The Netherlands, for example, all male 18-year-olds take a version of the Raven as part of a military induction requirement. The mean scores of those annual samples rose steadily between 1952 and 1982, gaining the equivalent of 21 IQ points in only 30 years! This amounts to a rate of no less than 7 points per decade—a figure confirmed by data from many other countries (Flynn, 1987). What can these increases mean?

One way to address this question is to look more closely at the abilities that the Raven requires. The example in Figure 1 makes some of them obvious: One must be able to "educe" abstract relations among

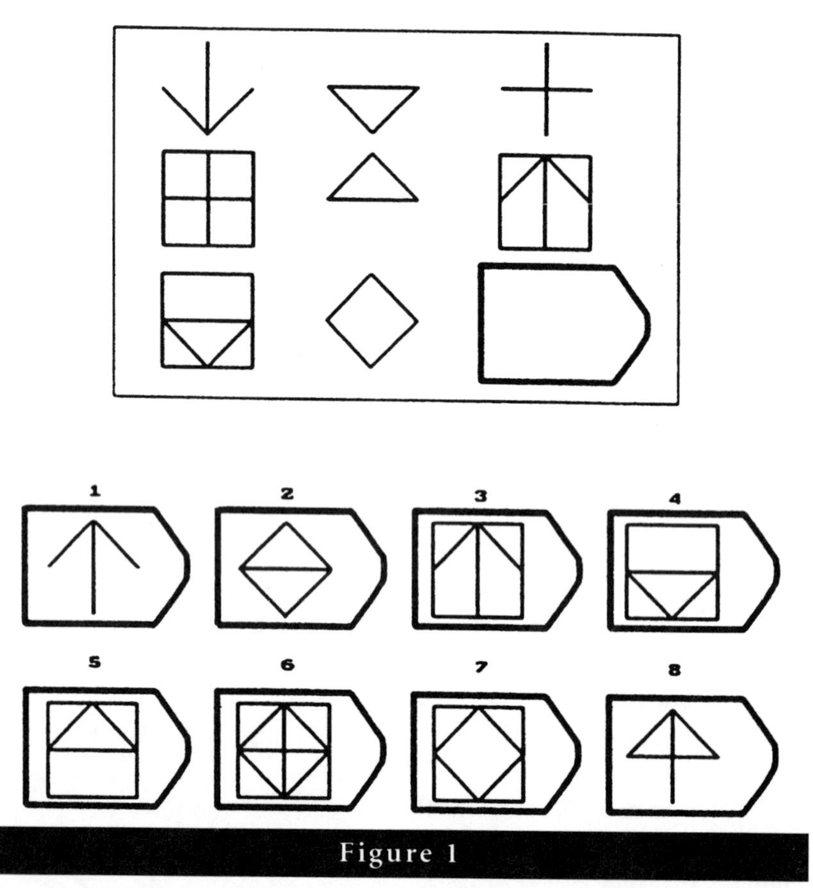

Figure 1

A problem similar to those in the Raven Advanced Progressive Matrices. The matrix at the top of the figure is missing one element; which of the eight pieces at the bottom would complete it appropriately? From "What One Intelligence Test Measures," by P. A. Carpenter, M. A. Just, and P. Shell, 1990, *Psychological Review, 97,* p. 409, Figure 4C. Copyright 1990 by the American Psychological Association. Reprinted with permission.

a series of meaningless figures and to keep track of several such series at once (Carpenter, Just, & Shell, 1990). A special form of visual analysis is also required: Each entry must be dissected into the simple line segments of which it is composed before the processes of abstraction can operate. Although this kind of seeing is rarely needed in the ordinary environment (where figures are usually seen as wholes), it may have become more familiar in recent years. As our exposure to movies, television, and other technical optical displays increases, our skills of visual analysis may be increasing too. In chapter 4 of this volume, Patricia Greenfield explores the hypothesis that those very changes are responsible for the Flynn rise.

STANDARD TESTS OF "IQ"

The other main strand in the history of testing begins with a more practical problem. Some children do not do well in school; how can they be helped? According to Alfred Binet, who addressed this question in the early 1900s, a good first step is to determine the child's level of intellectual maturity, that is, his or her mental age (MA). No theory of intelligence is necessary for this purpose; one only needs a pool of test items that have already been presented to samples of children of different ages. Binet's samples showed, for example, that the average French 5-year-old could copy a square, count four pennies, indicate the heavier of two cubes, and so on; 7-year-olds could copy a diamond and define familiar objects in terms of their use; 9-year-olds knew the day of the week and could make change in simple play-store transactions. A child's mental age can be established simply by seeing where he or she succeeds and begins to fail in such an age-graded series of items.

Binet was trying to help children learn, not to assign a fixed level of intelligence to any given child. His tests were so practical and their results so consistent, however, that others were soon using them for that very purpose. One need only divide a child's MA by his or her actual age (and multiply the result by 100) to get an intelligence quotient (IQ), which has useful metric properties. IQs predict school grades rather well (although far from perfectly); they are fairly stable throughout the developmental years (although large changes sometimes occur); and their distribution in a given age group roughly follows the bell-shaped normal curve (not quite, but I ignore the discrepancies here). IQ testing quickly became popular, especially in the United States, where Lewis M. Terman soon standardized a test based on Binet's principles. He called it the Stanford–Binet.

Since that time, many different kinds of intelligence tests have come on the market. In a particularly influential series developed by David Wechsler—including the Wechsler Intelligence Scale for Children (WISC) and the Wechsler Adult Intelligence Scale (WAIS)—each test is organized into a dozen subscales that each include a single type of item: vocabulary, comprehension, block design, and so on. In modern tests like these, IQs are defined in ways that do not involve quotients. Nevertheless, the definition always begins with the scores of a *standardization sample*, selected to represent various population age groups, that took the test at a certain point in time. The average score for a given age group in that sample defines an IQ of 100; its standard deviation (a statistical measure of spread around the mean) defines 15 IQ points.

IQ GAINS OVER TIME

At first glance it would seem that by this definition, the mean IQ must always be 100. Moreover, given the properties of the normal curve, about 2 ½% of the population should always have IQ scores above 130 (such individuals are called "very superior"), and another 2 ½% should score below 70 (called "intellectually deficient"). However, these implications hold only as long as the standardization sample represents the current population, which may not be the case if that population's competence has changed. It is partly to guard against this possibility that tests are restandardized from time to time. The WAIS, for example, was originally normed in 1953; its successor, the WAIS-R, was normed on a new sample in 1978. Wechsler (1981) then asked a group of people to take both tests; the result was that they scored 7.5 points higher on the (older) WAIS! Given that 25 years had elapsed between standardizations, this represents a gain of 3 points per decade.

Wechsler's (1981) result was no fluke. In almost every study in which the same group has taken two tests standardized at different times, scores have been higher on the earlier test (Flynn, 1984). Generally speaking, a person who has a given IQ score with respect to his or her contemporaries scores substantially higher when compared with earlier samples. These gains are steady and systematic: Performance on IQ tests has been going up 3 points per decade ever since the first Stanford–Binet was normed in 1932. Although this rise is smaller than the remarkable increases on the Raven (see Figure 2), it is still impressive.

The seven chapters in Part 1 of this volume are devoted to these remarkable gains. Flynn himself argues in chapter 2 that they are too large to be genuine; we cannot possibly be that much smarter than our grandparents! His is a minority position. All the other contributors to Part 1 believe that there have been real and substantial gains in at least some kinds of intelligence. But then, what is responsible for them?

A wide range of candidate explanations—sociological, psychological, and biological—is considered in these pages. The sociological analysis is presented by Carmi Schooler (chapter 3), who reviews existing evidence on the effects of environmental complexity and modernization as part of his general critique of Flynn's argument. In chapter 4, Patricia Greenfield focuses on technologically driven changes in the visual environment, changes that may well have had specific effects on the modes of thought required by mental tests. In chapter 5, Wendy Williams reviews a broad spectrum of school- and home-related variables that may have contributed to the rise in different ways. The last three chapters in Part 1 focus on the nutrition hypothesis: Perhaps people have been getting smarter in the last hundred years or so for the same biological reason—dietary improvement—that they have been getting taller! (On average, taller people tend to have larger heads and brains.) Richard Lynn supports this explanation strongly in chapter 8, whereas Reynaldo Martorell is skeptical (chapter 7); Marian Sigman and Shannon Whaley take an intermediate position in chapter 6.

Even this wide-ranging set of hypotheses does not exhaust the possibilities. Robert Zajonc (1976) has long argued that intelligence is affected by family size and sibling position. The relationship is complex, but in most cases early-born children have an advantage over later borns because they spend more time in more intelligent company (that of their parents). Therefore, the long-term demographic trend toward smaller families, which implies that fewer children are being born in

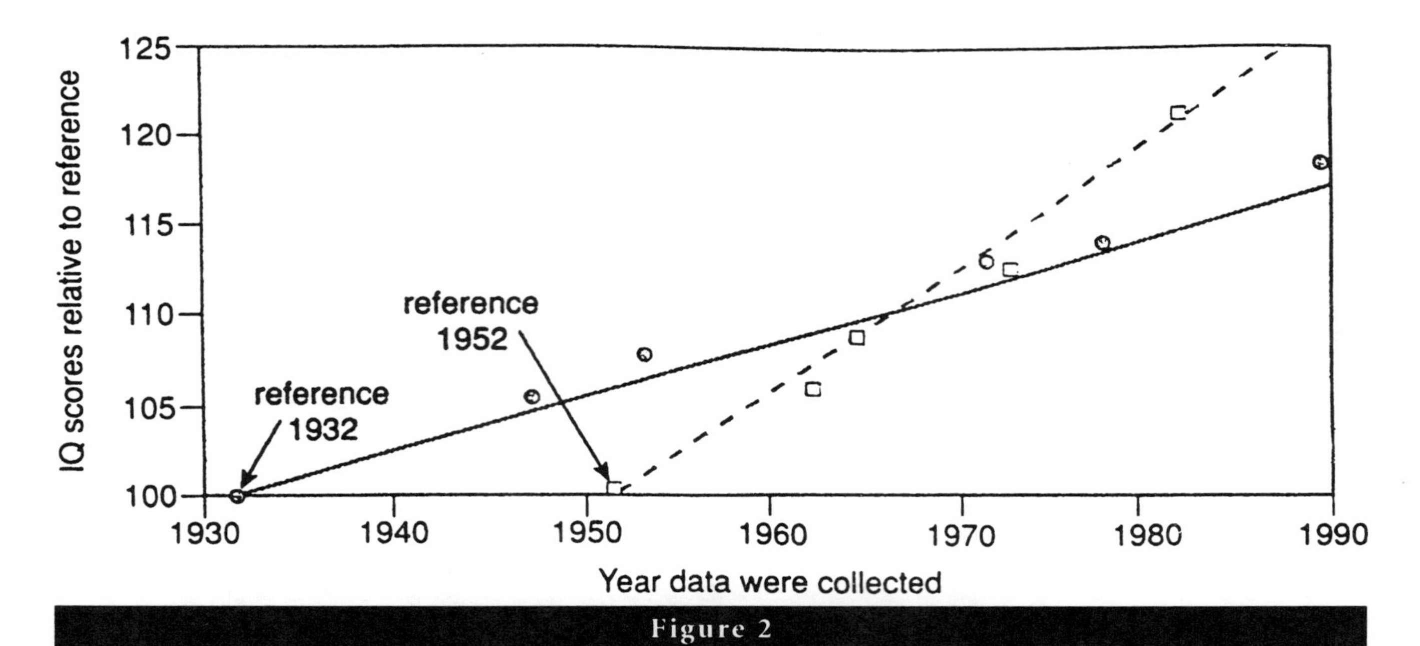

Figure 2

The rise in intelligence test scores since 1932. The solid line shows the adjusted mean performance of Americans on various Wechsler tests as well as the Stanford–Binet; the dashed line shows the mean performance of 18-year-old Dutch males on a version of Raven's Progressive Matrices. Data from Neisser (1997b).

later sibling positions, may have contributed to raising the average level of intelligence (Zajonc & Mullally, 1997).

GENETIC VERSUS ENVIRONMENTAL INFLUENCES

Terman and many of his successors were "hereditarians." They assumed both (a) that people's intellectual abilities—specifically, their IQ scores—are determined at or before birth by innate endowment; and (b) that *differences* in IQ scores of individuals within a given group—say, the adult White American male population of a given year—primarily reflect differences in those endowments. The early hereditarians did not distinguish between these two claims, but the development of modern behavior genetics has made it obvious that they are distinct. The first is false, but the second may not be far from the truth.

Every human trait develops through an interaction between genetic factors and the (internal and external) environment. Indeed, the environment must lie within a certain normal range if the trait is to exist at all. Intelligence is a case in point: No one could acquire cultural information, learn a language, or master any important cognitive skill without environmental support. Although this fact alone is enough to show that IQ scores are not determined exclusively by innate endowment, the existence of the Flynn effect makes that point in a more concrete way. However, one may choose to interpret it, the fact that (unknown) environmental factors are raising the mean IQ of Americans by 3 points per decade certainly shows that the environment matters!

The second proposition has quite a different status. Within a given population and a given range of environments (e.g., those that are characteristic for White American males in 1998), genetic factors do make a major contribution to individual differences. This has now been shown beyond a reasonable doubt by the methods of behavior genetics, a discipline that is primarily concerned with variability. The individuals in a given population differ on almost any measure one is likely to care about: their heights, weights, Raven scores, IQ scores, or anything else. Every such measure has a distribution, often a bell-shaped normal one.

Roughly speaking, there are two reasons for these individual differences. On the one hand, people differ in the sets of genes with which they are born (their *genomes*), and this may affect the trait in question. (Except for identical twins, no two people have the same genome.) On the other hand, people differ in the specific environments that they have encountered. An interesting question, then, is how much of the trait's total variability (technically, its variance) is due to genetic differences and how much to differences in environment.

The proportion of any trait's variance due to genetic differences is called its heritability (h^2). An h^2 of zero means that the genes make no contribution to the variance, whereas an h^2 of 1.00 means that genes account for all of it. In fact, it is hard to think of traits that fit either of those profiles. Most measurable characteristics are somewhat heritable, that is, have intermediate values of h^2. Heritabilities can be estimated on the basis of correlations between relatives: identical twins, fraternal twins, siblings, adopted children and their adoptive or natural parents, and so on. All these estimates have technical complexities that I cannot consider here. (One of the most interesting of those complexities is that some variables seem to straddle the genetic–environmental dichotomy and cannot readily be assigned to either side; an example is given below.) Nevertheless, the overall pattern of results is clear: The intelligence measured by test scores, like most other human traits, is at least moderately heritable.

It is now widely agreed that h^2 for IQ lies between .40 and .80 in the U.S. White population. In other words, genetic factors contribute substantially to individual differences in intelligence. (That is why some theorists, including the authors of *The Bell Curve*, have been so concerned about long-term dysgenic trends. We shall see in Part 3 that their concern may be misplaced.) But by the same token, there is also no doubt that the environment contributes to those differences: h^2 is well below 1.00. Unfortunately, no one knows what it is about the environment that makes this contribution to differences in IQ scores. Some obvious possibilities, such as the economic and intellectual quality of children's home situations, may be less important than was once believed. The surprising fact is that when biologically unrelated children are raised in the same home (as in many cases of adoption), the correlation between their IQ scores is unimpressive in childhood and near zero as they grow up! This finding is important, but it is still negative: The aspects of the environment that *do* matter for the development of intelligence have not yet been identified.

In fact, there are surprisingly many things about intelligence that no one presently understands. One example has just been noted: There are evidently aspects of the environment that affect individual differences, but we do not know what those aspects are or how they work. Another case in point, considered earlier, is the Flynn effect itself. The worldwide rise in test scores is surely driven by some kinds of environmental change, but what are they? A third example, to be discussed later in this chapter, concerns the difference between the mean IQ scores of Blacks and Whites. As we shall see, its causes are equally mysterious.

One further discovery of modern behavior genetics should be mentioned at this point. Several different observations, including the pattern of correlations between IQs of adopted children mentioned above, have led to an unexpected discovery: Heritability itself seems to increase with age. Genetic differences contribute relatively more to the variability of IQ among *adults*, whereas environmental differences contribute relatively more to its variability among *children*. What can this mean? One possibility is that the environment does matter but that adults are somewhat able to choose their own environments (on the basis of genetically influenced preferences), whereas children cannot. This sort of gene–environment correlation does not fit comfortably on either side of the genetic–environmental dichotomy! I cannot pursue these issues here; for more on behavior genetics, see Neisser et al. (1996) and the references cited there.

BLACK–WHITE DIFFERENCES

The Flynn gains have implications for a related problem, one that the authors of *The Bell Curve* regard as intractable. The 15-point "gap" between the IQ means of Blacks and Whites in the United States has persisted for many decades. This gap has serious social consequences:

The abilities measured by intelligence tests are important in the workplace as well as the school. Nevertheless, its cause still is not known. Some of the more obvious hypotheses have long been refuted: It does not result simply from racial bias in the language or cultural content of the tests. (In fact, the Black–White difference is somewhat larger on nonverbal tests.) Some currently plausible explanations refer to differences of caste and culture; there are also hypotheses based on early experience, nutrition, and the like. These hypotheses may seem plausible, but in this case, too, none of them are firmly established (for a review, see Neisser et al., 1996).

Some theorists have suggested that the persistent Black–White difference may have a genetic basis. Although the authors of *The Bell Curve* regard this hypothesis as plausible, there is little direct evidence to support it (Neisser, 1997a). In particular, the existence of genetic differences *within* both the White and the Black populations implies nothing one way or another for the difference between those populations. A well-known example given by Lewonton (1970) makes this point clear. Imagine that two fields of corn have been planted with the same strain of genetically varied seeds but that only one field is adequately watered and fertilized. The result will be an entirely environmental between-field difference, together with a large and entirely genetic within-field variance.

Whatever the merits of these various explanations, they may all soon be out of date. This is for two reasons. The first reason, of course, is the rise in test scores. As we have seen, the 3-point-per-decade gain documented by Flynn means that the test performance of Black Americans today is roughly equivalent to that of Whites in the 1940s. Even if the mean test scores are still 15 points apart, it is now clear that a gap of this size can easily result from environmental differences, specifically, from the differences between the general American environments of 1940 and 1990.

The second reason is that the gap itself may be closing. Where IQ scores themselves are concerned, the situation is not clear. Although some researchers have reported a recent reduction in Black–White IQ differences among children (Vincent, 1991), the samples have been small and there are also data to the contrary (Lynn, 1996). Meanwhile, a more impressive convergence has appeared in a related domain: measures of children's school learning. This trend is the subject of Part 2 of this volume.

SCHOOL ACHIEVEMENT

IQ scores predict school achievement fairly well; that is what they were designed to do. Nevertheless, what children actually learn depends on many factors other than their intelligence. Some of those factors are other characteristics of the child, such as effort, attitude, and the like, but many are characteristics of the school environment. Other things being equal, how much mathematics a child learns will depend on how much time his or her school devotes to instruction in that subject, how much the importance of that instruction is emphasized, and so on. These emphases can vary from time to time and classroom to classroom as well as from country to country and school to school.

The most widely known index of school achievement is probably the SAT (recently renamed the Scholastic Assessment Test), but this test has many disadvantages from a scientific point of view. The most important of these is that it does not involve a representative sample: Children do not take the SAT unless they expect to go to college. A much better measure is the National Assessment of Educational Progress (NAEP), which is regularly given to appropriate national samples of children. NAEP data show that the average reading and math achievement of African American children went up substantially in the 1970s and 1980s, while those of White children did not. The Black–White gap in the math scores of 17-year-olds (e.g.) was about 1.1 standard deviation units as recently as 1978; by 1990, it had shrunk to about 0.6 units. Verbal scores showed similar trends. These gains were not limited to African Americans: Hispanic children showed comparable though somewhat smaller increases.

Most of the discussion in Part 2 centers on the NAEP data. (In chapter 12, Huang & Hauser report similar gains on a vocabulary scale that is regularly administered as part of a social survey.) The focus is not simply on the existence of the gains, which is well established, but on their possible causes. What factors might have produced such substantial increases in so short a time? In chapter 10, David Grissmer et al. show that the gains partly reflect demographic shifts. Today's African American children tend to come from smaller families and to have better educated parents than was the case in earlier times; both of these factors are known to be associated with higher school achievement. Nevertheless, demographic shifts are not enough to account for all of the observed gains. Something else must be going on too, and these authors suggest a plausible candidate. During the 1960s and 1970s, the federal government adopted a wide range of policies designed to improve the education of Black children; these policies included, but were not limited to, school desegregation. Despite what is often assumed, those efforts seem to have been somewhat successful. The relative rise in Black children's school achievement is thus less mysterious than the general rise in IQ scores; we know at least some of its causes. Whether these phenomena are linked in any way remains to be seen.

THE HYPOTHESIS OF DYSGENIC TRENDS

Parts 1 and 2 of this book begin with established effects—the worldwide rise in IQ scores and the narrowing Black–White gap in school achievement—and consider their possible causes. That is, they start with results, and then speculate about "mechanisms." Part 3, in contrast, begins with a mechanism and goes on to speculate about its effects. The "mechanism" in question is the trend mentioned earlier: People of high intelligence (or at least of high IQ) tend to have fewer children than those of lower intelligence. Even though we are presently going through an environmentally driven rise in IQ scores, it remains possible that a genetic trend in the opposite direction is also taking place. This hypothesis is defended by Richard Lynn in chapter 13 and critically evaluated in chapters 14–16.

There are actually two steps in the argument. The first is to show that a "negative fertility differential" presently exists, that is, that people of low IQ have more surviving children than people of high IQ. Although Lynn presents a range of data to establish this point, such differentials can change rather quickly from one cultural era to the next. The second step is to prove that such a differential inevitably produces a decline in the average IQ of the whole population. Lynn, like Galton and others before him, simply takes this for granted. Samuel Preston shows in chapter 15, however, that it is a mistake to do so: Under many conditions, differential fertility rates are fully compatible with a stable population mean! Although the contributors to Part 3 do not reach a consensus on this and other issues, one can at least conclude that no dysgenic trend for intelligence has been conclusively demonstrated. If such a trend does exist, it is probably too small to be cause for alarm.

The Flynn effect, in contrast, is large indeed. It is also profoundly significant: The existence of 3-point-per-decade gains in IQ (and of even larger gains on *g*-loaded tests) will surely transform the intelligence debate. The authors of *The Bell Curve* assumed that the present (1990s) test scores of certain population groups set inevitable limits on what they and their children can attain. This assumption turns out to be false. The intellectual abilities of population groups are not carved forever in the genes; they can and do go up from one generation to the next. IQ, *g*, and school achievement are all massively affected by environmental factors. The next task for research and analysis is to understand what those factors are and how they have their effects. That understanding will set the stage for a new and more constructive debate, a discussion focused on the nature and meaning of the rising curve.

REFERENCES

Carpenter, P. A., Just, M. A., & Shell, P. (1990). What one intelligence test measures: A theoretical account of the processing in the Raven Progressive Matrices test. *Psychological Review, 97*, 404–431.

Cronbach, L. J. (1970). *Essentials of psychological testing* (3rd ed). New York: Harper & Row.

Flynn, J. R. (1984). The mean IQ of Americans: Massive gains. *Psychological Bulletin, 95*, 29–51.

Flynn, J. R. (1987). Massive IQ gains in 14 nations: What IQ tests really measure. *Psychological Bulletin, 101*, 171–191.

Herrnstein, R. J., & Murray, C. (1994). *The bell curve: Intelligence and class structure in American life.* New York: Free Press.

Jensen, A. R. (1980). *Bias in mental testing.* New York: Free Press.

Lewonton, R. (1970). Race and intelligence. *Bulletin of the Atomic Scientists, 26,* 2–8.

Lippmann, W. (1976). The mental age of Americans. In N. J. Block & G. Dworkin (Eds.), *The IQ controversy* (pp. 4–8). New York: Pantheon Books (Random House).

Lynn, R. (1996). Racial and ethnic differences in intelligence in the United States on the Differential Ability Scale. *Personality and Individual Differences, 20,* 271–273.

Neisser, U. (1997a). Never a dull moment. *American Psychologist, 52,* 79–81.

Neisser, U. (1997b). Rising scores on intelligence tests. *American Scientist, 85,* 440–447.

Neisser, U., Boodoo, G., Bouchard, T. J., Boykin, A. W., Brody, N., Ceci, S. J., Halpern, D. F., Loehlin, J. C., Perloff, R., Sternberg, R. J., & Urbina, S. (1996). Intelligence: Knowns and unknowns. *American Psychologist, 51,* 77–101.

Schaie, K. W. (1983). The Seattle Longitudinal Study: A twenty-one year exploration of psychometric intelligence in adulthood. In K. W. Schaie (Ed.), *Longitudinal studies of adult psychological development* (pp. 64–135). New York: Guilford Press.

Schaie, K. W. (1997). The course of adult intellectual development. *American Psychologist, 49,* 304–313.

Spearman, C. (1904). "General Intelligence" objectively determined and measured. *American Journal of Psychology, 15,* 201–293.

Tuddenham, R. (1948). Soldier intelligence in World Wars I and II. *American Psychologist, 3,* 54–56.

Vincent, K. R. (1991). Black/White IQ differences: Does age make a difference? *Journal of Clinical Psychology, 47,* 266–270.

Wechsler, D. (1981). *WAIS-R manual.* New York: Psychological Corporation.

Zajonc, R. B. (1976). Family configuration and intelligence. *Science, 192,* 227–236.

Zajonc, R. B., & Mullally, P. R. (1997). Birth order: Reconciling conflicting effects. *American Psychologist, 52,* 685–699.

PART ONE

Gains on Intelligence Tests

2

IQ Gains Over Time: Toward Finding the Causes

James R. Flynn

This chapter describes the magnitude, pattern, duration, and prevalence of IQ gains over time. Its ultimate purpose is to suggest and evaluate research strategies that might generate promising causal hypotheses. However, the phenomenon to be explained dictates the task of explanatory hypotheses, which poses a fundamental question: Granted that people are better at taking IQ tests, what other, related cognitive skills are they better at? Estimates of the size of this package might range all the way from doing better at IQ tests plus some related cognitive skills too trivial to have significant real-world effects, to doing better on IQ tests plus all of the cognitive skills that are usually enhanced when one goes from a student with an IQ of 75 to a student at the next desk with an IQ of 100 or 125. My purposes dictate covering the following topics: describing and evidencing the brute phenomenon of IQ gains over time, discussing what other cognitive skills have escalated in tandem, using that discussion to critique the attempts at causal explanation made thus far, and suggesting research strategies that might engender better hypotheses. In addition, I discuss theoretical and practical implications of IQ gains when these issues seem significant enough to justify a digression.

IQ GAINS OVER TIME

Data now are available for 20 nations, and there is not a single exception to the finding of massive IQ gains over time. These countries include the most advanced nations of continental Europe, that is, The Netherlands, Belgium, France, Norway, Sweden, Denmark, the former East and West Germany, Austria, and Switzerland. They include virtually all English-speaking nations, that is, Britain including Scotland, Northern Ireland, Canada, the United States, Australia, and New Zealand. They include two nations outside Europe but predominantly of European culture, namely, Israel and urban Brazil. They include two Asian nations that have adopted European technology, namely, Japan and urban China. The first pattern revealed is a correspondence between IQ gains and industrialization. Recent data show that IQ gains in Britain began no later than the last decade of the 19th century at a time when, paradoxically, IQ tests did not exist (see pp. 33–34). The time between the advent of industrialization and the beginning of IQ gains is probably short, and the two events may well coincide (Flynn, 1987a, 1994; Raven, Raven, & Court, 1993).

Recent IQ gains, those covering the last 60 years, are largest on tests that are supposed to be the purest measures of intelligence, tests of fluid intelligence (or fluid *g*) that are also culture reduced. The best example is the Raven Progressive Matrices, in which one identifies the missing parts of patterns that are presumed to be easily assimilated by people across a wide variety of cultures. It tests fluid intelligence because it measures the mind's ability to solve problems at the moment, which is distinguished from crystallized intelligence (or crystallized *g*). The latter represents the kind of knowledge an acute mind normally tends to acquire over time and is measured by tests like the Vocabulary, General Information, and Arithmetic subtests of the Wechsler verbal scale.

Raven data, and data for other tests as well, may be categorized as strong, fair, or weak. Strong data have come from military testing of comprehensive and nationwide samples of young adults or similar samples (comprehensive or random) of schoolchildren. Fair data have come from excellent local samples or nationwide samples of the quality of U.S. Wechsler standardization samples. Weak data have come from samples that fall short of U.S. Wechsler quality. Military data on the Raven, or matrices tests derived from the Raven, for The Netherlands, Belgium, Israel, and Norway are particularly valuable, not only for sample quality but also for the maturity of the subjects. All nations but Norway have shown gains at a rate of about 20 IQ points per generation (30 years). Norway was similar until 1968, but between that date and 1980, the gains ran at a generational rate of only 7.5 points. Raven data of only fair quality put British adults at 16 points (British children show much lower rates), Canada at 12 points, and Australia at 10. Strong data show that Sweden and Denmark have gained at a generational rate of about 10 points. However, although the tests measured fluid intelligence, none of the Swedish subtests and only one (of four) Danish subtests were matrices type or culture reduced. In 1985, Sweden may have become the first nation to register losses on a test of spatial visualization. It appears that gains in Scandinavia have been lower than in the low countries and Israel, but the fact that the tests differed is a confounding variable. Weak data mainly from the Raven have shown wide-ranging results for another six nations (Emanuelsson, Reuterberg, & Svensson, 1993; Emanuelsson & Svensson, 1990; Flynn, 1987a; J. Goldenberg, personal communication, March 4, 1991; Raven, et al., 1993, Graph G2; Teasdale & Owen, 1989).

Performance scale gains from Wechsler samples of schoolchildren have been similar to the results from tests of fluid intelligence. Only weak samples have shown gains above 20 points over a generation or below 9 points. However, there are two reasons for caution. First, the gains of White American children are the best evidenced, and they are at the lower end of this range at a generational rate of about 10 points from 1948 to 1972, perhaps a bit higher for 1972 to 1995. Second, tests like the Raven have generated much adult data, the Wechsler tests, very little. There is no obvious tendency for gains to diminish with age, but recent data from a small sample show that Japan might be an exception. Japanese schoolchildren have doubled the rate of gain of White American children, whereas Japanese and U.S. adults show similar rates (Flynn, 1987a, pp. 185–186; K. Hattori, personal communication, November 30, 1991; Wechsler, 1992, p. 198).

Verbal IQ gains vary from almost nil to 20 points per generation, with 9 as a rough median, and some of the evidence comes from adult data from military testing. Among the 11 countries that allow a comparison, there is not one in which verbal gains match the gains on Raven's type, performance, or nonverbal tests. Often the ratios run against verbal gains by two or three to one. Vocabulary gains have been similar to verbal gains in West Germany and Vienna, but they are lower in English-speaking countries, particularly in Northern Ireland and Scotland, where they are nil. British adults of all ages gained 27 points over 50 years on Raven's but gained only 6 points over 45 years on the Mill Hill Vocabulary Scale. Wechsler subtest data show negligible gains for general information and losses for arithmetic reasoning (Flynn, 1984b, p. 46, 1987a, pp. 185–186, 1990, p. 47; Lynn, 1990, p. 139; Raven, Raven, & Court, 1994, Table MHV3; Raven et al., 1993, Graphs G2, G6; Schallberger, 1987, p. 9; Schubert & Berlach, 1982, p. 262; Wechsler, 1992, p. 198).

IQ AND OTHER COGNITIVE SKILLS

I want to contrast the significance of IQ differences between people who belong to the same generation with the significance of IQ differences between people who belong to different generations, that is, generations separated by time. Take three schoolchildren seated in a row with IQs of 75, 100, and 125. As one goes from one student to another, one would expect a certain escalation of associated cognitive skills. As IQ rose, one would expect the child to be better at arithmetic reasoning, as distinct from the mere mechanics of calculations, and to have a larger nonspecialized vocabulary and fund of general information. If within the school, one found a gifted class with a mean IQ of 125, one would expect them to learn more quickly and be more original and one would predict that in adulthood they would be inventive and make original contributions to their society. If one found a special-needs class with a mean IQ of 75, one would expect them to show a more limited participation in everyday life.

For example, Arthur Jensen related an interview with a young man with a Wechsler IQ of 75. Despite the fact that this man volunteered baseball as his chief interest and attended or viewed games frequently, he was vague about the rules, did not know how many players composed a team, could not name the teams his home team played, and could not name any of the most famous players. Jensen later put the same questions to someone with a high IQ, a learned colleague who disdained the sport and had never attended a baseball game in his life. He answered them all and was puzzled as to how he knew so much about something he enjoyed so little (Jensen, 1981, p. 65).

Reverting to the three schoolchildren, if the child with an IQ of 100 were the first in the school to get a home computer or happened to belong to a chess club, he or she would be likely to have certain advantages over the child with an IQ of 125. For example, she might know how to word process, or have committed the rules of chess to memory, or know something about openings and endgames. However, that would not engender doubts about their comparative IQs because no matter what theory of intelligence one holds, one distinguishes among learning, memory, and intelligence. In my youth, the Irish, Italian, and Polish children who attended Catholic schools took more science and math, at least in the Southern and mid-Atlantic states, than children in public (or state) schools. Therefore, Catholic children demonstrated a higher competence in these subjects. But that would not lead one to expect them to have a higher mean IQ; if it did, one would be mistaken, as Thomas Sowell's data show (Flynn, 1991, p. 30).

I do not mean to imply that distinguishing intelligence from learning is always easy in practice. Richard Lynn (1987) argued that a substantial part of recent IQ gains over time really do represent intelligence gains, roughly a gain of a full standard deviation over 50 years. As evidence, he cited the enormous increase in scientific output between the current generation and the last, plus the great increase in the number of schoolchildren passing secondary school exams and going on to universities.

I remain unconvinced, for one thing, that the argument proves too much for its own credibility. Just as there are more scientists living today than in all previous history and more students at higher levels, that has been true for every generation since the industrial revolution. If this sort of data evidence a standard deviation of intelligence gain every 50 years, there must have been two or three such gains in recent history, which simply does not seem plausible. More to the point, using the fact that the present generation has more scientists than the last generation to evidence enhanced intelligence rests on a false analogy with other types of group differences. When Chinese Americans produce more doctors than White Americans, there is at least a *prima facie* case that intelligence is a factor but only because both groups are competing for a limited number of places in medical schools at the same place at the same time. The progress of science does not show one generation *competing* more successfully than the last, but one generation *building* on the achievements of the last. It no more signals a group intelligence advance than do other generational differences that represent cumulative trends, such as increased numbers of accountants from one generation to another or of cordon bleu cooks.

Ken Vincent believes that the complexity of the modern world both causes and proves massive intelligence gains. He has said that people in industrialized countries live in a world with daily stimuli "beyond the wildest dreams of their grandparents" and that "our grandparents, because of a lack of environmental stimulation, were simply not bright enough as a group to have run the modern world" (Vincent, 1993, p. 62). He was careful to note that he does not deny that the current generation's grandparents had the potential to run the modern world, if only they had enjoyed the necessary environmental advantages.

No doubt these grandparents raised without video recorders, word processors, and computer games would (and do) find it difficult to cope if plunged into the modern world. Do we call this new ability our generation has developed "enhanced intelligence" or "acquired learning"? I have developed my mind by focusing on something far more complex than the modern world, namely, Plato's later dialogues. Those who have done this cope with something that the philosophically naive would find daunting without decades of study, no matter what generation they belong to. Does that mean I am more intelligent than a nonphilosopher with the potential to understand Plato or that I am more learned? If I am merely more learned, why call the present generation more intelligent than the last?

This argument may not be fair to Vincent's point. Consider an analogy from sports: When athletes train to be pole vaulters, they do not just learn skills about manipulating the pole; they also develop new musculature, that is, alter their bodies in a way that confers an ability the untrained lack. The modern world has not only taught us how to "finger" modern contrivances, it has also altered our minds in a way that confers an ability to cope with such contrivances, an ability previous generations lacked. Does one call this new ability an addition to "crystallized intelligence" or a new "achievement"? Is the label arbitrary, just so long as one is clear about what has happened? If so, one would be free to choose either label on rational grounds and might argue that morality should guide the choice. Few would want to label Australian Aborigines "dumb" simply because they have been conditioned to cope with the Australian outback rather than the modern world.

I think it is counterproductive to become obsessed with labels like *intelligence, learning,* and *achievement.* The dominant theme is far more important: What package of enhanced cognitive skills does it make sense to bundle up with IQ gains over time, preparatory to seeking causal explanations? Here there is some unfinished business: Have IQ gains over time been accompanied by enhanced ability to participate in everyday life? To assess this, one needs some everyday activity shared by both the present world and the world of previous generations. Jensen has already provided readers with something. Within the present generation, take people with IQs below 75 when scored against today's norms: They often have difficulty coping with the complex rules of sports like baseball. What of previous generations, would those with IQs below 75 when scored against today's norms have had similar problems? I call this facet of intelligence, ideally suited to bridge distance between generations, "understanding-baseball intelligence."[1]

To progress, one needs an estimate of what proportion of some previous generation would have low IQs against current norms; this brings me to the British Raven data. In 1942, J. C. Raven administered the first standardization of the Raven Progressive Matrices. For ages 20 to 30, he selected soldiers in army camps whose education matched that of the total population of British men of similar age. For older ages, he tested large samples from a private firm and from a government department in which the majority of employees joined as youths and remained until retirement. They gave him a curve of performance from one age cohort to another, and he grafted that curve onto his military sample, thereby deriving norms covering all ages (Foulds & Raven, 1948; J. C. Raven, 1941). In 1992, Raven's son John restandardized Raven's on a representative sample of the adult population of Dumfries in Scotland, selected as typical of an area whose norms matched those of Britain as a whole (J. Raven, 1981, p. RS1.25). John Raven then took the test scores of all of these participants, those aged 25 to 65 tested in 1942 and those aged 25 to 65 tested in 1992, and plotted them by birth date. This gave him scores for people born all the way from 1877 to the 1970s (Raven et al., 1993, Graph G2).

The birth-date method of estimating trends over time entails the assumption that performance is constant between maturity and old age; that is, Raven assumed that the 65-year-olds tested in 1942 would have received much the same scores if they had been tested as 25-year-olds in 1902. There have been no longitudinal studies of the effects of aging

[1]Jensen's assessment of the limitations of low IQ is not, of course, based on a single person. He said that overwhelming evidence shows the man he interviewed was typical: "Adults with an IQ below 75 can seldom manage their own affairs; they often need assistance from their families or from social agencies" (Jensen, 1981, p. 12). I hope it is plain that my own use of "understanding-baseball intelligence" is meant simply to convey how dysfunctional a society would be if 70%, or even 40%, of its members could not participate fully and autonomously in everyday life. No doubt, people can be found who understand baseball despite low IQ, but *if* they are being scored against current norms, there will be significant limitations somewhere. This is not to belittle those whose programs have enhanced the ability of persons with low IQ to cope. However, such programs hardly operated on a mass scale in previous generations. Even recently, Spitz (1986, p. 215) offered a word of caution: "We have no prescription that will change their capacity . . . to solve real-life challenges of some complexity."

on the Raven, but Pat Rabbitt (personal communication, September 19, 1997) is measuring the effects on Cattell's (Individual or Group Culture Fair Intelligence Test, 1960 ed.), which is a similar matrices test of fluid *g*. His results have suggested a drop of no more than 10 IQ points. Another source of error: The Raven was administered without a time limit to both the 1942 and 1992 standardization samples, but the 1942 sample took the test under supervision, whereas the 1992 sample took it unsupervised in their own homes. Raven and Gudjonsson have debated whether unsupervised administration may have inflated the 1992 scores. Comparative data, plus data from a short test each participant completed under supervision, suggest that if score inflation occurred, it was primarily among the top 10%, not extending below the 50th percentile (Raven, 1995). Therefore, I use the 5th percentile from 1992 to compare the two standardization samples.

Figure 1 shows that the bottom 90% of Britons born in 1877 fall below the 5th percentile of those born in 1967, which is to say below an IQ of 75. Assume that members of the 1877 cohort deserve an extra 10 points to compensate for the effects of aging, and throw in another point or two for good measure. This would put 70% of late-19th-century Britons below an IQ of 75 when scored on current norms. To identify IQ gains with understanding-baseball intelligence, one would have to assume that 70% of Britons could not, even if it became their chief interest, understand cricket in 1897. Even if one put the percentage at 60%, or down to 50%, would such a thing be plausible?

Moreover, data whose quality cannot be challenged have posed the same question. The Dutch military data, like those of Israel, Norway, and Belgium, are near exhaustive; but even better, Vroon compared a sample of the total population of Dutch examinees with the scores of their own fathers. There is simply no doubt that Dutch men in 1952 had a mean IQ of 79 when scored against 1982 norms. Has the average person in The Netherlands ever been near mental retardation? Does it make sense to assume that at one time almost 40% of Dutch men lacked the capacity to understand soccer, their most favored national sport?

One could argue that rather than scoring 1952 Dutch men against

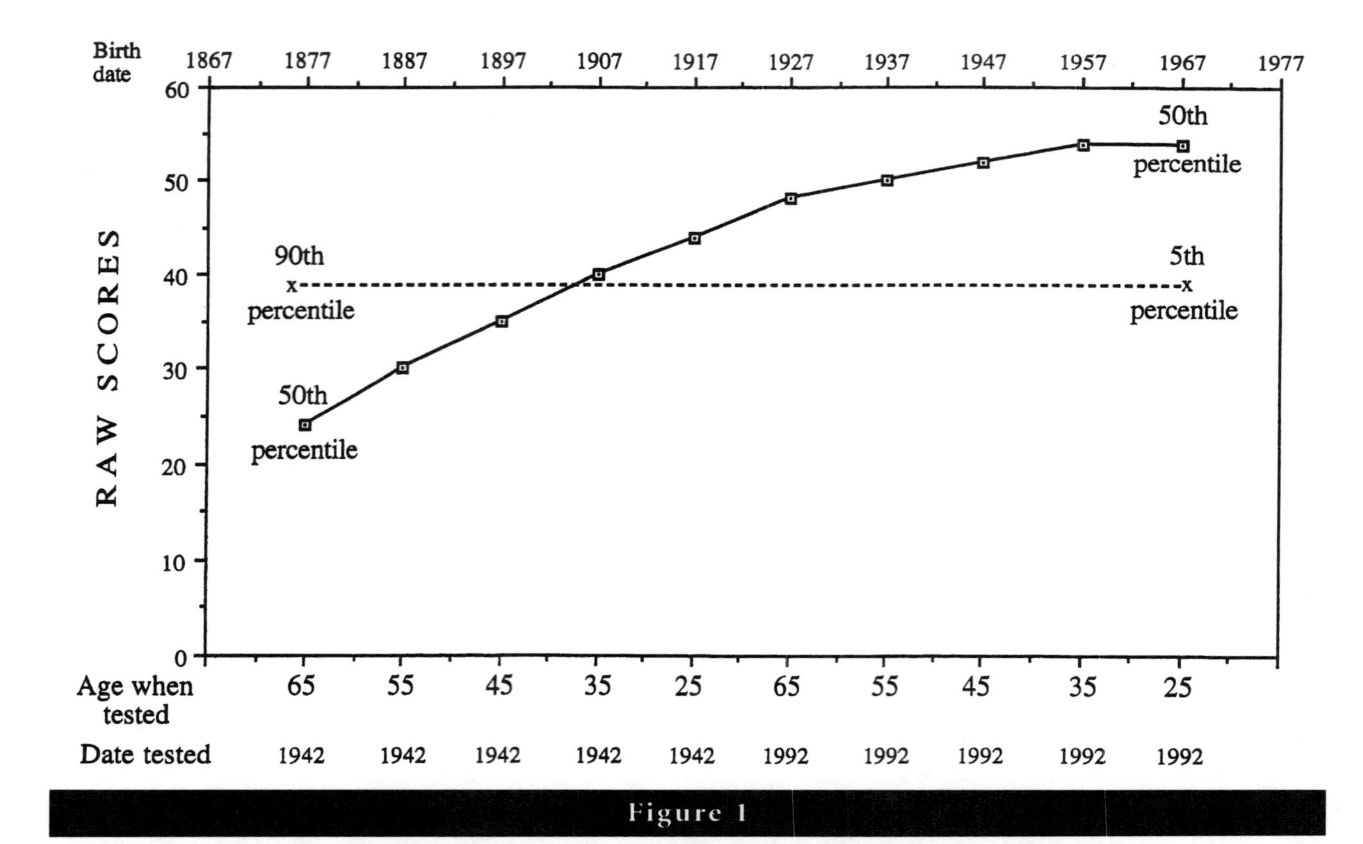

Figure 1

Great Britain and the Raven Progressive Matrices. The bottom 90% born in 1877 fall within the bottom 5% born in 1967 (IQ 75 or below). Data from Raven, Raven, and Court (1993).

later norms, one should score 1982 men against the earlier norms, giving a 1982 mean IQ of 121. But this would entail a new problem, looking for evidence of widespread giftedness in the present generation. Dutch teachers in 1982 should have enjoyed classrooms in which over 25% of children had IQs of 130 or above, children who should have flashed through ordinary schoolwork. The number of people with IQs above 140 would have increased from 1 in 260 to 1 in 10, the sort of people whose adult achievements are so clear that they fill the pages of *American Men of Science* and *Who's Who*. However, a careful survey of serious Dutch publications revealed not a single reference to a dramatic increase in cognitive ability or escalating giftedness among schoolchildren. The number of inventions patented in fact showed a sharp decline over the last generation (Flynn, 1987a, pp. 172, 187).

These scenarios are derived from gains on tests of fluid intelligence. Jensen's participant had a Wechsler IQ of 75, and Wechsler tests measure a mix of fluid and crystallized intelligence. It may be said that such tests are a better measure of mental retardation and, therefore, that my scenarios are suspect. The next step, therefore, is to examine U.S. data from Wechsler and Stanford–Binet samples, in which all participants took tests that measure a mix of fluid and crystallized intelligence. These samples, although carefully chosen, cannot match the quality of either comprehensive or random samples. For example, the most recent measures of U.S. gains are based on the Wechsler Intelligence Scale for Children (WISC-III) and the Wechsler Adult Intelligence Scale (WAIS-III) standardization samples tested in 1989 and 1995, respectively. Participants who took both tests show that even if these two samples had been selected at the same time, the latter would have performed about 1.70 IQ points below the former. Results from the first sample suggest a recent rate of gain of just over 0.30 points per year, and results from the second, about 0.20 points per year. When sampling error of this sort produces two estimates, one can do little but split the difference and put post-1972 gains at about 0.25 points per year.

Nonetheless, the U.S. data clearly show massive gains from 1932 to 1995, and they strongly suggest that these gains began no later than 1918. Every study from the interwar era shows large gains, and they are supported by a comparison of performance on the Stanford–Binet by soldiers in 1918 with that of the standardization sample of 1932 (Flynn, 1984b, 1993; Terman & Merrill, 1937, p. 50; Wechsler, 1992, p. 198; 1997; Yerkes, 1921, pp. 654, 789). As Figure 2 shows, White Americans gained almost 25 points on Wechsler–Binet tests between 1918 and 1995.[2] This means that in 1918, when scored against today's norms, Americans had an average IQ of 75 on tests in which the crystallized component is at least as great as that of the Wechsler tests. Does that mean that during World War I about half of White Americans lacked the capacity to understand the basic rules of baseball?

It is now possible to make some decisions about what package of IQ gains plus other enhanced skills poses a problem of causal explanation. Some enhanced skills one expected to accompany IQ gains can be excluded simply because the enhancement has not occurred. Recall the three schoolchildren sitting in a row who belong to the same generation. The escalation of cognitive skills associated with IQ differences, going from a child with IQ of 75, to one with IQ of 100, to one with IQ of 125, is not in evidence when one compares generations over time. Wechsler subtests, going back as far as 1948, show no gain for arithmetic reasoning, some gains for nonspecialized vocabulary in West Germany and Vienna (but such gains were small or nil in English-speaking nations), and negligible gains for nonspecialized general information. If one regards the present generation as if it were a gifted class with a mean IQ of 125, where are the reports from teachers of long experience that children today surprise them with their speed of learning and sheer

[2]The estimate for 1918–1932 is based on a comparison between the 1918 White draft and the 1932 Stanford–Binet standardization sample, mental ages of both calculated in terms of 1916 Stanford–Binet norms, plus regional studies from that period; all of which show gains greater than those represented in Figure 2. Storfer (1990, pp. 89–94) analyzed Stanford–Binet data, military data covering the period between World War I and World War II, and longitudinal studies and concluded that substantial IQ gains began in America in the early 1890s. The estimates for the decades 1930–1970 are based on numerous comparisons of Wechsler and Binet standardization samples, as noted above. The post-1970 estimates, terminating in 1995, are based on the WISC-R (1972) to WISC-III (1989) estimate and the WAIS-R (1978) to WAIS-III (1995) estimate. These overlap so much, and cover so great a portion of this period, that they were both treated as comprehensive and averaged.

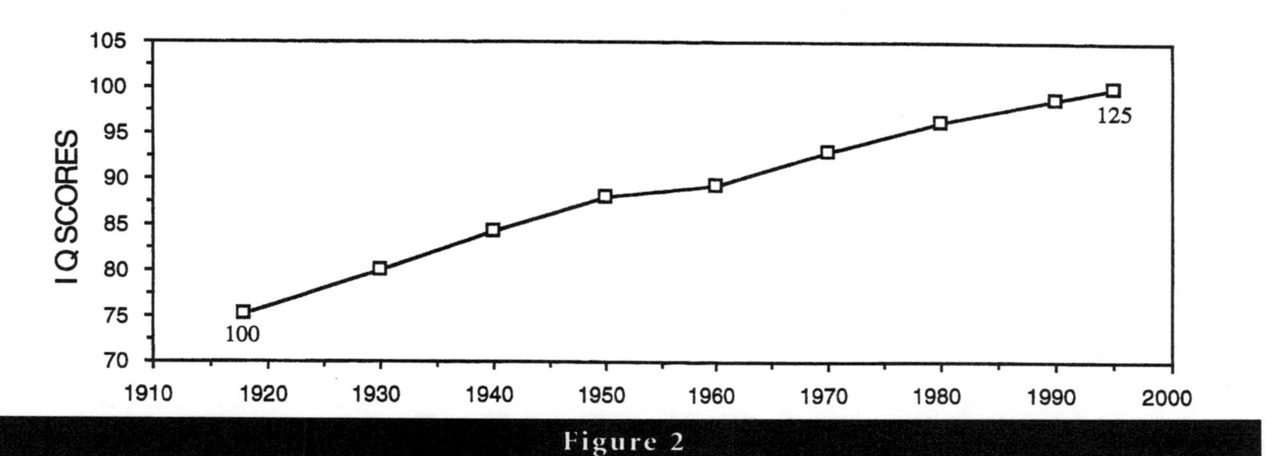

Figure 2

The United States and Wechsler–Binet tests: Whites gained 25 IQ points from 1918 to 1995. A previous graph was revised and extended to accommodate the WAIS-III standardization sample. American gains on Wechsler–Binet tests do not break down into discrete time periods like Raven's gains; for example, in The Netherlands, virtually the total 18-year-old cohort was tested every 10 years. The data are compiled primarily from studies in which what appear to be representative samples (the draft, standardization samples) can be compared as to the norms set. I have estimated rates of gain decade by decade, but this must not conceal the fact that any time periods chosen involve merging various estimates and taking averages. For example, if a rate of gain from a pair of standardization samples overlaps with 7.5 years of a decade, its rate is multiplied by 7.5 and a rate from another pair of samples might be multiplied by 2.5; then a weighted average could be calculated for the full decade. The mechanics are spelled out in Flynn (1984b). From "Get Smart, Take a Test," by J. Horgan, 1995, *273*(5), p. 14. Copyright 1995 by *Scientific American.* Adapted with permission from publisher and artist.

intelligence? The present generation at adulthood, at least in The Netherlands, patents fewer inventions. If one treats past generations as if they were a special-needs class with a mean IQ of 75, one finds no evidence of widespread problems in coping with everyday life. The frustration of our within-generation expectations, concerning enhanced cognitive skills, seems to imply that IQ gains have not been accompanied by intelligence gains.

I do not deny the existence of other important trends: the rise of computer skills, the larger number of people going on to universities and learning more advanced subjects, the larger number of scientists. In my view, these trends no more lead to an expectation of enhanced intelligence than they do when certain people are advantaged over others within a generation; recall the advantages of the first child to get a home computer or those who attended Catholic schools. I would label such trends as enhanced *achievement* and *learning*. However, to insist on my preferred labels would be to violate my fundamental purpose. Call the new skills and educational and scientific progress enhanced *intelligence*, if you will. The labels are not important, but what is important are the contents of the package. The fundamental question is this: Will it be productive or counterproductive to bundle up such enhanced skills with IQ gains *for the purpose* of seeking causal explanation?

After all, over the past 50 years, historians and social scientists have provided detailed explanations of educational and scientific progress. But not one of them used their hypotheses to deduce a prediction of IQ gains over time or even suggested looking for such. This seems odd, if the two explanatory tasks have much in common. Bundling together the wrong phenomena for causal explanation can be a distracter. It made sense to seek a common origin of the movement of the tides and the motions of Mercury, but it would have been a distracter to include the problem of why a stick looks bent when it is half in and half out of water. Granted that this example could be used to suggest that one is wise only after the event. Even though dogmatism is not in order, I believe that bundling up IQ gains with scientific and educational progress is a bad bet.

THEORY AND PRACTICE

It would be a mistake to think that the only important thing about massive IQ gains is providing causal explanation. Therefore, I interrupt the analysis to say a few words about theoretical and practical implications.

Massive IQ gains over time pose a direct threat to the Spearman–Jensen theory of intelligence. That theory is based on *g*, the general intelligence factor derived from the tendency of the same people to excel on a wide range of IQ tests and items. The Raven is the test that best operationalizes *g*, that is, shows that *g* refers to a coherent set of mental abilities (Flynn, 1987b). Jensen (1987, pp. 380–381; 1991, pp. 59–60, 68–69) himself has accepted that IQ gains over time are too great, even on tests whose credentials as measures of *g* are impeccable, to be equated with intelligence gains. The huge gains on the Raven are, of course, especially troublesome. Some years ago, Jensen (1980) envisaged tests running from the detour problem (finding one's way around a barrier) through an adapted form of the Raven, which would allow one to measure the intelligence of cats and chickens, Kalihari Bushmen and polar Eskimos—even extraterrestrials. Today, we have reason to suspect that the Raven cannot compare the Dutch of 1982 and the Dutch of 1952 for intelligence, perhaps not even the Dutch of 1982 and 1972.

Problems remain even for those who abandon *g* but who accept the distinction between the Raven as a measure of fluid intelligence and the Wechsler–Binet tests as adding the dimension of crystallized intelligence. Theory posits a functional relationship between these two, so that a problem that afflicts one transfers to the other. Take the problem of putting past generations at a mean IQ of 70 or 80 on tests of fluid intelligence. Such populations should not be capable of soaring much above that for crystallized intelligence, the skills they need to deal with the real world. It is quite possible that people whose fluid intelligence did not decline until old age should retain the information and vocabulary they acquired earlier, at a time when their fluid intelligence was normal. The evidence of many studies suggests that this is true (Horn, 1989). However, it is quite another thing to imagine people acquiring normal levels of knowledge and vocabulary if their fluid intelligence never, during their entire lives, rose much above the level of mental retardation.

Flynn, (1992, in press) gave an overview of the full range of practical implications of massive IQ gains over time, but I highlight a few examples here. Between 1948 and 1972, the period between standardizations of the WISC, IQ gains lowered the number of American children eligible to be classified as mentally retarded from 8.8 million to 2.6 million. There is no evidence in the literature that clinical psychologists were aware of this (Flynn, 1985). The recent WISC-III manual gives a criterion for learning disabilities or reading disorders in terms of differential performance on four subtests (Wechsler, 1992, pp. 212–213). Thanks to differential gains over time on those subtests, perfectly normal children are in danger of misdiagnosis.

Vernon (1982) analyzed studies of Chinese Americans in which they were scored not against their White contemporaries but against the lower performance of Whites from previous decades. This made Chinese Americans appear to be an IQ elite when, in fact, they had no higher mean IQ than White Americans. Despite this, Chinese Americans have outperformed White Americans academically and occupationally by huge margins (Flynn, 1991). Massive IQ gains add viability to an environmental hypothesis about the IQ gap between Black and White Americans. It appears that the former have enjoyed a slightly higher rate of gain than the latter (Herrnstein & Murray, 1994, pp. 276, 292). This implies that since 1945, Blacks have gained at an average rate of over 0.30 points per year and gained a total of 16 points over 50 years. So the Blacks of 1995 should have matched the mean IQ of the Whites of 1945. Therefore, an environmental hypothesis need assume only that the average Black environment of 1995 matches the quality of the average White environment of 1945.

CAUSAL HYPOTHESES

Scholars are accustomed to providing causes of within-generation IQ differences, such as the IQ differences that separated the three children attending school at the same place and time. They are tempted to regard the problem of explaining between-generation differences as similar. However, the two phenomena are radically differentiated by the fact that the link between IQ and other cognitive abilities, so firm within generations, has snapped between generations. The breaking of that link, the fact that IQ gains are simply not attended by the enhancement of most of the real-world cognitive skills one might expect, creates a peculiar criterion for evaluating the plausibility of causal hypotheses. Throughout this section, I use the label *intelligence* to refer to the missing real-world cognitive skills, but it has only that significance, nothing more.

Factors that are evidentially weak, such as test sophistication or personal irresponsibility (see **The Brand Hypothesis**), qualify as plausible simply because they might raise IQ scores without raising intelligence. Factors that prima facie look evidentially strong, such as nutrition, are labeled implausible simply because they could not possibly raise test scores except through the vehicle of enhanced intelligence. Then there are factors like socioeconomic status (SES) and urbanization that must be given a peculiar formulation to qualify as plausible. Higher SES is thought to be correlated with higher IQ scores for two reasons: Competition within a generation for high status ranks people for intelligence without enhancing the overall level of intelligence; the better environment high SES provides perhaps enhances intelligence. The first rationale does not apply between generations because past and present generations do not compete. The second rationale would have to be modified to read that SES environmental gains between generations raise IQ scores without raising intelligence. The case for urbanization as a factor would have to be reformulated in exactly the same way. The plausibility of education is less affected, that is, less affected by the lack of association between IQ and real-world cognitive skills.

I review the cases for the factors just named, trying to strike a balance between citing the broken link when it is relevant and ignoring it when it gets in the way of evaluating evidence on its merits. The factors reviewed are environmental rather than genetic: Only a fanatic eugenics program could have made a significant contribution to IQ gains, and if anything, mating trends have been dysgenic (Lynn, 1996).

Increased outbreeding, as local communities became less isolated between 1850 and 1950, may have been a quasi-genetic factor, but it is unlikely that it could be used to explain post-1950 gains in advanced European nations.

Test Sophistication

The 20th century has seen a change from totally unaware participants to people bombarded by standardized tests, and undoubtedly, a small portion of gains in most nations is explained by test sophistication. However, its role must be relatively modest. Gains antedate the period when testing became common and have persisted into an era when IQ testing, owing to its unpopularity, has become less frequent. More to the point, even when naive participants are repeatedly exposed to a variety of tests, IQ scores rise by only 5 or 6 points, and the rate of gain reduces sharply after the first few exposures. It would be difficult to put British gains at less than 30 IQ points, and some nations, like The Netherlands, show the rate of gain escalating decade after decade.

The Brand Hypothesis

Brand (1987a, 1987b) argued that the permissive society advantages the present generation as test takers. He considered it significant that IQ gains are correlated with increasing rates of sexual promiscuity, illegitimacy, divorce, irreligiousity, cigarette consumption, accidents, and crime, as well as with Britain's leadership in the field of popular music. The last generation was scrupulous and painstaking; the present generation tends toward personal liberalism. The former wasted time trying to get every item correct; the latter are prone to intelligent guessing and finish more items within the time allotted. The former, even when tests were untimed, became demoralized when they could not answer a hard item and did not persist to answer subsequent easy items; the latter skim hard items and persist.

This hypothesis is theoretically ideal. It explains IQ gains in terms of something that implies no intelligence gains and cites environmental factors independent of mere exposure to tests. However, it has now been proved false. John Raven analyzed his own test and reported that responsible test takers persist to the end. In addition, people do not get items right or wrong by guessing but by mastering or not mastering rules that govern the orderliness of the matrix (Raven, Raven, & Court, 1995, p. G59). On the test Brand chose to evidence his hypothesis, the Verbal scale of the WISC (a test exerting little time pressure), Scottish children in fact made a generational gain of fully 13 IQ points (Flynn, 1990). Flieller, Jautz, and Kop (1989, pp. 11–12) analyzed a Binet-type test with a fairly even balance of timed and untimed items. They found that the last generation left more questions unanswered on both kinds of items and that worse performance on items completed accounted for virtually all of the last generation's score deficit.

Nutrition

The nutrition hypothesis cannot be bettered as an example of the peculiar problem of explaining IQ gains over time. Better nourished brains would function better in the test room but only because they also functioned better in everyday life. Therefore, if improved nutrition has caused IQ gains of 20 or 30 points, one would be driven to posit huge understanding-baseball intelligence gains.

Richard Lynn (1987, 1989) enhanced the plausibility of this hypothesis by ascribing only a portion of IQ gains to nutrition. He ascribed the remainder to other causes such as defective tests. For example, the Raven is held to measure increased arithmetic skills as well as intelligence gains and therefore to overestimate intelligence gains. The critique of the Raven poses many evidential problems. Norwegian draftees made matrices gains while suffering losses on a test modeled on the Wechsler adult arithmetic subtest. Military samples from Israel showed comparable male and female performance on the matrices, which runs counter to most gender data concerning mathematics. As for the magnitude of matrices gains, they are larger than those of other nonverbal tests in Britain but equivalent in Belgium and smaller in Australia, Canada, and Scotland (Flynn, 1987a, pp. 173–174, 176, 1990; J. Goldenberg, personal communication, March 4, 1991).

Even if given a diminished explanatory role, the nutrition hypothesis has its own peculiar evidential problems. Lynn (1987, p. 467) focused on Britain. He cited a height gain of 1 *SD* over the last 50 years; this equals his estimate of British intelligence gains over that period, that is, 15 points. However, some European countries have been reporting height gains for fully a century or two, and these amount to more than 1 *SD*, sometimes to 2 or 3 *SD*s (Floud, Wachter, & Gregory, 1990, pp. 16, 23, 26). If height gains are truly accompanied by intelligence gains, they pose a familiar question: Did the Dutch in 1864 really have the same intelligence as people who today score 65 on IQ tests? Did Norwegians in 1761 really resemble those who today score 62?

The best experimental study of the effects of vitamin–mineral supplements on IQ showed that in California, a modest supplement had little effect, a moderate one had a significant effect, and a large one had little effect (Schoenthaler, Amos, Eysenck, Peritz, & Yudkin, 1991, pp. 357–358). That every nation has continuously enhanced nutrition just the right amount, neither too little nor too much, for decade after decade, seems unlikely. Moreover, historic nutritional fluctuations have sometimes proved "impotent." The Netherlands almost certainly provided children born after World War II with better nutrition than it provided those born during the great wartime famine. The effect on IQ gains of the fluctuating nutritional quality over time was nonexistent (Flynn, 1992, p. 346).

The experimental data concerning dietary supplements also have shown that 75% of participants enjoy very modest gains, whereas 25%, presumably persons who are subclinically malnourished, make large gains. The latter tend to have lower IQs than the former, which means that if enhanced nutrition is a factor, IQ gains over time should come disproportionately from those with below-average IQs. Denmark's data fit that pattern, but the data of most nations do not. A good sign that IQ gains extend to every IQ level is that score variance remains unchanged over time or diminishes only because of clear ceiling effects. Military samples or samples of equivalent quality show this pattern for Belgium, Norway, Sweden, Israel (men), Canada, and New Zealand. Dutch Raven's data and U.S. Wechsler data provide the full IQ curves and allow one to verify gains at all levels (Bouvier, 1969, pp. 4–5; Clarke, Nyberg, & Worth, 1978, p. 130; Elley, 1969, p. 145; Emanuelsson

& Svensson, 1990; Flynn, 1985, p. 240; J. Goldenberg, personal communication, March 4, 1991; Rist, 1982, p. 47; Teasdale & Owen, 1989, pp. 258–259; P. A. Vroon, personal communication, November 5, 1984).

The Storfer Hypothesis

Storfer (1990) made an attempt at causal explanation that has much in common with that of Lynn, although he added an unusual twist at the end. Once again, only a portion of IQ gains are to be identified with intelligence gains, this time 22 points (rather than 15). Once again, the remainder is ascribed to defective tests. This time the Raven is held to be a spatial test measuring a peripheral rather than a core component of intelligence. I believe that the Raven may have a spatial memory component if administered with great time pressure, but the component is small. Jensen (1980, pp. 646–647) called the notion of a significant spatial component a "common misconception" and emphasized that, factorially, the Raven measures fluid *g* and little else. Storfer cast doubt on the concept of so-called fluid intelligence and argued that only tests heavily weighted toward crystallized intelligence are true measures of intelligence. The theoretical price to be paid for downgrading fluid intelligence has already been discussed. The plausibility of even a 22-point intelligence gain is not directly confronted. That is, rather than discussing the consequences of putting the mean IQ of this generation's grandparents at 78, in terms of today's norms, the usual within-generation evidence for the validity of IQ tests is cited.

Storfer (1990) supplemented improved nutrition as a cause with factors like the eradication of childhood diseases and improvements in the cognitive quality of the preschool home environment. He argued that in the United States, when acting purely as environmental variables, these factors could explain an 11-point intelligence gain since 1900. The analysis takes within-generation data and applies them across generations by making certain assumptions, such as that half of American infants were in unfavorable home environments in 1900 compared with only 20% today. As I have demonstrated, the assumption that the within-generation potency of a factor holds between generations is suspect, particularly when extended to factors like favorable versus unfavorable home environments.

The 11 points supposedly explained fall well short of the 22-point intelligence gain Storfer posited. He doubled the explanatory potential of his factors by formulating a new Lamarckian theory of evolution, citing cholinergic neurons as the vehicle for the inheritance of an acquired characteristic. These neurons might convey an environmentally induced change in the brain cells to the testes, allowing that change to be passed from one generation to the next. There is no harm in this sort of speculation, but acceptance awaits anatomical and biochemical evidence. Until then, there is no substantial body of evidence to assess.

SES and Urbanization

Enhanced SES over time should capture something of the improved home environment Storfer and many others have posited. Whatever role SES played in the first half of the 20th century, the eternally puzzling Dutch data imply little impact since 1950. de Leeuw and Meester (1984, pp. 14, 16, Figures 5, 7) provide data that allow an estimate of SES gains from 1952 to 1962 as measured by father's occupation. When projected over 30 years, this amounts to 1.18 *SD*s. The correlation between father's occupation and son's IQ is .33 (de Leeuw & Meester, 1984, pp. 13, 16); therefore, SES gains might appear to account for 5.84 of the 20-point Dutch generational gain ($1.18 \times .33 = 0.3894$ *SD* units; $0.3894 \times 15 = 5.84$ points). However, the correlation between father's occupation and son's IQ may not represent a causal link: When P. A. Vroon controlled for father's IQ and father's education level, variables with a high genetic loading, the path correlation between father's occupation and son's IQ was .02, or virtually zero (P. A. Vroon, personal communication, October 9, 1984). A generous estimate for SES, as an environmental variable, would be that it caused a 3-point IQ gain in the current generation.

The best data on urbanization, or the migration of people to cities, come from Flieller, Saintigny, and Schaeffer (1986), covering the years 1944 to 1984. They estimated that French 8-year-olds gained 24 IQ points (1.6 *SD*s) over those 40 years on a Binet-type test. Their occupational breakdown evidences the IQ deficit of the children of farmers and other rural workers. Using algebra, one can show that even if urbanization totally eliminated these occupations, only 3.2 points of the 24-point gain would be explained. Calculation of the IQ gain of children whose parents had urban occupations gave a similar result: Only 3 points of the total sample's gain disappeared. The authors also provided data that show more preschooling accounting for 3.75 points, test sophistication for 1.81 points, and enhanced SES for 1.37 points. Simply adding the values for these four variables accounts for 10 of the 24 points explained, but the results would be heavily confounded. The shift from rural to urban living would in itself account for much of the rise in preschooling, test sophistication, and SES. No precise estimate of the effect of the total package is possible, but an estimate of 6 points is plausible and tallies with the Dutch data. There, 5 points out of 20 were explained by a package of SES plus test sophistication plus more education (Flynn, 1987a, pp. 188–189).

Education

Education seems an obvious cause because, at its best, it awakens the mind and teaches students to analyze and criticize. During the 20th century, semiformal and formal education have been extended down into the preschool years and upward into adulthood. Using Stanford–Binet data for 1932 to 1971–1972, Thorndike (1977) concluded that American children aged 6 and younger have made greater IQ gains than older children. Therefore, he sought causal factors likely to affect preschoolers more than others such as TV in general and educational TV in particular. Flynn (1984a) compiled a wider array of data that showed that the atypical gains of young children were either an artifact of sampling error or antedated 1947, ruling out TV as an age-specific factor. Moreover, he used the WISC standardization sample to compare American IQ gains from 1932 to 1947–1948 with those from 1947–1948 to 1972, the periods immediately before and after the introduction of TV. The rates of gain for both periods were roughly equal. Lynn (1987) has hypothesized that children who grew up during the Great Depression and World War II may have had their IQs depressed. If so, the WISC standardization sample of 1947–1948 would have had an atypically poor performance, which would deflate estimated gains prior to the introduction of TV and inflate the estimate for the period thereafter. In other words, TV might have lowered IQ gains, an effect concealed by the depressed performance of the WISC sample. However, there is ample international data that show massive gains for people born after World War II, which counts against the depression–World War II hypothesis (Flynn, 1988, p. 349, 1990; J. Goldenberg, personal communication, March 4, 1991; Lynn, 1990; Wechsler, 1992, p. 198). There is no reason to believe that TV either increased or reduced the rate of IQ gains in the United States.

Every one of the 20 nations evidencing IQ gains shows larger numbers of people spending longer periods of their life being schooled and examined on academic subject matter. IQ gains in Denmark appear highly correlated with increased years of schooling and more people attaining higher credentials (Teasdale & Owen, 1989). However, the reverse is true in The Netherlands, where matching across generations to hold educational level constant eliminated only 6.5% of a massive gain (Flynn, 1987a, p. 188). Gains among schoolchildren, on the basis of comparing 6th or 12th graders with their counterparts of a generation ago, cannot be influenced by years of schooling because the number of years is by definition the same.

As for quality of schooling, some educational reforms may actually have impeded IQ gains. Rist (1982, pp. 56–58, 63) noted that when students trained in the new math reached military age, Norwegian gains on a math test, modeled on the Wechsler Adult Arithmetic subtest, turned into losses. Setting that aside, those who endorse quality of schooling as a factor must argue the following: either that better teaching of the learned content of an academic curriculum has raised IQ or that better teaching of decontextualized problem-solving skills has raised IQ (Cole & Means, 1981; Scribner & Cole, 1981; Sharp, Cole, & Lave, 1979). I now examine those two subhypotheses.

The first subhypothesis, concerning better teaching of school-learned content, has already been falsified by the pattern of IQ gains over time. As I have shown, gains drop as one goes from Raven's type tests to performance tests to verbal tests to Wechsler subtests like

Arithmetic, Information, and Vocabulary. This implies that the gains tend to disappear when material closer to the learned content of the school curriculum is tested.

This leaves the second subhypothesis, namely, that schools are teaching better decontextualized problem-solving skills. Perhaps they are, but the hypothesis is empty unless (a) these school-taught skills are identified; (b) they are linked to the problem-solving skills used on IQ tests, particularly culture-reduced tests of fluid *g*; and (c) they are linked to some kind of real-world problem solving or, the greatest puzzle, it is explained why there is no such link. The very fact that children are better and better at IQ test problems logically entails that they have learned at least that kind of problem-solving skill better, and it must have been learned somewhere. However, simply to assert that the enhanced IQ test skill can be equated with some enhanced school skill is arbitrary and vacuous.

The fact that education cannot explain IQ gains as an international phenomenon does not, of course, disqualify it as a dominant cause at a certain place and time. Particular countries are sometimes influenced by a factor that is culture specific. Comparing age cohorts has suggested that the urban Chinese gained 22 IQ points on the Raven Progressive Matrices between 1936 and 1986 (Raven & Court, 1989, p. RS4.8). Learning to read Chinese characters involves memorizing complex symbols, combining them to alter meaning and signal pronunciation, and taking such tasks seriously. The literacy that follows urbanization might be an important cause of matrices gains peculiar to China.

Evaluation of Causal Hypotheses

It is logically possible that peculiar factors dominate in each and every one of the 20 nations: years of schooling in Denmark, urbanization in China, perhaps test sophistication in Brazil, and so forth. However, the universal pervasiveness of massive IQ gains and the fact that there are such striking counterexamples to these factors make this highly unlikely. I believe it is fair to say that up to now, efforts to identify the environmental factors that have caused IQ gains have not come to much. The tendency has been to take the massive IQ gains since the last century, carve out a small portion to be treated as an intelligence gain, try to explain that portion by familiar within-generation factors, and treat the remainder as a nonintelligence gain caused by faulty tests. The tests castigated, it is worth noting, were hitherto considered reliable and central to the theory of intelligence. If that is a fair summary, new departures—new research strategies—are necessary.

RESEARCH STRATEGIES

I intend to discuss five new research strategies. First, Arthur Jensen is experimenting with using behavioral and physiological variables to measure intelligence. He hopes these will not generate the theoretical and practical problems that IQ gains pose for IQ tests. Second, if there are national differences in terms of rates of gain, one might find something present in a nation with a high rate but absent in a nation with a low rate. Third, it might be possible to "rerun history" since the industrial revolution by studying a nation whose locales range all the way from untouched by industrialization to fully modern. Fourth, if IQ gains differ in terms of age, one might find something present at one age but absent at another. Finally, perhaps one can add to the package of enhanced cognitive abilities. Clarifying the effects to be explained may well provide a clue as to causes. There will be more on this last possibility at the end of this chapter.

Jensen and Physiology

Jensen believes that IQ tests normally measure intelligence, but that when used to compare generations, they are sensitive to factors that distort the measurement. He called on the analogy of measuring height by using shadows (Bower, 1987). At a particular time and place, shadows can rank people for height with considerable accuracy. But if one made comparisons over time, compared shadows in summer with shadows in winter, the latter would be longer and would register height gains that were spurious. The distorting factor to which shadows are sensitive is, of course, the angle of the sun's rays.

Therefore, Jensen (1988, 1989) is experimenting with behavioral and physiological variables that show correlations with IQ, the electrical response of the cerebral cortex to sights and sounds, how quickly people can react to stimuli, the time taken for an injection of glucose to reach and be absorbed by the brain. The logical culmination of this process would be the replacement of IQ tests by a battery of chronometric and physiological tests. Although the prospect of such a revolution is exciting, the new battery would have no real value unless it could outperform existing tests as a measure of intelligence; therefore, it must be assessed in terms of external validity, and that assessment may lead to ambiguous results.

I take up Jensen's analogy. It replaces IQ as a measure of intelligence with shadows as a measure of height. Because shadows have proved unreliable measures over time, from summer to winter, the search for physiological correlates begins. Imagine that someone discovered a physiological correlate of shadow differences, such as pulse rate. Would it make sense to replace shadows with pulse as a measure of height? It would not if pulse were as sensitive to the seasons as shadows are; for example, the pulse rate may rise in winter. But even if it is less sensitive, it might be worse than shadows as a measure of height in most cases. At a particular place and time, the relationship between height and shadows is strong. The correlation between height and pulse might be low: The hearts of tall people may tend to work harder but not that much harder. In addition, to make the situation truly analogous to IQ, one must assume that these people have no direct measure of height, just as IQ tests give no direct measure of intelligence. All people have are the criteria of external validity that gave them confidence in shadows in the first place. Shadows usually predict who can reach a given shelf without a stepladder, who can vault a particular fence, and so on. If pulse gave worse predictions, they would reject it as a crude measure of height, even for the purpose of comparing height differences between summer and winter. Would they assign it the more modest role of providing a warning sign that seasonal shadow differences cannot be equated with height differences? They already have plenty of warning signs: People can no more dispense with stepladders in winter than in summer, and they are no better at vaulting fences.

To return to the world of IQ, one does not know whether scores on chronometric and physiological tests would remain unchanged if one tested representative samples of two generations. But even if they did, that would not establish their credentials as measures of intelligence. Achievement test scores correlate with IQ and have been relatively stable between the generations; yet, they are almost certainly worse measures of intelligence than IQ. There should be no presumption in favor of physiological correlates as such, so long as people are ignorant of the physiological processes that establish the correlation with IQ. As for the physiological tests playing the role of providing warning signs that IQ gains cannot be equated with intelligence gains, people already have the only warning sign they need. Despite massive IQ gains, this generation does not exhibit the enhanced arithmetic reasoning, vocabulary, creativity, and speed of learning one would normally expect.

If chronometric and physiological tests are to be accepted as better measures of intelligence than IQ tests, they will have to be assessed against the usual criteria of external validity, which means walking the long hard path blazed by Binet. Do their scores rise with age at least to maturity? How do they correlate with teacher estimates of ability, achievement test scores, SES, upward mobility of sibling versus cosibling, and so forth? I suspect that they will fall below Wechsler–Binet tests in these assessments and below the Raven itself. Here the example of the Raven is instructive. Jensen (1980) argued that the Raven does not predict achievement as well as Wechsler–Binet tests because it is too "factor pure": It measures intelligence alone and screens out other factors that contribute to achievement such as motivation and education. The new tests are being groomed to replace the Raven as measures of intelligence or *g*. If they are even less predictive of achievement, they will pose a choice: Are they less adequate measures of intelligence because of physiological processes of which people are ignorant, or are they better measures of intelligence because intelligence has less causal potency for achievement than suspected hitherto?

The theoretical problems surrounding physiological measures are great (Flynn, 1987b), the practical problems of their administration are even greater. Gumming electrodes onto children's scalps and injecting glucose may provoke consumer resistance. If IQ tests remain as the mainstream measures of intelligence, no physiological research strategy can substitute for knowledge of the causes of IQ gains. Somewhere out there, environmental variables of enormous potency are creating IQ differences: score differences that seem to have little more to do with intelligence than the score differences caused by test sophistication. It seems likely that these variables operate to some degree within generations as well as between generations. If so, some children are getting inflated or deflated estimates of their intelligence compared to the child sitting next to them. Therefore, the causal factors responsible are well worth knowing.

Looking for National Differences

The strategy of analyzing different rates of gain between nations makes sense only if the data allow for true comparisons. Only the military samples from The Netherlands, Israel, Norway, and Belgium qualify. They alone provide reliable samples (near saturation) from portions of the same period (1952–1982) for persons of the same age (young adults) taking equivalent tests (derivatives of the Raven).

Figure 3 compares the slopes of the rates of gain of those nations plus Britain. Britain only half belongs; I included it because it yields adult data for the Raven for the whole period from 1942 to 1992. Using all ages from 18 to 67, I compared the top half of the 1942 curves with the bottom half of the 1992 curves, thus discarding the top percentiles from 1992. I discarded the top percentiles from 1992 because their scores were suspect: On the one hand, they were depressed by an obvious ceiling effect; on the other hand, they were the ones, it will be recalled, who may have profited from taking the test unsupervised. The comparison yielded a gain of 27 IQ points between 1942 and 1992. Because all ages are represented for both years, this is a better estimate than was obtained using birth dates to project back to the 19th century. At any rate, taking Figure 3 as a whole, perhaps someone can find

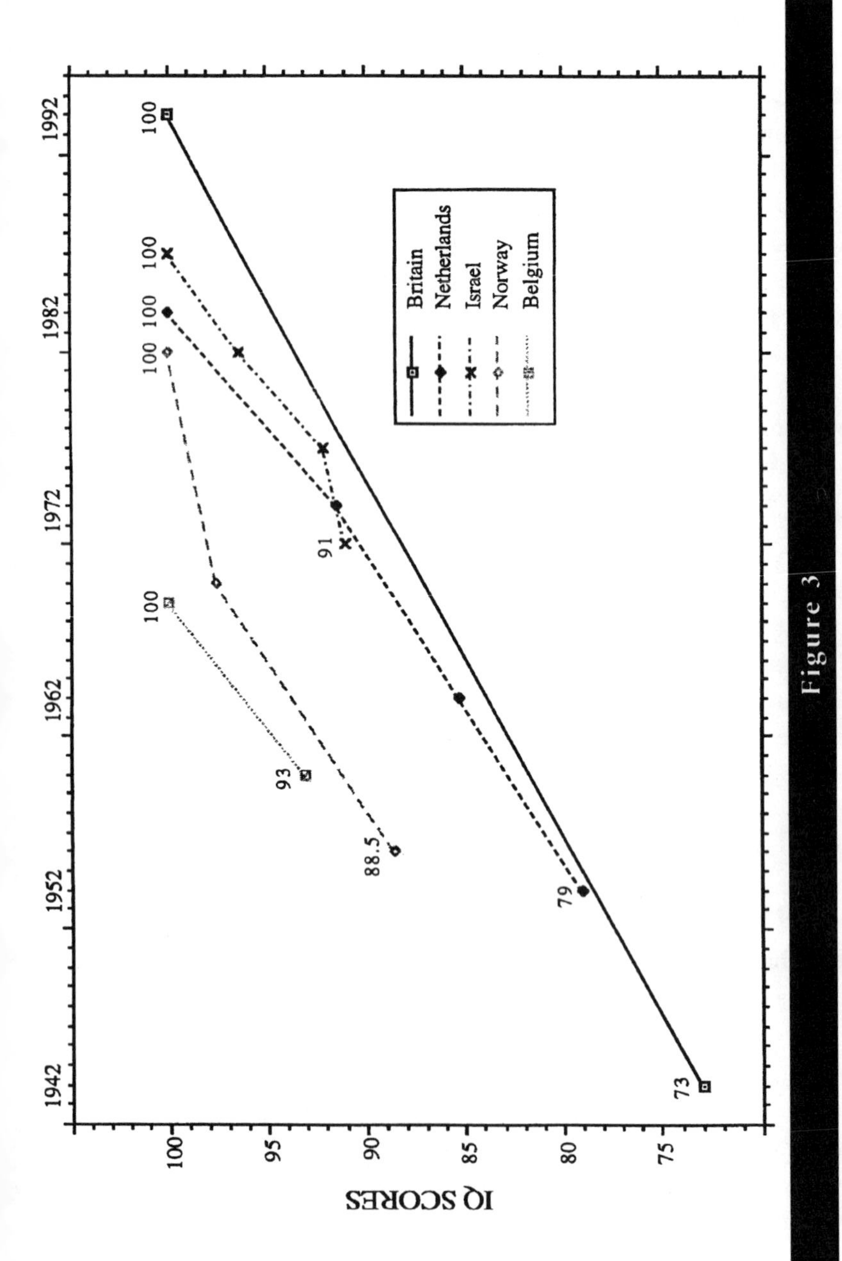

Figure 3

Five nations and matrices tests: A comparison of rates of IQ gain. Every nation is normed on its own samples. Therefore, even though nations can be roughly compared in terms of different rates of IQ gain, they cannot be compared in terms of IQ scores. That is, the fact that the mean IQ of one nation appears higher than another at a given time is purely an artifact. Data from Flynn (1987a, pp. 172–174), J. Goldenberg (personal communications, March 4, 1991), and Raven, Raven, and Court (1993, Graph G2).

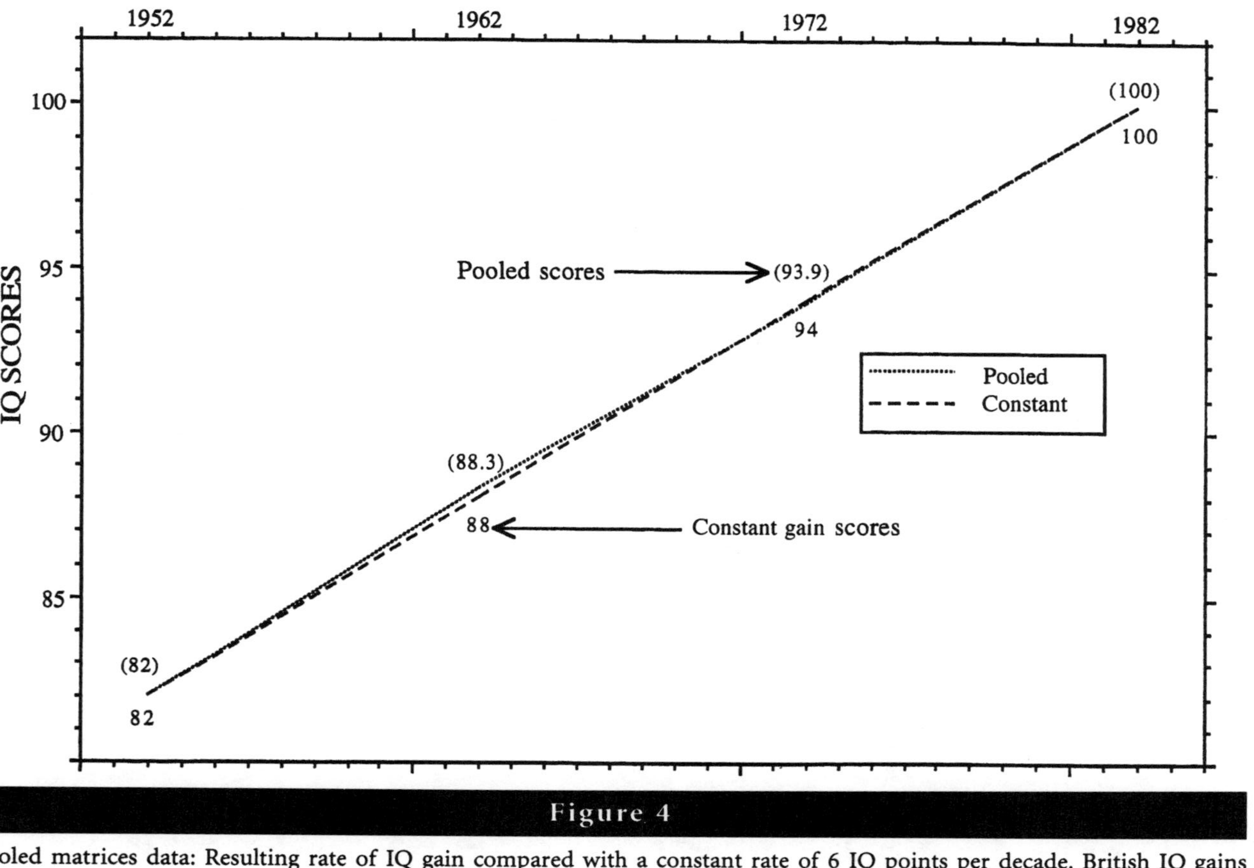

Figure 4

Pooled matrices data: Resulting rate of IQ gain compared with a constant rate of 6 IQ points per decade. British IQ gains were not measured between 1952 and 1982, which means that they had to be "uniform" within that period as a result. Therefore, whereas they were used to calculate the magnitude of the overall rate of gain, they were not allowed to reduce variations in the rate of gain from decade to decade.

significant differences among these five nations in terms of rates of gain, that is, differences that suggest causes present in some nations but absent in others. To me, the slopes are depressingly similar, the rates of gain range only from a low of 5.40 points per decade for Britain to a high of 7.78 points for Belgium.

Therefore, I decided to pool the data to see if the five nations collectively showed different rates of gain over time, that is, from one decade to another. I hoped that significant differences would emerge, for example, between 1952–1962 and 1962–1972, that would signal causal factors waxing or waning. As described at the bottom of Figure 4, the British data, which necessarily show a uniform rate transcending decades, were not allowed to have a leveling effect. Figure 4 shows that the pooled rates of gain came astonishingly close to a constant rise over the whole 30 years. With no variation in the rates of gain, it was of course impossible to look for a correlation with some variation in possible causal factors. The pattern is unsettling: It is as if some unseen hand propelled scores upward at an unvarying rate between 1952 and 1982, a rate of 6 IQ points per decade, with individual nations scattering randomly around that value.

Trying to Rerun History

IQ gains, or better, the factors causing IQ gains, appear to have been triggered by the industrial revolution. One cannot get into a time machine, go back and test one's ancestors, and watch them evolve as they enter the modern world. However, a nation like India shows great diversity from one locale to another, some virtually untouched by the industrial revolution, some strongly influenced, and others intermediate. One might target 10 areas so as to create a simulation of the history of the last 100 years. A report published in June 1995 put the mean IQ of Indian schoolchildren at 125 ("Children Working," 1995, p. 34). Clearly, IQ gains are occurring in India: A mean of 125 can only be a product of scoring contemporary children against the performance of lower scoring children tested some years before. Ten testing teams could provide a Raven's score map of the targeted areas, and these areas could be studied to see what causal factors kick in going from lower to higher

IQ areas. No one would fund such a project as a stand-alone, but it might be attached to a larger project studying the impact of industrialization on Indian society.

Looking for Age-Specific Differences

If IQ gains over time are age specific and show their greatest magnitude at a certain age, this might indicate either that the causal factors are more prominent at that age or that people are more receptive to them at that age. Imagine that IQ gains were greatest for children under 10. That would suggest factors like better parenting, more preschooling, and educational TV, the factors Thorndike proposed when he mistakenly thought IQ gains were evident mainly among young children. Or imagine that Raven gains escalated dramatically at 20, 30, or 40 years of age. That would suggest that over the last few generations, something had happened to enhance the on-the-spot problem-solving ability of adults. Perhaps something societal stimulated people in a different way from previous generations of adults, or something physiological made them more receptive to stimulation despite growing older. Isolating the factors responsible for unusually large age-specific gains would not, of course, explain the large gains evidenced for all ages. However, if we were lucky, the two trends—shifts over time in age receptivity and generalized gains over time—would have some causal overlap.

Figure 5 illustrates the point. Cohort 1, born in the earlier year of 1950, is susceptible to a factor escalating IQ at a constant rate of 5 points per decade, but only until the age of 20, at which point it develops an "immunity." Cohort 2, born in the later year of 1960, remains receptive to the factor beyond the age of 20. Therefore, the measured IQ gap between the two cohorts, only 5 points at ages 20 and under, dramatically jumps to 10 points at age 30. Do data exist that mimic Figure 5? The British Raven data are tantalizing: They show IQ gains between 1942 and 1992 of "only" 20 points for ages 18–32 but almost 30 points for ages 33–67. Before one gets too excited, remember that the British Raven data come from two cross-sections of all ages tested 50 years apart. Therefore, the rate of gain over time and age-specific

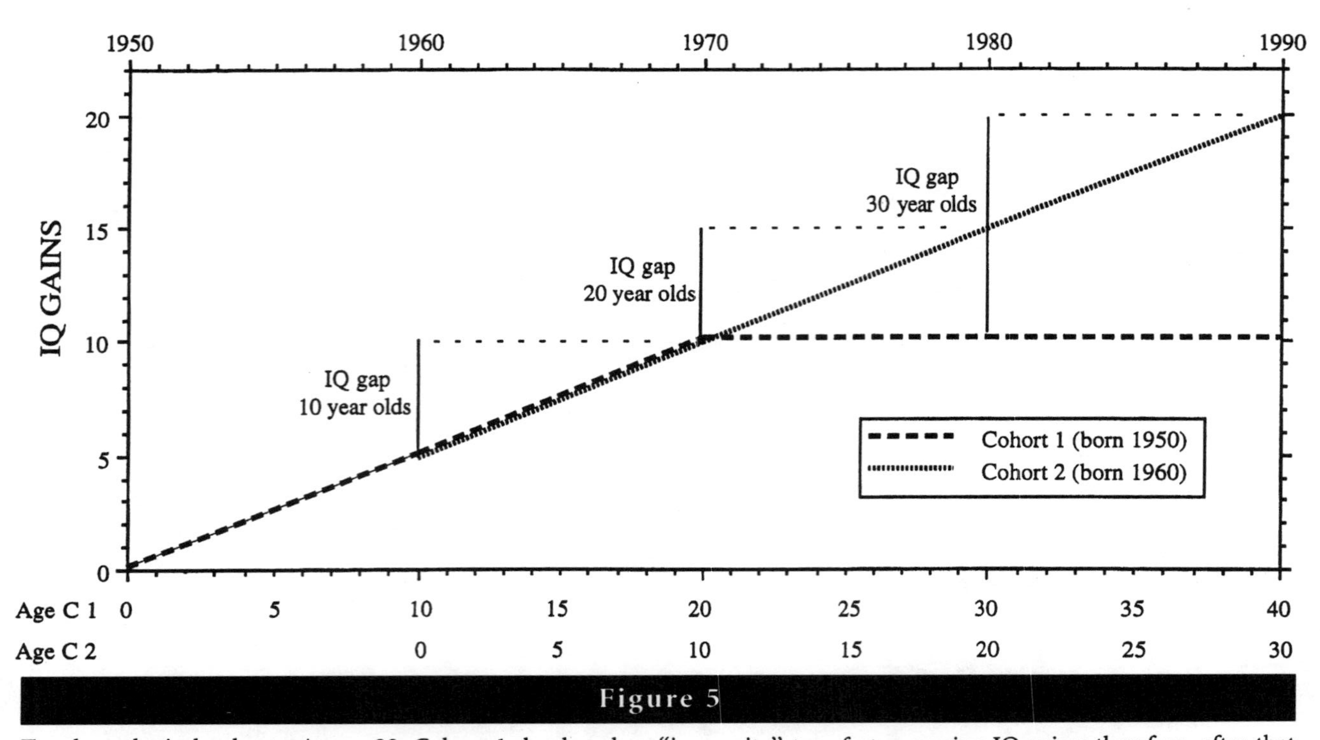

Figure 5

Two hypothetical cohorts: At age 20, Cohort 1 developed an "immunity" to a factor causing IQ gains; therefore, after that age, cohort comparisons show a widening IQ gap.

peak performance are hopelessly confounded in a way that only genuine longitudinal studies can cure.

Figure 5 can help to correct a common misconception. It shows that uniform presence of a causal factor at all ages produces an IQ gap fully intact in early childhood. In other words, one must not overestimate the importance of a pattern in which IQ gains do not escalate as people age (see chapter 8, this volume). Such a pattern means that the causal factors are undeniably present in early childhood but not necessarily that they impact exclusively or evenly mainly there. This is apparent when one considers the fate of many intervention programs. Factors that boost IQ are applied in early childhood and create an IQ gap between experimental participants and controls, but then the gap disappears as the children mature because no factors are applied after early childhood. Of course, the "IQ gains over time" boosting factor might be different; that is, it might create an IQ gain in early childhood that persists, even though the factor itself disappears and is succeeded by no later factor. However, one would know that the "IQ gains over time" boosting factor was of this special sort only after identifying it. With two possible causal explanations of an IQ gap beginning in early childhood and persisting until old age, either an early-childhood factor with permanent effects or factor(s) operating throughout life, the IQ gap itself cannot choose between them. All that such an IQ gap can tell scholars is that a causal explanation that omits factors operating in early childhood is unlikely to be true.

Back to Packaging Effects

It may appear that my purpose has been to reduce the package of cognitive trends to be explained to one, that is, nothing but IQ gains over time. In fact, I hope to expand the package, but I believe this must be done carefully and evidentially, without preconceptions based on the cognitive skills that normally accompany IQ.

I remain convinced that neither giftedness (the capacity to learn more quickly and make creative leaps) nor understanding-baseball intelligence (the capacity to absorb the usual rules of social behavior) has increased significantly. But even I believe that enhanced problem solving in the test room must signal some kindred gain in problem solving in the real world, however subtle. Identifying these two effects and comparing them could provide a priceless guide from effect to cause. Let me use a sports analogy. Assume that juggling skills have dramatically escalated over time, and this takes everyone by surprise because no one has been training people to be jugglers. At first, people find no carryover to a socially significant sport, so we are baffled. Then they find that although people seem no better at football, or baseball, or basketball, they are better at the minor sport of archery. The link between juggling skills and archery skills and lack of link to other sports would identify the effect to be explained: Perhaps because of its highly repetitive technique, better juggling does not involve improved reflexes or hand–eye coordination, much less greater speed or strength; all it really involves is (like archery) a steadier hand, better concentration, lower distractibility, and more patience. So one looks for plausible causes of the latter traits rather than the former.

Fortunately, thanks to the work of Carpenter, Just, and Shell (1990), it is now known what cognitive skills improve performance on the Raven, namely, mastering five rules that collectively determine the element needed to complete the matrix pattern. The five rules are (a) The same value occurs across a row but alters down a column; (b) there is a quantitative pattern between adjacent entries, for example, the number of black squares in each entry increases across a row from one to two to three; (c) a figure from one column added to (or subtracted from) another produces a third; (d) three values from something like a figure type are distributed across a row, for example, diamond, square, and triangle run across rows according to a distributive rule; and (e) two values are distributed across a row, but the third value is null. The problem is how to link this test-room mastery to real-world cognitive tasks. The U.S. military has attempted to correlate mental test scores with job performance on tasks that include land navigation, use of a night-vision device, and so forth, some of which look cognitively demanding (Wigdor & Green, 1991). Perhaps they would be willing to study personnel, separated by 10, 15, or 20 years of age, and determine whether the profile of performance on such tasks varies from one cohort to another.

There is no guarantee one would perceive links between enhanced Raven skills and enhanced cognitively demanding job-performance skills. However, IQ gains as an isolated phenomenon are a dead end. So long as one is locked in the test room, the failure of explanations like test sophistication and the Brand hypothesis will leave one baffled. All my instincts tell me that a better understanding of the test skills plus discovery of associated real-world skills will produce the package of effects needed to identify probable causes.

CONCLUSION

Massive IQ gains began in the late 19th century, possibly as early as the industrial revolution, and have affected 20 nations, all for whom data exist. No doubt, different nations have enjoyed different rates of gain, but the best data do not provide an estimate of the differences. Different kinds of IQ tests show different rates of gain: Culture-reduced tests of fluid intelligence show gains of as much as 20 points per generation (30 years); performance tests show 10–20 points; and verbal tests sometimes show 10 points or below. Tests closest to the content of school-taught subjects, such as arithmetic reasoning, general information, and vocabulary, show modest or nil gains. More often than not, gains are similar at all IQ levels. Gains may be age specific, but this has not yet been established and they certainly persist into adulthood. The fact that gains are fully present in young children means that causal factors are present in early childhood but not necessarily that they are more potent in young children than among older children or adults.

IQ gains have not been accompanied by an escalation of the real-world cognitive skills usually associated with IQ. Going from past generations to the present, one does not see an evolution from widespread retardation to normalcy, or from normalcy to widespread giftedness, take your choice. Therefore, causal explanations were divided into first, those that would explain IQ gains as an isolated phenomenon in the test room and second, those that implied an escalation of real-world cognitive skills as well. The first kind of explanation included test sophistication and altered test-taking strategies, and these were evidentially weak. The second kind included nutrition, SES, urbanization, eradication of childhood diseases, historical trauma such as the Great Depression and World War 2, upgrading of the preschool home environment, educational TV, and education in general. It was acknowledged that these variables must have had some impact in the first half of the 20th century. However, since 1950, massive IQ gains have occurred in nations where only higher SES, urbanization, and enhanced education seemed to persist as significant factors. Since 1950, higher SES and urbanization, plus other trends with which they are confounded, probably account for only 5 or 6 points. Education appears to have been a potent factor in some countries but feeble in others.

Finding physiological correlates of IQ cannot substitute for better causal hypotheses. Comparative analysis is inhibited by the lack of well-evidenced national or age differences in rates of gain. The best way forward is through an analysis of what cognitive skills enhance performance in the test room and an attempt to identify similar real-world skills. The history of science shows many instances in which causal explanation awaits clarification of the package of effects to be explained.

REFERENCES

Bouvier, U. (1969). *Evolution des cotes a quelques tests* [Evolution of scores from several tests]. Brussels, Belgium: Belgian Armed Forces, Center for Research Into Human Traits.

Bower, B. (1987). "IQ's generation gap." *Science News, 132*, 108–109.

Brand, C. R. (1987a). Bryter still and bryter? *Nature, 328*, 110.

Brand, C. R. (1987b). Keeping up with the times. *Nature, 328*, 761.

Carpenter, P. A., Just, M. A., & Shell, P. (1990). What one intelligence test measures: A theoretical account of the processing in the Raven Progressive Matrices Test. *Psychological Review, 97*, 404–431.

Children working have higher IQs, study shows. (1995, June 14). *Otago Daily Times*, p. 34.

Clarke, S. C. T., Nyberg, V., & Worth, W. H. (1978). *Technical report on Edmonton grade III achievement: 1956–1977 comparisons.* Edmonton, Canada: University of Alberta.

Cole, M., & Means, B. (1981). *Comparative studies of how people think: An introduction.* Cambridge, MA: Harvard University Press.

de Leeuw, J., & Meester, A. C. (1984). Over het intelligente—Onderzoek bij de militarie keuringen vanaf 1925 tot heden [Intelligence—As tested at selections for the military service from 1925 to the present]. *Mens en Maatschappii, 59,* 5–26.

Elley, W. B. (1969). Changes in mental ability in New Zealand schoolchildren. *New Zealand Journal of Educational Studies, 4,* 140–155.

Emanuelsson, I., Reuterberg, S-E., & Svensson, A. (1993). Changing differences in intelligence? Comparisons between groups of thirteen-year-olds tested from 1960 to 1990. *Scandinavian Journal of Educational Research, 3,* 259–277.

Emanuelsson, I., & Svensson, A. (1990). Changes in intelligence over a quarter of a century. *Scandinavian Journal of Educational Research, 34,* 171–187.

Flieller, A., Jautz, M., & Kop, J.-L. (1989). Les réponses au test Mosaïque à quarante ans d'intervalle [Answers on the Mosaique test after an interval of forty years]. *Enfance, 42,* 7–22.

Flieller, A., Saintigny N., & Schaeffer, R. (1986). L'evolution du niveau intellectuel des enfants de 8 ans sur une periode de 40 ans, 1944–1984 [The evolution of the intellectual level of 8-year-old children over a period of 40 years, 1944–1984] *L'orientation scolaire et professionnelle, 15,* 61–83.

Floud, R., Wachter, K., & Gregory, A. (1990). *Height, health and history.* Cambridge, England: Cambridge University Press.

Flynn, J. R. (1984a). IQ gains and the Binet decrements. *Journal of Educational Measurement, 21,* 283–290.

Flynn, J. R. (1984b). The mean IQ of Americans: Massive gains 1932 to 1978. *Psychological Bulletin, 95,* 29–51.

Flynn, J. R. (1985). Wechsler intelligence tests: Do we really have a criterion of mental retardation? *American Journal of Mental Deficiency, 90,* 236–244.

Flynn, J. R. (1987a). Massive IQ gains in 14 nations: What IQ tests really measure. *Psychological Bulletin, 101,* 171–191.

Flynn, J. R. (1987b). The ontology of intelligence. In J. Forge (Ed.), *Measurement, realism, and objectivity* (pp. 1–40). Dordrecht, The Netherlands: Reidel.

Flynn, J. R. (1988). Japanese intelligence simply fades away: A rejoinder to Lynn. *The Psychologist, 9,* 348–350.

Flynn, J. R. (1990). Massive IQ gains on the Scottish WISC: Evidence against Brand et al.'s hypothesis. *Irish Journal of Psychology, 11,* 41–51.

Flynn, J. R. (1991). *Asian Americans: Achievement beyond IQ.* Hillsdale, NJ: Erlbaum.

Flynn, J. R. (1992). Cultural distance and the limitations of IQ. In J. Lynch, C. Modgil, & S. Modgil (Eds.), *Education for cultural diversity: Convergence and divergence* (pp. 343–360). London: Falmer Press.

Flynn, J. R. (1993). Skodak and Skeels: The inflated mother-child IQ gap. *Intelligence, 17,* 557–561.

Flynn, J. R. (1994). IQ gains over time. In R. J. Sternberg, S. J. Ceci, J. Horn, E. Hunt, J. P. Matarazzo, & S. Scarr (Eds.), *Encyclopedia of human intelligence* (pp. 617–623). New York: Macmillan.

Flynn, J. R. (in press). The schools: IQ tests, labels, and the word "intelligence." In W. Tomic & J. Kingma (Eds.), *Conceptual issues in research on intelligence.* Greenwich, CT: JAI Press.

Foulds, G. A., & Raven, J. C. (1948). Normal changes in the mental abilities of adults as age advances. *Journal of Mental Science, 94,* 133–142.

Herrnstein, R. J., & Murray, C. (1994). *The bell curve: Intelligence and class in American life.* New York: Free Press.

Horgan, J. (1995). Get smart, take a test. *Scientific American, 273*(5), 12–14.

Horn, J. L. (1989). Cognitive diversity: A framework for learning. In P. L. Ackerman, R. J. Sternberg, & R. Glaser (Eds.), *Learning and individual differences: Advances in theory and research* (pp. 61–114). New York: Freeman.

Jensen, A. R. (1980). *Bias in mental testing.* London: Methuen.

Jensen, A. R. (1981). *Straight talk about mental tests.* New York: Free Press.

Jensen, A. R. (1987). Differential psychology: Towards consensus. In S. Modgil & C. Modgil (Eds.), *Arthur Jensen: Consensus and controversy* (pp. 353–399). Lewes, Sussex, England: Falmer Press.

Jensen, A. R. (1988). *The Oddman-Out Test: A chronometric anchor for psychometric tests of* g. Unpublished manuscript.

Jensen, A. R. (1989). Raising IQ without increasing *g*? A review of *The Milwaukee Project: Preventing mental retardation in children at risk. Developmental Review, 9,* 234–258.

Jensen, A. R. (1991). Speed of cognitive processes: A chronometric anchor for psychometric tests of *g*. *Psychological Test Bulletin, 4,* 59–70.

Lynn, R. (1987). Japan: Land of the rising IQ. A reply to Flynn. *Bulletin of the British Psychological Society, 40,* 464–468.

Lynn, R. (1989). Positive correlation between height, head size and IQ: A nutrition theory of the secular increases in intelligence. *British Journal of Educational Psychology, 59,* 372–377.

Lynn, R. (1990). Differential rates of secular increase of five major primary abilities. *Social Biology, 38,* 137–141.

Lynn, R. (1996). *Dysgenics: Genetic deterioration in modern populations.* New York: Praeger.

Raven, J. (1981). *Manual for Raven's Progressive Matrices and Mill Hill Vocabulary Scales* (Research Supplement No. 1). London: H. K. Lewis.

Raven, J. (1995). Methodological problems with the 1992 standardization of the SPM: A response. *Personality and Individual Differences, 18,* 443–445.

Raven, J., & Court, J. H. (1989). *Manual for Raven's Progressive Matrices and Vocabulary Scales* (Research Supplement No. 4). London: H. K. Lewis.

Raven, J., Raven, J. C., & Court, J. H. (1993). *Manual for Raven's Progressive Matrices and Vocabulary Scales* (Section 1). Oxford, England: Oxford Psychologists Press.

Raven, J., Raven, J. C., & Court, J. H. (1994). *Manual for Raven's Progressive Matrices and Vocabulary Scales* (Section 5A). Oxford, England: Oxford Psychologists Press.

Raven, J., Raven, J. C., & Court, J. H. (1995). *Manual for Raven's Progressive Matrices and Vocabulary Scales* (Section J, General Overview). Oxford, England: Oxford Psychologists Press.

Raven, J. C. (1941). Standardization of progressive matrices. *British Journal of Medical Psychology, 19,* 137–150.

Rist, T. (1982). *Det intellektuelle prestasjonsnivaet I befolkningen sett I lys av den samfunns-messige utviklinga* [The level of the intellectual performance of the population seen in the light of developments in the community]. Oslo, Norway: Norwegian Armed Forces Psychology Service.

Schallberger, U. (1987). HAWIK and HAWIK-R: Ein empirischer Vergleich [HAWIK and HAWIK-R: An empirical comparison]. *Diagnostica, 33,* 1–13.

Schoenthaler, S. J., Amos, S. P., Eysenck, H. J., Peritz, E., & Yudkin, J. (1991). Controlled trial of vitamin-mineral supplementation: Effects on intelligence and performance. *Personality and Individual Differences, 12,* 351–362.

Schubert, M. T., & Berlach, G. (1982). Neue Richtlinien zur interpretation des Hamburg Wechsler-Intelligenztests für Kinder (HAWIK) [New guidelines for the interpretation of the Hamburg Wechsler Intelligence Tests for Children (HAWIK)]. *Zeitschrift für Klinische Psychologie 11,* 253–279.

Scribner, S., & Cole, M. (1981). *The psychology of literacy.* Cambridge, MA: Harvard University Press.

Sharp, D., Cole, M., & Lave, C. (1979). Education and cognitive development: The evidence from experimental research. *Monographs of the Society for Research in Child Development, 44,* 1–92.

Spitz, H. H. (1986). *The raising of intelligence.* Hillsdale, NJ: Erlbaum.

Storfer, M. D. (1990). *Intelligence and giftedness: The contributions of heredity and early environment.* San Francisco: Jossey-Bass.

Teasdale, T. W., & Owen, D. R. (1989). Continued secular increases in intelligence and a stable prevalence of high intelligence levels. *Intelligence, 13,* 255–262.

Terman, L. M., & Merrill, M. A. (1937). *Measuring intelligence.* London: George G. Harrap.

Thorndike, R. L. (1977). Causation of Binet IQ decrements. *Journal of Educational Measurement, 14,* 197–202.

Vernon, P. E. (1982). *The abilities and achievements of Orientals in North America.* New York: Academic Press.

Vincent, K. R. (1993). On the perfectability of the human species: Evidence using fixed reference groups. *Texas Counseling Association Journal, 22,* 60–64.

Wechsler, D. (1992). *WISC-III: Wechsler Intelligence Scale for Children,* (3rd ed., Australian adaptation). New York: The Psychological Corporation and Harcourt Brace Jovanovich.

Wechsler, D. (1997). *WAIS-III: Wechsler Intelligence Scale for Children* (3rd ed.). New York: The Psychological Corporation and Harcourt Brace Jovanovich.

Wigdor, A., & Green, B. F. (Eds.). (1991). *Performance assessment in the workplace* (Vol. 1). Washington, DC: National Academy Press.

Yerkes, R. M. (1921). Psychological examining in the United States Army. In *Memoirs of the National Academy of Sciences* (Vol. 15). Washington, DC: U.S. Government Printing Office.

3

Environmental Complexity and the Flynn Effect

Carmi Schooler

I have two major disagreements with James Flynn's chapter (chapter 2, this volume), neither of which has to do with his empirical conclusions. Instead, my qualms center on his unwillingness to accept the implications of what he has found. First, I question Flynn's reasons for doubting that there actually has been a major increase in the general level of intellectual functioning in industrialized societies. Second, I believe that he neglects several bodies of relevant research, much of which appeared in the sociological literature. This research not only gives credence to the view that such an increase in intellectual functioning has occurred, but provides a glimpse of the mechanisms involved producing it.

Common to both of my disagreements with Flynn is my belief that he downplays the possible importance of changes in environmental complexity in explaining his results. There is a substantial amount of evidence and considerable theoretical rationale indicating that increases in environmental complexity increase intellectual functioning and that the complexity of the environment has generally been increasing since the start of the industrial revolution.

PLATO'S LATER DIALOGUES, THE PERPLEXED BASEBALL FAN, AND THE MISSING INVENTIONS

Three examples illustrate the rhetorical arguments that Flynn uses to cast doubt on whether there has been any basic change in the level of intellectual functioning. The first is Plato's later dialogues. Flynn uses these dialogues to raise questions about the nature of intelligence. In chapter 2, this volume, he argues that

> I have developed my mind by focusing on something far more complex than the technologies of the world, namely, Plato's later dialogues. Those who have done so cope with something that the philosophically naive would find daunting without decades of study, no matter what generation they belong to. Does that mean I am more intelligent that a nonphilosopher with the potential to understand Plato or that I am more learned? (pp. 30–31)

Several basic questions about the nature of intelligence are raised by this allusion to Plato's later dialogues. One is whether the concept *intelligence* refers to the individual's potential or present level of intellectual functioning. My strong preference is to limit the use of the term to the presently observable. Although an individual's present level of functioning may be a reasonably accurate predictor of his or her future level, a good argument can be made against defining intelligence in terms of some presently unmeasurable potential for future performance. As seen later, there is substantial evidence that the environmental conditions to which one is exposed continue to affect one's level of intellectual functioning throughout the life span. A person's future intellectual functioning will definitely be affected by the nature of the environments to which he or she will be exposed.

A related question raised by Flynn's allusion to Plato's later dialogues is whether something that is learned counts as mere "knowledge" as opposed to intelligence. This question boils down to whether there are types of learned knowledge that can affect the individual's present level of intellectual functioning. I believe that the answer is *yes*. For example, if one has not learned about probability theory, which was unknown in Plato's day, one cannot use it to solve gaming problems. As Hacking (1975) suggested, anybody with a 17th-century knowledge of probability theory would likely have become very rich in classical Greece.

At a less dramatic level, a wide range of cognitive studies have shown that not only the content but the nature of thinking processes can be affected by experiences such as training in and practice with various cognitive strategies. The likelihood of the occurrence of such practice and training is a function of both the society in which the individual lives and the individual's particular place in that society. This is true even for types of thinking apparently closely linked to fluid intelligence, which supposedly entails the "abstract" capacity for problem solving unaffected by practical experience. Spatial thinking (e.g., Regian, Shute, & Pellegrino, 1985), mathematical thinking (e.g., Schoenfeld & Herrmann, 1982), and logical thinking (e.g., Cheng, Holyoak, Nisbett, & Oliver, 1986) have now all been demonstrated to be affected by training or practice. (For an extensive review of how learning and experience affect the nature of thinking processes, see Schooler, 1989.)

Flynn clearly believes that the answer to the question of whether learning affects intelligence is *no*. He asserts in chapter 2, this volume, that "no matter what theory of intelligence one holds, one distinguishes among learning, memory, and intelligence" (p. 29). Flynn carefully avoids formally defining intelligence. Nevertheless, he seems to imply that intelligence is an inborn, or at least mainly biologically determined, potential to understand daily life and creatively solve complex problems.

In his rhetorical use of Plato's later dialogues, as in many places in his chapter, Flynn's removal by fiat from the definition of intelligence of any possible effect of learning, and consequently of experience, leads him to doubt the meaning of his findings. If, however, intelligence is defined in terms of the level of complexity of the problems that the individual is currently able to solve, the question of the biological or socioenvironmental source of that ability can be freely pursued. Furthermore, this can be done without having to make the dubious assumption that tests such as the Raven matrices or nonverbal parts of the Wechsler Adult Intelligence Scale (WAIS) represent accurate assessments, unaffected by the socioenvironmental circumstances under which they have been made, of some basic intellectual ability that is essentially unaffected by experience.

Even within present day U.S. society, it is possible that the socioenvironmental circumstances under which complex tasks are performed are so different from those envisaged by the developers of intelligence tests that there is no relationship between complex problem solving of a kind that would probably match even Flynn's views of intelligence and performance on a standard test. For example, Ceci and Liker (1986), examining racetrack betting, found that "expert handicapping was a cognitively sophisticated enterprise, with experts using a mental model that contained multiple interaction effects and nonlinearity" (p. 255). The IQs of expert and nonexpert handicappers as measured by the WAIS were found to be equivalent. Ceci and Liker concluded "that (a) IQ is unrelated to skilled performance at the racetrack and (b) IQ is unrelated to real-world forms of cognitive complexity that would appear to conform to some of those that scientists regard as hallmarks of intelligent behavior" (p. 255).

In terms of the way intelligence is formally assessed, the most important kind of learning may be the understanding of the nature of hypothetical problems. Such understanding involves both the knowledge of how to approach "nonreal" problems that are framed in terms of concepts and situations far removed from one's daily concerns and the belief that such problems should be taken seriously. Research reviews by Greenfield (chapter 4, this volume) and Ceci (1990) have shown that this knowledge and such beliefs have a profound effect on intelligence test performance, are affected by culturally determined experiences, and are especially readily learned in school. Furthermore, as I have mentioned, many experimental cognitive psychology findings strongly suggest that individuals' ways of thinking about relatively abstract problems can be affected by their society and their place in it. Thus, abstract tests that are supposedly made "culture free" by being purged of specific knowledge are in reality highly culture bound.

Equally relevant are the findings of Claude Steele and colleagues regarding the effects of such culturally induced factors as stereotype threat on the cognitive test performance of individuals from stigmatized social groups (cf. Steele & Aronson, 1995).

In chapter 2, this volume, Flynn uses his second rhetorical case, Jensen's (1981) description of the befuddlement of a baseball fan with a measured IQ of 75, for two purposes. The first is again to cast doubt on the meaningfulness of the steep rise in intelligence that his analyses have disclosed. He finds it unimaginable that the average person 100 years ago, whose IQ would measure about 75 by present standards, would have been as unable as Jensen's present-day fan with a 75 IQ to cope intellectually with such a popular game. It seems likely, however, that the paths leading to these apparently low intellectual levels would have been different than they are now. Given the rising levels of environmental factors linked to better intellectual performance, such as complexity, schooling, and nutrition, someone with a 75 IQ now is more likely to be biologically than environmentally handicapped than was someone with the "equivalent" IQ of 100 a century ago. The latter would most likely have had quite different capabilities than the former. Such a person may well have had a form of intelligence that permitted him or her to understand baseball rules better than someone whose IQ measures 75 now.

The possibility that an IQ of 75 implies different types of intellectual functioning depending on the decade in which it was measured raises the more general problem of whether there are different forms of intelligence. Flynn alludes to this problem with his second use of the befuddled baseball fan example when he distinguishes between what he sees as two basic types of intelligence: (a) "understanding-baseball intelligence—the capacity to absorb the usual rules of social behavior" (p. 59) and (b) "giftedness—the capacity to learn more quickly and make creative leaps" (p. 59). There have, of course, been more definitions of more types of intelligence and more theoretical and empirical explorations of how they might be interrelated than could possibly be reviewed in a chapter 10 times the assigned length of this one. It is far from obvious, however, that Flynn's distinction represents the best way to cut the pie.

Flynn uses his second definition of intelligence to argue further against the meaningfulness of his findings. In doing so he disregards some important theoretical insights and empirical findings from sociology. He suggests that present-day societies do not exhibit the high levels of inventiveness and creativity that he believes should result from the supposedly high levels of intellectual functioning of their individual members. He uses the third of the rhetorical examples cited above—missing inventions—as evidence that the actual level of intellectual functioning in The Netherlands could not have risen as dramatically between 1952 and 1982 as his analyses indicate that Dutch IQ norms have risen. Flynn cites as evidence that the number of inventions patented had actually shown a sharp decline during that period. His assumption that changes in individual-level phenomena are necessarily reflected in changes in social-level phenomena is not one that a sociologist could comfortably make.

SOME LESSONS FROM SOCIOLOGY

A basic lesson of sociology is that what happens within a social system is not completely determined by the individual characteristics of its members. The nature of the system's social structure, particularly its institutions, affects not only the psychological characteristics of its members, but also how these characteristics are expressed behaviorally. (For a full theoretical discussion of these issues, see Schooler, 1994; for a review of relevant cross-cultural studies, see Schooler, 1996.) In the case of inventions in The Netherlands, many socioeconomic, legal, and institutional factors might have affected the number of patents granted between 1952 and 1982. Unless one knows that all of these factors remained constant, the change in the number of patents is not a particularly good index of the intellectual level of individual members of Dutch society. It is possible, for example, that changes in the laws governing what is patentable (conceivably passed to deal with a flood of patent applications) resulted in such a decrease.

On the other hand, just as there is not necessarily a perfect correspondence between the levels of technical and cultural productivity of a society and the average level of intellectual functioning of its members, there is no necessary reason why the general level of intellectual functioning of individuals in a society has to meet the cognitive demands their society places on them. Thus, although as seen shortly, socioenvironmental changes coinciding with the advent of industrialization most probably increased general levels of intellectual functioning, it is not necessarily the case that the level of this increase has to match the level of the increased intellectual demands society imposes on the individual.

The relevance of the industrial revolution—another potential lesson from sociology—is suggested by Flynn's conclusion that "the . . . advent of industrialization and the beginning of IQ gains . . . may well coincide" (p. 26). The pioneer quantitative, sociological study examining the effects of industrialization and modernization was by Alex Inkeles and colleagues. The research was conducted in six industrializing nations: Argentina, Bangladesh, Chile, India, Israel, and Nigeria; almost 6,000 men served as survey respondents. Most directly relevant to the present discussion is that in each country studied, exposure to social-structural conditions associated with industrialization was empirically correlated with "being relatively open-minded and cognitively flexible" (Inkeles & Smith, 1974, p. 291). Although Inkeles and his associates described several mechanisms through which such changes may take place (i.e., reward and punishment, modeling, exemplification, and generalization), they did not isolate which aspects of the modernization experience have these effects.

One such mechanism is suggested by the research program on the psychological effects of occupational conditions that Melvin Kohn and I developed. In terms of intellectual functioning, its central finding, based on structural equation modeling of longitudinal data from a representative sample of American men, is that job conditions providing intellectual challenge and an opportunity for doing self-directed substantively complex work act to increase intellectual flexibility. By the same token, work conditions that limit intellectual challenge and self-direction on the job decrease intellectual flexibility (Kohn & Schooler, 1983). These causal connections between work conditions and intellectual functioning have been confirmed in women in a variety of cross-cultural settings (e.g., Japan, Poland, the Ukraine), in other types of work (i.e., schoolwork, women's housework), and at different stages of the life span (i.e., students, younger workers, and older workers). (For fuller reviews of these extensions of the Kohn–Schooler findings, see Schooler, 1990b, 1996.)

There is also long-standing evidence in the sociological literature that exposure to a complex environment during childhood has effects on cognitive functioning similar to those of later exposure to complex occupational conditions (Schooler, 1972). In my 1972 article, I linked, on theoretical and empirical grounds, complexity of childhood environment to being young; having a well-educated father; and being brought up in an urban setting, in a liberal religion, and in a region of the United States far from the South. Each of these background characteristics had an independent effect in raising the level of adult intellectual performance.

The Kohn–Schooler occupational studies and their extensions fit together with a large body of research from a wide range of disciplines, including animal-based neurobiology (e.g., Bennett, Rosenzweig, Morimoto, & Herbert, 1979; Kempermann, Kuhn, & Gage, 1997), indicating that exposure to complex environments increases intellectual functioning (Schooler, 1984, 1990b). I have also presented a rough-hewn theory of this effect (Schooler, 1984). According to this theory, the complexity of an individual's environment is defined by its stimulus and demand characteristics. The more diverse the stimuli, the greater the number of decisions required, the greater the number of considerations to be taken into account in making these decisions, and the more ill-defined and apparently contradictory the contingencies, the more complex the environment. To the degree that such an environment rewards cognitive effort, individuals should be motivated to develop their intellectual capacities and to generalize the resulting cognitive processes to other situations.

If technical and economic development lead to more complex environments, such increased environmental complexity should result in higher levels of intellectual functioning (as well as increased individualism; for empirical reviews and theoretical discussions of this relationship, see Schooler, 1989, 1996). Such an increase in environmental complexity almost certainly occurred in the moves from rural to urban settings and from premodern agricultural to commercial and manufacturing occupations. There is also evidence that the substantive complexity of work continued to increase at later levels of development (Attwell, 1987; Form, 1987; Spenner, 1988; but see Head, 1996, who suggested that the world may now be at the beginning of a downward trend in such complexity). Furthermore, in every nation in which the issue has been empirically examined, social-structurally determined differences in exposure to complex environments affect the degree to which various segments of society exhibit intellectual flexibility (Kohn, Naoi, Schoenbach, Schooler, & Slomzynski, 1990; Schooler, 1996). Given the increasing (at least until now) complexity of life in the industrial world and the apparent causal link between levels of environmental complexity and intellectual functioning, it seems highly plausible that the ongoing gains in intelligence test performance do represent a real increase in intellectual functioning, one caused in some large part by the way the world has become more complex.

Flynn does consider, but rejects, the possibility voiced by Ken Vincent, "that the complexity of the modern world . . . causes . . . massive intelligence gains" (p. 30). He does so as an introduction to the rhetorical arguments he gives, which are based on Plato's later dialogues, against any meaningful modern increase in intelligence. He further argues that even if "the modern world . . . has also altered our minds in a way that confers an ability to cope with [modern] contrivances" (p. 31), because this new ability could arbitrarily be labeled either "'crystallized intelligence' or . . . 'achievement'[,] . . . morality should guide the choice. Few would want to label Australian Aborigines 'dumb' simply because they have been conditioned to cope with the Australian outback rather than the modern world" (p. 31). Leaving aside the fact that complex environmental demands are not limited to the modern world (for a discussion of the possible effects of environmental complexity in different types of hunting–gathering societies, see Schooler, 1990a), I find both Flynn's Platonic and his Australian Aboriginal arguments neither necessary nor unconvincing.

The effects on intellectual functioning of complex environmental conditions such as self-directed and intellectually demanding work are mediated and sustained not only by social structure but also by culture. An example of such a long-term effect of culture on intellectual functioning in the United States is found in the evidence that people from ethnic groups with a recent history of serfdom exhibit not only the non-self-directed orientations and values, but also the relatively low levels of intellectual functioning characteristic of those who work under conditions limiting the individual's opportunity for self-direction (Schooler, 1976). This finding holds true even when a wide range of potentially confounding variables are controlled (i.e., age, father's education, rurality, religion, region of the country, occupational self-direction). Every link in the causal chain cannot be confirmed. Nevertheless, a model emphasizing the lagged effects on an ethnic group's culture of historical conditions that restricted the individual's autonomy and deemphasized the importance and effectiveness of intellectual concerns seems a probable and parsimonious explanation of these ethnic differences.

CONCLUSION

Cultures and social structures clearly affect the cognitive functioning of the individuals who are part of them. They similarly affect the likelihood that potential products of cognitive functioning such as inventions, philosophies, or works of art will be produced and remain visible for others to see. Cultures and social structures also both affect and are affected by the environmental conditions faced by those in them. At least one of these conditions, environmental complexity, has a demonstrated effect on both the content of a culture and the intellectual functioning of its participants. Individuals' positions in the social structure also both affect and are affected by their intellectual functioning. These potential effects of culture, social structure, and environmental complexity are often overlooked by those who focus only on the biological aspects of intellectual functioning or limit their studies to particular samples of subjects (e.g., college sophomores; see Schooler, 1989; Sears, 1987). A notable part of what seems to be biologically based intellectual stability may be a reflection of lack of environmental change. What seem to be universal laws of cognitive functioning may simply reflect the mode of operation of individuals in a particular social status of a particular culture.

Flynn, of course, can hardly be accused of narrowing his focus to the biological or of limiting his samples in terms of social status and culture. In these ways, his work provides an exemplary model for others to follow. Furthermore, given the scope of the changes in test norms that he has documented, Flynn has every reason to be concerned that his findings are somehow artifactual. He is right to be skeptical of simplistic, single-cause explanations. Nevertheless, the changes in culture, social structure, and environmental complexity that have taken place in the last century are at least as rapid and as massive as the changes in intelligence test norms that Flynn has found. Flynn, I believe, errs in downplaying the role of these socioenvironmental changes in explaining the effect he discovered.

REFERENCES

Attwell, P. (1987) The deskilling controversy. *Work and Occupations, 14,* 323–346.

Bennett, E. L., Rosenzweig, M. R., Morimoto, H., & Herbert, M. (1979). Maze training alters brain weights and cortical RDA/DNA ratios. *Behavioral Neural Biology, 26,* 1–22.

Ceci, S. J. (1990). *On intelligence . . . more or less: A bio-ecological treatise on intellectual development.* Englewood Cliffs, NJ: Prentice-Hall.

Ceci, S. J., & Liker, J. K. (1986). A day at the races: A study of IQ, expertise and cognitive complexity. *Journal of Experimental Psychology: General, 115,* 255–266.

Cheng, P., Holyoak, K., Nisbett, R., & Oliver, L. (1986). Pragmatic versus syntactic approaches to training deductive reasoning. *Cognitive Psychology, 18,* 293–328.

Form, W. (1987). On the degradation of job skills. *Annual Review of Sociology, 13,* 29–47.

Hacking, I. (1975). *The emergence of probability: A philosophical study of early ideas about probability, induction and statistical inference.* London: Cambridge University Press.

Head, S. (1996). The new, ruthless economy. *The New York Review of Books, XLIII (4),* 47–52.

Inkeles, A., & Smith, D. H. (1974). *Becoming modern: Individual changes in six developing countries.* Cambridge, MA: Harvard University Press.

Jensen, A. R. (1981). *Straight talk about mental tests.* New York: Free Press.

Kempermann, G., Kuhn, H. G., & Gage, F. H. (1997). More hippocampal neurons in adult mice living in an enriched environment. *Nature, 386,* 493–495.

Kohn, M. L., Naoi, A., Schoenbach, C., Schooler, C., & Slomzynski, K. M. (1990). Position in the class structure and psychological functioning: A comparative analysis of the United States, Japan, and Poland. *American Journal of Sociology, 95,* 964–1008.

Kohn, M. L., & Schooler, C. (1983). *Work and personality: An inquiry into the impact of social stratification.* Norwood, NJ: Ablex.

Regian, J., Shute, V., & Pellegrino, J. (1985). *The modifiability of spatial processing skills.* Paper presented at the 26th Psychonomic Society Conference, Boston.

Schoenfeld, A., & Herrmann, D. (1982). Problem perception and knowledge structure in expert and novice mathematical problem solvers. *Journal of Experimental Psychology, 8,* 484–494.

Schooler, C. (1972). Social antecedents of adult psychological functioning. *American Journal of Sociology, 78,* 299–322.

Schooler, C. (1976). Serfdom's legacy: An ethnic continuum. *American Journal of Sociology, 81,* 1265–1286.

Schooler, C. (1984). Psychological effects of complex environments during the life span: A review and theory. *Intelligence, 8,* 259–281.

Schooler, C. (1989). Social structural effects and experimental situations: Mutual lessons of cognitive and social science. In K. W. Schaie & C. Schooler (Eds.), *Social structure and aging: Psychological processes* (pp. 129–147). Hillsdale, NJ: Erlbaum.

Schooler, C. (1990a). Individualism and the historical and social-structural determinants of people's concern over self-directedness and efficacy. In J. Rodin, C. Schooler, & K. W. Schaie (Eds.), *Self-directedness and efficacy: Causes and effects throughout the life course* (pp. 19–44). Hillsdale, NJ: Erlbaum.

Schooler, C. (1990b). Psychosocial factors and effective cognitive functioning through the life span. In J. E. Birren & K. W. Schaie (Eds.), *Handbook of the psychology of aging* (pp. 347–358). Orlando, FL: Academic Press.

Schooler, C. (1994). A working conceptualization of social structure: Mertonian roots and psychological and sociocultural relationships. *Social Psychology Quarterly, 57,* 262–273.

Schooler, C. (1996). Cultural and social structural explanations of cross-national psychological differences. *Annual Review of Sociology, 22,* 323–349.

Sears, D. O. (1987). Implications of the life-span approach for research on attitudes of social cognition. In R. P. Abeles (Ed.), *Life-span perspectives and social psychology.* Hillsdale, NJ: Erlbaum.

Spenner, K. (1988). Occupations, work settings and the course of adult development: Tracing the implications of select historical changes. In P. Baltes, D. Featherman, & R. Lerner (Eds.), *Life-span: Development and behavior* (Vol. 9, pp. 243–286). Hillsdale, NJ: Erlbaum.

Steele, C. M., & Aronson, J. (1995). Stereotype threat and the intellectual test performance of African Americans. *Journal of Personality and Social Psychology, 69,* 797–811.

4

The Cultural Evolution of IQ

Patricia M. Greenfield

"Successful adaptation to its own niche marks an animal form as intelligent" (Scheibel, 1996). This definition leads to the following question: What constitutes specieswide adaptation to the human niche? Three features that reach their apex of complexity in the human species stand out: (a) technology, (b) linguistic communication, and (c) social organization. The ability to acquire competence in each one through development, learning, and socialization is my definition of *panhuman genotypic intelligence*: All normal members of the human species have that ability. However, particular languages, communication conventions, technologies, forms of social organization, and social norms vary from culture to culture. Therefore, *phenotypic intelligence varies from culture to culture*. Figure 1 presents this model of intelligence. The "bottom line" of the model is that although the range of within-culture genotypic variation of intelligence may be the same from culture to culture

I thank Ioakim Boutakidis and Helen Davis for their help with the library research for this chapter. Special thanks to Abraham Seidman for drawing the illustration of Tetris used in Figure 3.

Partial support for preparation of this chapter was provided by a grant from the Fogarty International Center of the National Institutes of Health to UCLA (Steven Lopez, Principal Investigator) and by the Colegio de la Frontera Sur (ECOSUR), San Cristobal de las Casas, Chiapas, Mexico.

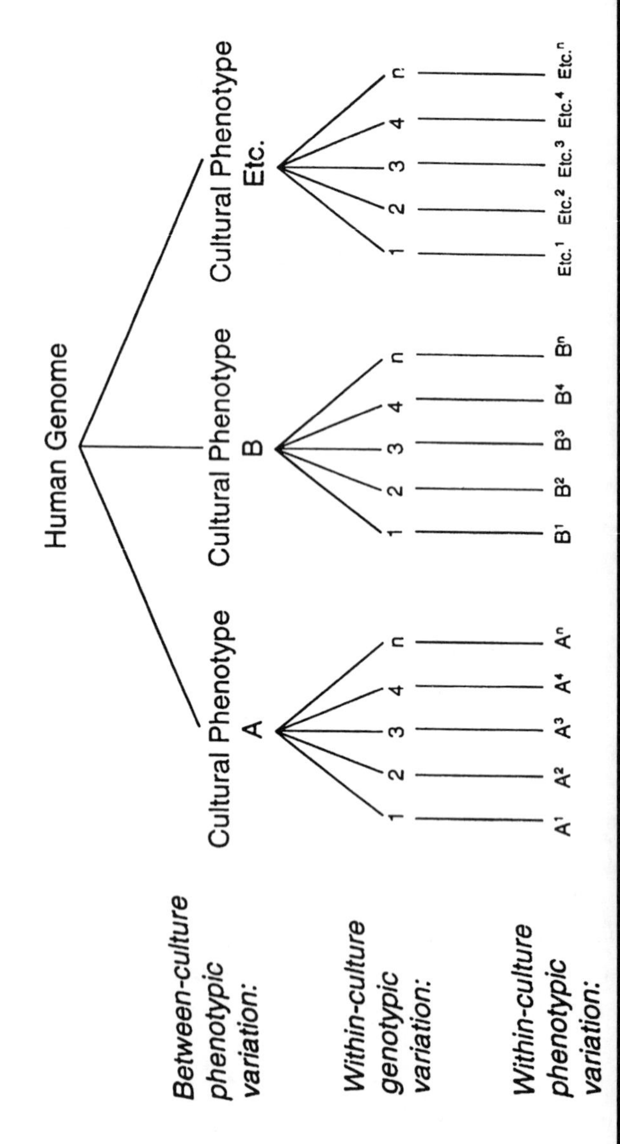

Figure 1

Model of the variability in human intelligence: Within- and across-culture variation in the paths of intelligence.

(1 through *n*, repeated three times in the diagram), a given genotype will be expressed differently (as a different phenotype) in a different cultural milieu (e.g., A^1 vs. B^1 in the diagram).

LANGUAGE, INTELLIGENCE, AND ADAPTATION

Because they possess the panhuman genotype, normal infants are born with the ability to learn any language. However, in most cases, they actually learn one (e.g., Cultural Phenotype A or B, see Figure 1). The universal, innate language-learning capacity is applied in a specific linguistic environment, and the resultant language learned is therefore adapted to this linguistic environment. If one tries to test the "innate linguistic ability" of a child exposed to Language A using stimuli from Language B, the child will look stupid, as though lacking in "innate ability." Of course, this is obvious, and it is something that no one would dream of doing, yet it is done with "intelligence" tests. The cartoon in Figure 2 parodies this practice.

This cartoon makes the important point that each phenotypic form of intelligence is adapted to a particular ecocultural niche. With respect to the top line of the model presented in Figure 1, Cultural Phenotype A would be adapted to Ecocultural Niche A, Cultural Phenotype B would be adapted to Ecocultural Niche B, and so forth. Two important points then follow:

1. Cultures define intelligence by what is adaptive in their particular ecocultural niche.
2. Definitions of intelligence are as much cultural ideals as scientific statements.

The logical necessity of cross-cultural differences in definitions of intelligence follows from these two points, and indeed such differences do exist. For example, in Africa in the 1970s, it was found that there was an emphasis on social (rather than technological) intelligence (Mundy-Castle, 1974), on behavior that leads to respect for and compliance with society's ways (Wober, 1975), and on slowness, that is, deliberation (Wober, 1974). These definitions contrast directly with pre-

Figure 2

The teepee test: An example of cultural bias in testing. The Native American boy on the left has scored 100 on his teepee test, whereas the European American boy on the right has flunked. From *Psychological Tests and Social Work Practice* (p. 39), by M. L. Arkava and M. Snow, 1978, Springfield, IL: Charles C Thomas, Copyright 1978 by Charles C Thomas. Reprinted with permission.

suppositions about intelligence in the Western world: that it involves understanding of the physical world more than the social world, that it involves "being able to think for yourself" rather than compliance, and that it involves speed rather than slowness.

Traditional African definitions of intelligence are adapted to the traditional ecocultural niches of small, face-to-face groups and kin-oriented societies. The Western definition of intelligence is adapted to a different set of ecocultural niches. These niches feature urbanization, high technology, and schooling as the major modes for socializing the young for adulthood. Because schooling is the main method of socialization for adulthood in this type of society, it is not surprising that IQ tests have focused on forms of intelligence that are cultivated in school, what Neisser (1976) calls "academic intelligence." Indeed, Binet, the originator of intelligence tests, explicitly developed his test to identify children who needed special education in school.

Serpell (1993) and Dasen (1984) have shown that when schooling (the origin of which is a colonial European influence) is introduced to an African community, a school-influenced definition of academic intelligence also develops. African school attendees combine the traditional and academic definitions of intelligence. Nonetheless, these are contrasting definitions with contrasting cultural roots.

In essence American intelligence tests measure what Americans think intelligence is, not what Africans think it is. (Remember, however, that definitions are not static but change as a function of historical time as well as place.) In the words of Cole and Cole, "intelligence cannot be tested independently of the culture that gives rise to the test" (1993, p. 502). The point here concerns the cultural relativity of intelligence tests. Contrary to traditional psychometric theory, intelligence tests are not universally applicable. Instead, they are cultural genres, relevant to a particular definition of intelligence. This is the conceptual context I use to explain the worldwide rise in IQ that has been termed the *Flynn effect.*

THE FLYNN EFFECT

James Flynn has focused on the socially and theoretically important phenomenon of historical IQ rise in many countries over the period of the 20th century and even before (Flynn, 1984, 1987, 1994). At the same time, he has called attention to the seeming paradox of a decline in scores on the verbal Scholastic Aptitude Test (SAT) in the United States during the same period in which IQ has been rising. Finally, Flynn has observed that IQ has risen fastest on nonverbal tests such as the Performance battery of the Wechsler Adult Intelligence Scale (WAIS) or the Raven Progressive Matrices.

In summary, the patterning of the Flynn effect is as follows: a worldwide rise in IQ scores, particularly nonverbal or performance IQ, with a concurrent drop (in the United States) in verbal SAT scores.

I argue that only an explanation that focuses on cultural history can account for the particular patterning of changes Flynn described. The strategy I use to construct this argument goes as follows:

1. Identify historical changes in the ecocultural niche that could account for these changes in test performance. (These must be widespread trends.)
2. Cite both traditional and "natural" experiments to demonstrate a causal link.
3. Develop theory and evidence regarding the mechanisms behind these causal links.

HISTORICAL CHANGES IN THE ECOCULTURAL NICHE

My thesis is that ecocultural changes in three areas—technology, urbanization, and formal education—account for a major portion of the changes in test performance. These areas are closely allied with and complementary to other areas that may also contribute to historical changes in IQ test performance: changing patterns of mother–child interaction, the ubiquitous presence of certain artifacts in the cultural environment, and improved nutrition (see chapters 5, 6, 7, and 8, this volume). From a historical and ecocultural perspective, changes in one of these factors—urbanization—has often been accompanied by changes in the other two—technology and formal education. For example, urban environments often provide higher levels of formal education for their populations and are more technologically sophisticated as well.

Coordinated Development in Technology, Urbanization, and Education

Wheeler (1942/1970) made a study of East Tennessee mountain children in 1930 and 1940, administering the same IQ tests at a 10-year interval. In 1930, approximately 1,000 children in Grades 1 through 8 in 21 mountain schools were tested with the Dearborn IA and IIC

Intelligence Tests. A subset of these children between Grades 3 and 8 were also given the Illinois Intelligence Test. In 1940, children from the same schools, "from the same areas and largely from the same families" (Wheeler, 1942/1970, p. 122), were given the same test. Two thousand additional children from 19 other mountain schools were also tested in 1940. Between 1930 and 1940, Wheeler found that IQs rose an average of 11 points across grade levels from 1 to 8.

What happened in East Tennessee that could account for the 10-point rise in IQ performance? Indeed, a series of coordinated changes in the three hypothesized areas had taken place.

1. *Technology*:
   a. *An excellent new road system.* This "developed transportation facilities for schools and industry" (Wheeler, 1970, p. 122).
   b. *Agricultural income was supplemented by industrial income.* By 1940, the rapid growth of industry in the area enabled about 60% of the families in one county and 40% in another to have one or more members working in industrial plants (Wheeler, 1942/1970).
2. *Access to urban areas.* The road system gave "every community access to progressive areas outside of the mountains" (Wheeler, 1970, p. 122). As part of this trend, new houses tended to be built on or near the new main highways.
3. *Formal education*:
   a. *A rise of 32% in average daily attendance and a rise of 17% in enrollment.* These changes were in large part due to the introduction of a system to transport thousands of children to and from school, which itself depended on the road system, an aspect of technological development.
   b. *Movement from one-room schoolhouses toward larger, better equipped schools.* This change was made possible by the road system and the school transportation system. The improved equipment included a state- and county-provided circulating library.
   c. *Improvement in teacher training.* In 1930, the average teacher

training included less than 2 years of college work. By 1940, all new teachers were required to have 4 years of college training.

d. *Hot lunches served in the larger schools.* Thus, the development of formal education was linked with improved nutrition, another potential factor in IQ rise (see chapters 6, 7, and 8, this volume).

It is notable that the rate of change in this sample was about double that of the typical Flynn effect (Flynn 1984, 1987; chapter 2, this volume). Environmental changes that occurred slowly elsewhere seemed to have occurred rapidly in East Tennessee, with a correspondingly rapid rise in IQ.

The fact that coordinated changes in technology, urbanization, and formal education were correlated with IQ rise over a 10-year period is suggestive of a causal relation but not definitive. Caution in drawing causal conclusions is warranted because we are dealing with a historical correlation rather than an experimental effect. It is possible that still other factors were the motor behind the IQ rise. Another limitation of this study is that even assuming that the three factors of technology, urbanization, and education were of causal importance, it does not reveal what role each factor played. Therefore, I turn now to more specific experiments that isolate a single factor at a time.

Technological Development

This factor was isolated in a field experiment carried out by McFie (1961) in Uganda. He "gave 26 boys entering a training school verbal tests, and a Block Designs (adapted from Koh's Blocks) test. After two years' technical training, scores on Block Designs and related tests had improved, while verbal tests had not similarly shown gains" (Wober, 1975, p. 64). This study is particularly significant because it simulates the *patterning* of the Flynn effect—selective improvement on nonverbal tests—and indexes an ecological factor—technological development—that has been going on in all of the Flynn-effect countries (e.g., France, The Netherlands) throughout the 20th century.

Urbanization

A number of studies (summarized by Cole & Cole, 1993) have indicated that urbanization has a positive effect on IQ. For example, Klineberg (1935) found that after people moved from rural areas to the city, their IQs rose. This demographic factor could be quite relevant to the Flynn effect in the United States, where increases in population density and urbanization have been documented throughout the 20th century (*Encyclopaedia Britannica*, 1938, p. 684, 1972, p. 689, 1990, p. 740, 1994, p. 169). Indeed, a concomitant increase in population density and urbanization is a worldwide phenomenon.

Formal Education

An array of studies (summarized by Wober, 1975) have provided a comparison between schooled and unschooled persons in Africa and showed that schooled persons do better on various intelligence tests, particularly nonverbal ones. One such study using the Raven Progressive Matrices was carried out in the Belgian Congo by Ombredane, Robaye, and Robaye (1957, as reported in Wober, 1975). The Raven is a nonverbal performance test that has provided a substantial proportion of the evidence for the Flynn effect (Flynn, 1984, 1987). The cognitive requirements of this test are examined in a later section of this chapter. In the Congo, Ombredane et al. (1957) found that school experience bore a considerable relationship to scores on the Raven (Wober, 1975). Lynn (chapter 8, this volume) also argues that gains in Raven scores result from schooling.

When variations in school experience are compared with variations in age, schooling is more predictive of test performance than is age. This has been established both in Africa (Schmidt, 1960) and in Israel (Cahan & Cohen, 1989). In other words, actual performance on intelligence tests is more closely related to years of schooling than it is to chronological age. (This research has taken the form of natural experiments in societies where age and and schooling vary more independently than they do in the United States or Europe.)

In Western societies, the role of schooling in producing gains in tested intelligence was probably most important hundreds of years ago when school attendance first became widespread. A similar situation has prevailed in the last 40 years in Africa, where there may well have been a large "Flynn effect" owing to increased schooling. However, the relevant research has not been done.

Even in the United States, adult illiteracy rates have gone down steadily during the 20th century (*Encyclopaedia Britannica*, 1938, p. 684, 1990, p. 740), and high school attendance took a quantum leap between 1900 and 1937 (*Encyclopaedia Britannica*, 1972). This trend might be particularly relevant to the Raven, which may depend heavily on secondary education (Ombredane, 1956, reported in Wober, 1975).

The Connection Between Formal Education and Maternal Behavior

One of the mechanisms by which school works is to change how mothers interact with their children, even after just a few years of schooling. The mother comes to reflect the teacher as a model for maternal behavior (Uribe, LeVine, & LeVine, 1994). For example, schooling provides experience with "known-answer questions" (in which the questioner already has the requested information). Minimally schooled mothers, like teachers, ask their children such questions; unschooled mothers do not (Duranti & Ochs, 1986). Responding to known-answer questions is the most basic convention on an intelligence test. In general, there is a connection between education and the kinds of maternal behavior posited by Ramey (1996) to make a positive contribution to children's IQ (e.g., Laosa, 1978).

The Factors Across Space and Time

Note that although some general principles may explain the Flynn effect across cultures, particular factors, their instantiation, and their strength vary from country to country and epoch to epoch. Focusing particularly on the United States and on symbolic technologies, the remainder of this chapter provides a closer look at the mechanisms that mediate the operation of these factors.

THE SEARCH FOR MECHANISMS TO EXPLAIN THE FLYNN EFFECT

The key to understanding these mechanisms is to be able to explain the patterning of the Flynn effect. The patterning in the United States, as described earlier, is as follows: There has been a modest rise in verbal IQ, in concert with a considerable decline in verbal SAT scores. Over the same period, a much larger rise in nonverbal, or performance, IQ has taken place. In the United States, a rise of at least 21 IQ points took place between 1918 and 1989 (Flynn, 1984, 1994), with a spurt (close to half a point a year) between 1972 and 1989 (chapter 2, this volume). The next step in the argument being developed is to provide evidence concerning the causal factors and mechanisms for each element in the pattern.

A Relatively Large Rise in Nonverbal IQ Scores

My search for the mechanisms to explain the relatively large rise in nonverbal IQ focuses on communication and information technologies: film, TV, video games, computers. Ever since film became popular in the 1920s, the spatial and iconic imagery featured by such media has been getting increasingly important. This has been a gradual change, involving the diffusion as well as invention of new communication and information technologies. This development of the technological environment is not a matter of a single medium at a particular time but involves many media and a long-term trend. In the next section of this chapter, I use experimental data to demonstrate that the spatial and iconic imagery characterizing this technological environment has effects on the mental processes of the users. External modes of representation enhance and develop the corresponding internal modes of representation required to handle them. These internal modes are required by performance IQ tests.

I start by discussing video games as an example pertinent to recent gains in nonverbal IQ. Video games are interesting from the perspective of the Flynn effect because they are a mass medium with over 90% penetration in the United States. In turn, I demonstrate how video games develop skills in visual–spatial representation and iconic imagery and how these same skills are utilized in important nonverbal portions of major IQ tests. Although my examples come mainly from action video games, many other popular electronic games (e.g., Scriven, 1987) and computer applications involve the same cognitive and perceptual skills.

Visual-Spatial Skills

Tetris is a dynamic spatial puzzle. It involves the putting together of pieces that fall from the top of a computer screen (see Figure 3). Using a joystick or keyboard, the player must turn and horizontally position descending pieces to make a solid wall at the bottom of the screen. Okagaki and Frensch (1994/1996) investigated the effects of Tetris on the mental manipulation of spatial imagery, tasks common on nonverbal IQ tests. In their experiment, 6 hr of playing Tetris enhanced performance on several paper-and-pencil tests that are similar to nonverbal IQ measures. One example was a puzzle assembly task (see Figure 4). In this task, the participant had to figure out which pieces were required to compose the large triangle shape on the left. Note, however, that the same skills enhanced by playing the video game Tetris are tested in performance, or nonverbal, IQ tests. For example, the Object Assembly subtest of the Wechsler intelligence scales for both children and adults (Wechsler, 1981, 1991) involves the manipulation and fitting together of puzzle pieces. Figure 5 shows a simulated item from the Object Assembly subtest of the Wechsler Intelligence Scale for Children (WISC-III).

Block design tests, found on many major IQ tests including the Wechsler scales for adults and children and the Stanford–Binet children's intelligence scale, also involve puzzle assembly skills related to those enhanced by Tetris. An example is the Pattern Analysis test from the Stanford–Binet. The point of these examples is clear: If practice on a video game can affect puzzle task performance in an experiment, why not on IQ tests?

It is interesting that the experimental effects of playing Tetris on paper-and-pencil tests, such as the one shown in Figure 4, occurred in male participants only (Okagaki & Frensch, 1994/1996), particularly because most of the Flynn data involve male army recruits (e.g., Flynn, 1984, 1987). (However, female participants in the Okagaki and Frensch study did show a positive effect of playing the video game on computer-tested [rather than paper-and-pencil] spatial skills more similar to Tetris itself.)

The effect of Tetris play on the mental manipulation of spatial imagery indicates, on a theoretical level, that external forms of repre-

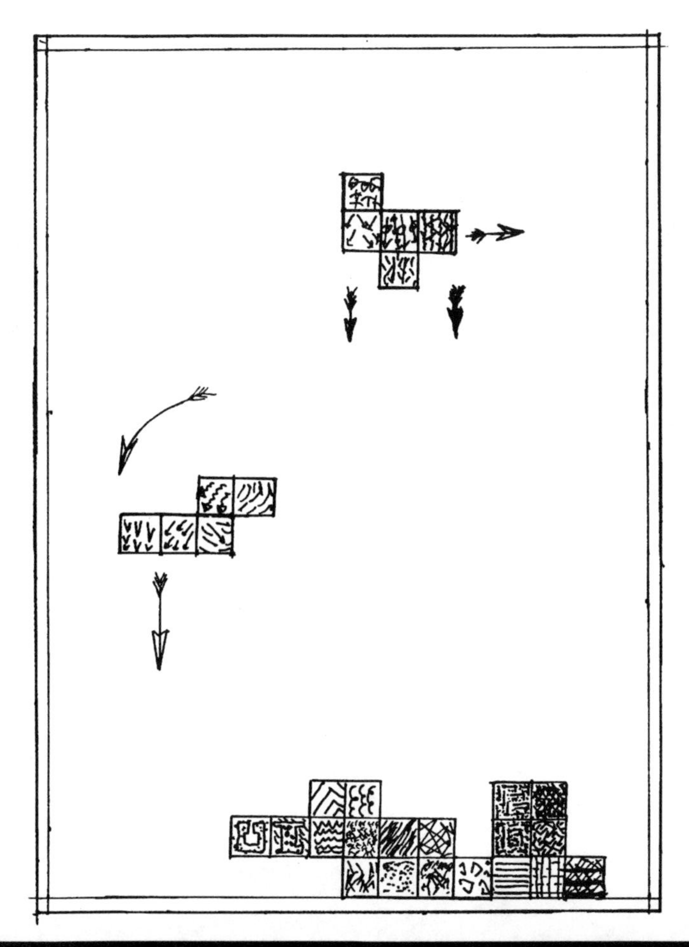

Figure 3

Schematic drawing of a screen from the popular video game Tetris, with arrows to indicate the direction of the movement of the pieces. In Tetris, puzzle pieces fall from the top of the screen (indicated by downward facing arrows). The player uses a joystick or keyboard to control horizontal position (indicated by a horizontal arrow) and to rotate pieces (indicated by a curved arrow). The goal is to fill all the spaces in each row of squares at the bottom of the screen. Given the location of the pieces at the bottom of the illustration, this can no longer be done in Row 1, but it is still possible to fill in Rows 2 and 3. Drawing by Abraham Seidman, 1996.

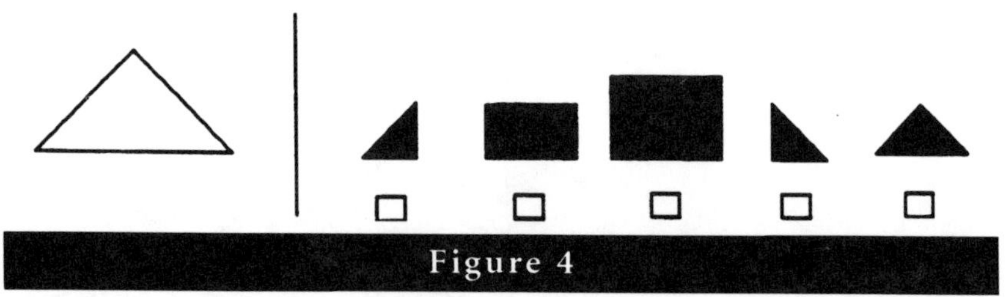

Figure 4

Performance on this puzzle assembly task was enhanced after participants played 6 hr of the computer game Tetris. From "Effects of Video Game Playing on Measures of Spatial Performance: Gender Effects in Late Adolescence," by L. Okagaki and P. A. Frensch, 1994, *Journal of Applied Developmental Psychology, 15*, p. 40. Copyright 1994 by Ablex. Reprinted with permission. (See also Greenfield & Cocking, 1996.)

sentation stimulate internal forms of representation. That is, the spatial manipulation of puzzle pieces on a computer screen both requires and develops internal modes of spatial representation that can be subsequently generalized to paper-and-pencil tests. These modes of spatial representation are among those assessed in performance IQ tests.

Other popular video games, including older ones, utilize still other skills in spatial representation that are also assessed on nonverbal IQ

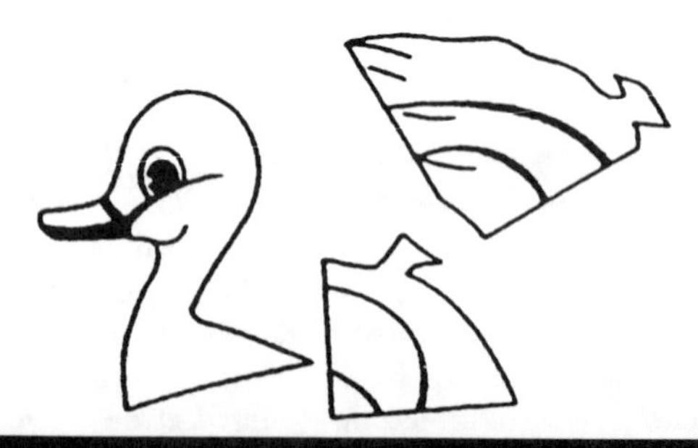

Figure 5

Simulated item similar to item in the Wechsler Intelligence Scale for Children: Third Edition. The pieces are asembled to form a duck. Copyright © 1990 by The Psychological Corporation. Reproduced by permission. All rights reserved. "Wechsler Intelligence Scale for Children" and "WISC-III" are registered trademarks of The Psychological Corporation.

tests. Whereas the video game Tetris and tests of puzzle assembly involve the manipulation of two-dimensional spatial representations, other games and performance tests involve the mental manipulation of three-dimensional representations presented on a two-dimensional screen or piece of paper. This is a more complex level in the mental representation of space.

For example, some genres of action video games require navigation through a two-dimensional representation of three-dimensional space. Mental paper-folding tests (see Figure 6 for an example) also require active mental manipulation of a two-dimensional representation of three-dimensional space. In the test shown in Figure 6, the participant must mentally fold the designs into a cube, noting which side of the two-dimensional design would touch the side of the hypothetical cube marked with an arrow.

Would a video game requiring the navigation through a two-dimensional representation of three-dimensional space require the same skills as mental paper folding? In a first study of one such game, *The Empire Strikes Back*, my students and I found a significant positive correlation between expertise in this game and performance on the mental paper-folding task shown in Figure 6 (Greenfield, Brannon, & Lohr, 1994/1996). In a second, experimental study, we explored whether experience with the game has a causative influence on the spatial skills required by mental paper folding (Greenfield, Brannon, & Lohr, 1994/1996). Our results indicated that an experimental treatment involving relatively brief exposure to *The Empire Strikes Back* game had no effect on mental paper folding. However, we did find, through structural equation modeling, that the initial level of accumulated expertise on *The Empire Strikes Back* was causally related to mental paper-folding skill (Figure 7). The model in Figure 7 showed a good fit to our data. It indicates that initial skill level with *The Empire Strikes Back* game (itself strongly influenced by gender) traces a significant causal path to mental paper-folding performance.

The importance of these findings in contributing to an explanation of the Flynn effect is that mental paper folding can be part of the measurement of nonverbal IQ. For instance, a version of mental paper

Below are drawings each representing a cube that has been "unfolded." Your task is to mentally refold each cube and determine which one of the sides will be touching the side marked by an arrow.

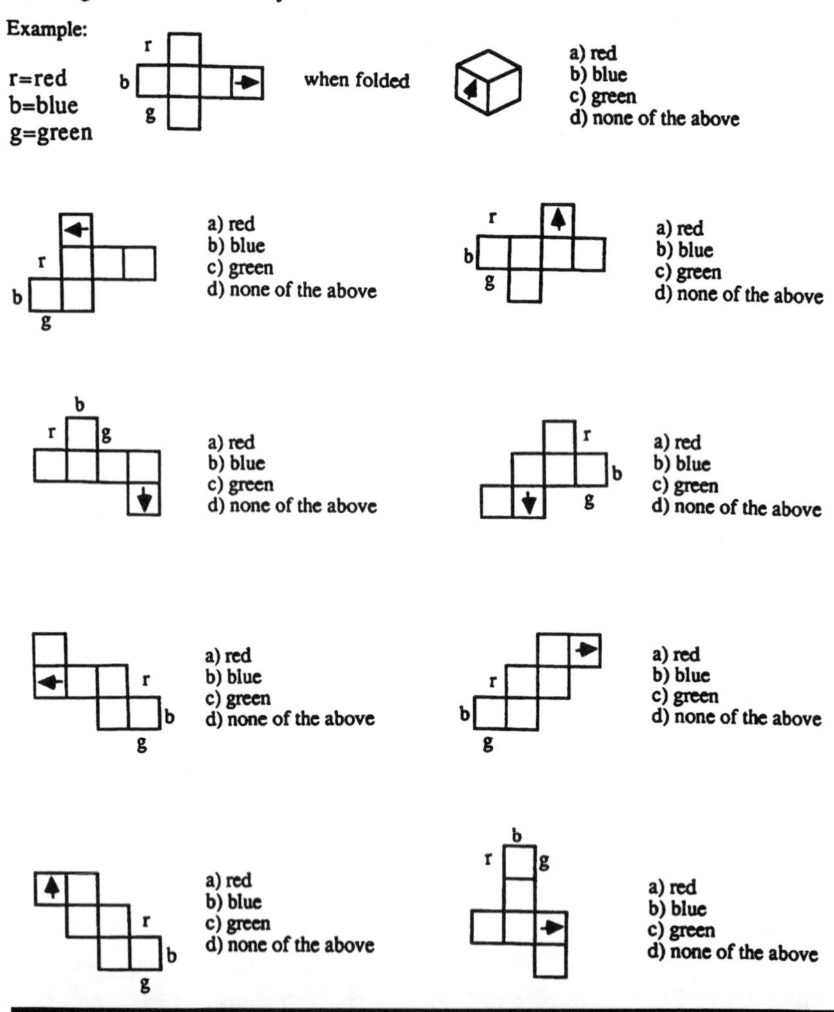

Figure 6

A mental paper-folding test, like a video game, requires active manipulations of a two-dimensional representation of three-dimensional space. From "Two-Dimensional Representation of Movement Through Three-Dimensional Space: The Role of Video Game Expertise," by P. M. Greenfield, C. Brannon, and D. Lohr, 1994, *Journal of Applied Developmental Psychology, 15*, p. 91. Copyright 1994 by Ablex. Reprinted with permission. (See also Greenfield & Cocking, 1996.)

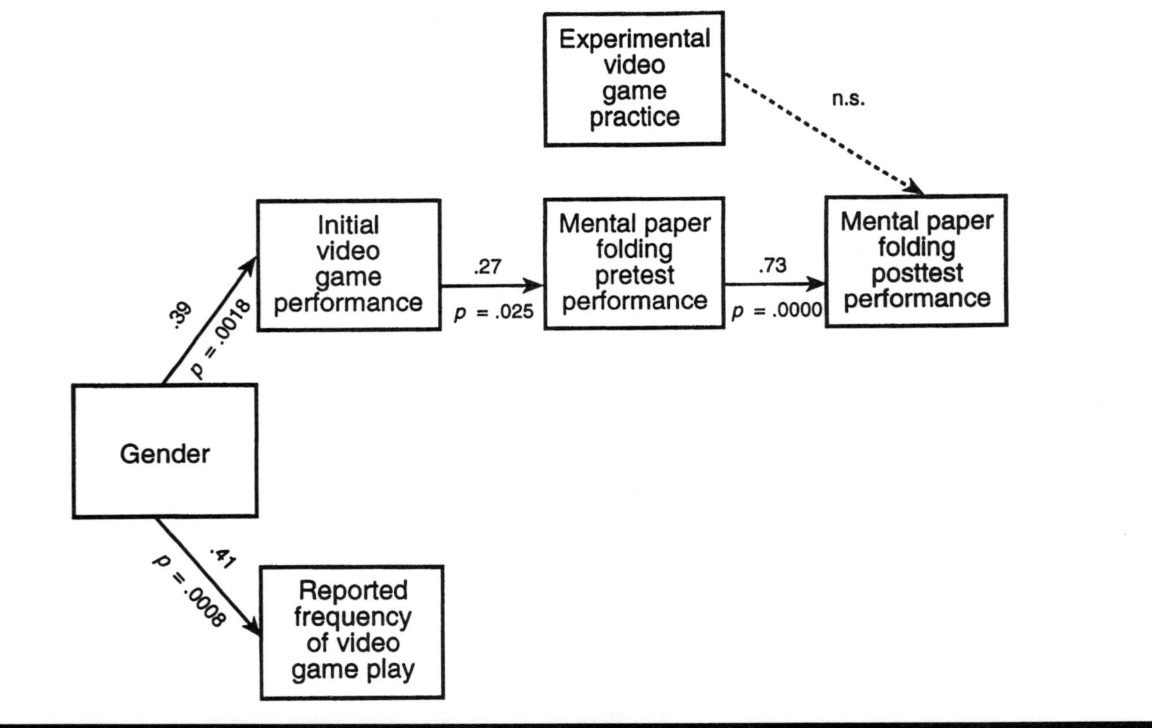

Figure 7

Relationship between video game playing and performance on a mental paper-folding task. From "Two-Dimensional Representation of Movement Through Three-Dimensional Space: The Role of Video Game Expertise," by P. M. Greenfield, C. Brannon, and D. Lohr, 1994, *Journal of Applied Developmental Psychology, 15,* p. 97. Copyright 1994 by Ablex. Reprinted with permission. (See also Greenfield & Cocking, 1996.)

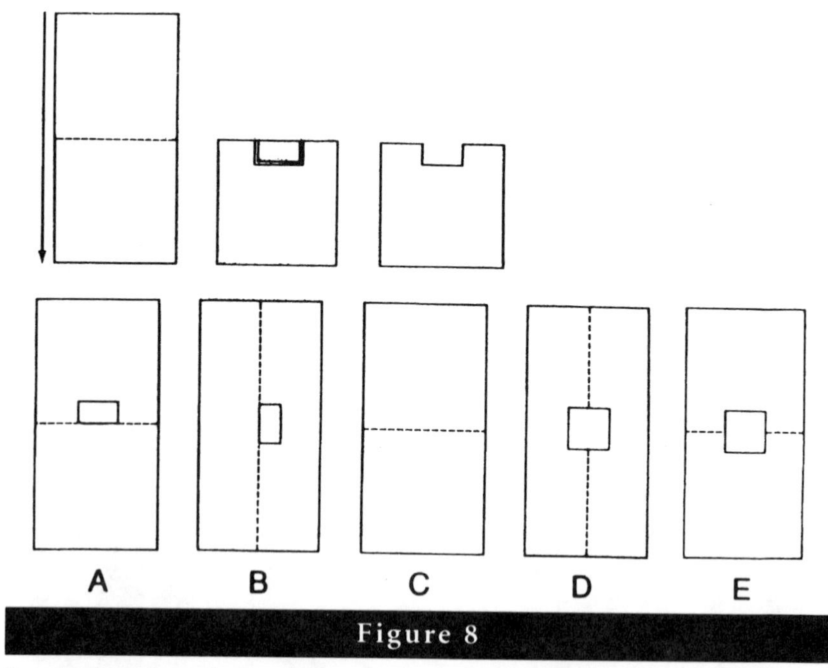

Figure 8

Practice item from the Stanford–Binet Mental Paperfolding Test. From *Stanford–Binet Intelligence Scale, Fourth Edition, Item Book 4* (p. 108), by Robert L. Thorndike, Elizabeth P. Hagen, and Jerome M. Sattler, 1986, Chicago: Riverside Publishing Company. Copyright 1986 by The Riverside Publishing Company. Reprinted with permission. All rights reserved.

folding is one of the subtests on the Stanford–Binet Intelligence Scale (see Figure 8 for a sample item). In this test, the experimenter first folds and cuts the sample (top row of the figure); the participant must look at the folded sample and select which of several choices (bottom row of figure) the unfolded sample will look like.

The message is the following: If video game expertise develops mental paper-folding skill as measured in an experimental situation, it should logically have the same effect on mental paper folding when this type of task appears on an IQ test.

In sum, there is a suite of visual–spatial skills developed by video games that are also tested in nonverbal IQ assessment. It follows that the historical ascendance of popular video games would contribute to the historical rise in nonverbal IQ. Indeed, this ascendance began within the span of years in which Flynn found a spurt in the rate of IQ increase in the United States (from 1972 to 1989).

Iconic Imagery

Film, television, video games, and computers all privilege iconic, or analog, representation over symbolic, or digital, representation. That is, they privilege image over word. This distinction between iconic and symbolic representation has a strong background in studies of cognitive development (Bruner, Olver, & Greenfield, 1966). For purposes of the present argument, the importance of iconic imagery in these communication media is that nonverbal IQ tests are also iconic in nature. Has experience with the iconicity of these media contributed to the historical rise in nonverbal IQ?

A cross-cultural study carried out in Los Angeles and Rome has provided some experimental results that are relevant to answering this question (Greenfield et al., 1994/1996). In fact, my colleagues and I demonstrated that a computer game can shift representational style from verbal to iconic. University students were required to write down their answers to a paper-and-pencil test of comprehension of an animated video display of the logic of computer circuitry. Figure 9 presents an example of one particular sequence in the animated displays; the sequences were taken from a piece of software called Rocky's Boot (Robinett, 1982). The two frames from this animated sequence, shown to participants on a video screen, demonstrate the operation of an *and*-circuit. An *and*-gate must receive energy input at both of two points to be activated. Activation causes energy to flow through the whole circuit; energy flow and activation are represented by the broader lines and lighter color and the switching of the sign from off to on in the figure. The rectangle shown in the figure represents an energy source. None of this was explained to the participants; they were simply asked to watch carefully and try to figure out what was going on so that they could answer questions about it afterward. After the animated simulation was shown to them, participants were given a paper-and-pencil test to assess their comprehension of the meaning of what they had just seen. Figure 10 shows some sample test items. One of these items

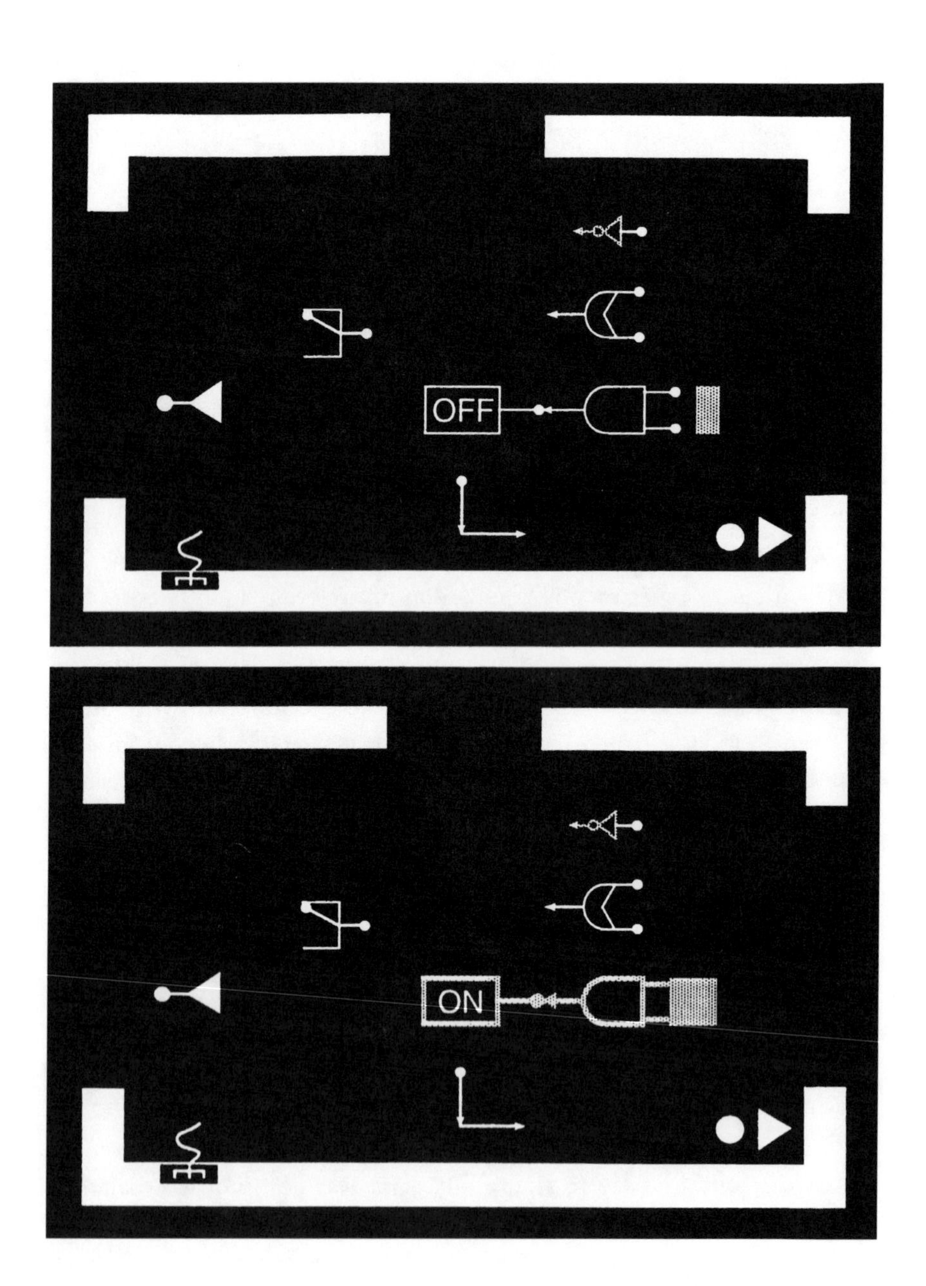
OFF
ON

(middle of Figure 10) assesses understanding of the simulated *and*-circuit shown in Figure 9. The results showed that participants used three different modes of representation in communicating their answers to the experimenter. These three modes were verbal, iconic, and mixed; examples are shown in Figure 11.

The paper-and-pencil comprehension test was given before and after a number of conditions. Of special relevance for the present argument are two conditions: the memory game of Concentration played on a computer screen and the same memory game played on a board. The point of the game is to find identical pairs, in this case, pairs of numbers. In both versions, numbers hidden behind "doors" are arranged in a grid.

In the *computer version,* virtual doors are "opened" by a joystick controlling a cursor in the shape of a hand (see Figure 12). In the board version, doors (shown in Figure 13) are opened manually; for this reason, this was termed the *mechanical version.*

Most relevant to the present argument is the fact that communication about the animated computer simulation became more iconic and less symbolic after participants played the computer game, but not after they played the same game on the board. The computer medium was the decisive factor. This finding is important for explaining the historical rise in performance IQ: Iconic images and diagrams are basic to all of the nonverbal performance tests. If modern computer tech-

Figure 9

(*Facing page*) Sequence from an animated video display of the logic of computer circuitry (Rocky's Boot; Robinett, 1982). In the top frame of the animated video display, the energy source is not touching the two input contacts of the *and*-gate; therefore, the circuit is not activated and the sign is switched off. In the bottom frame, the energy source has been moved to touch the two input contacts of the *and*-gate, thereby activating the circuit and turning the sign on. From "Cognitive Socialization by Computer Games in Two Countries: Inductive Discovery or Mastery of an Iconic Code," by P. M. Greenfield, L. Camaioni, P. Ercolani, L. Weiss, B. A. Lauber, and P. Perucchini, 1994, *Journal of Applied Developmental Psychology, 15,* p. 69. Copyright 1994 by Ablex. Reprinted with permission. (See also Greenfield & Cocking, 1996.)

NAME:

DATE: **FORM B**

What does this represent? 

What is its function?

How would you get the orange color to flow through the following game elements?

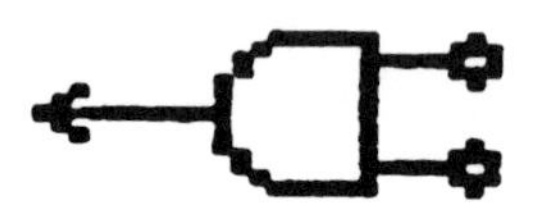

Figure 10

(*Facing page*) Sample items from the test of comprehension of the animated video display of the logic of computer circuitry. From "Cognitive Socialization by Computer Games in Two Countries: Inductive Discovery or Mastery of an Iconic Code," by P. M. Greenfield, L. Camaioni, P. Ercolani, L. Weiss, B. A. Lauber, and P. Perucchini, 1994, *Journal of Applied Developmental Psychology, 15*, p. 71. Copyright 1994 by Ablex. Reprinted with permission. (See also Greenfield & Cocking, 1996.)

nology is making people more iconic in their style of representation, it follows logically that people will do better on nonverbal IQ tests.

The cross-cultural findings were also relevant to the argument. Before being exposed to any experimental treatments, university students in Rome were predominantly symbolic in their representations,

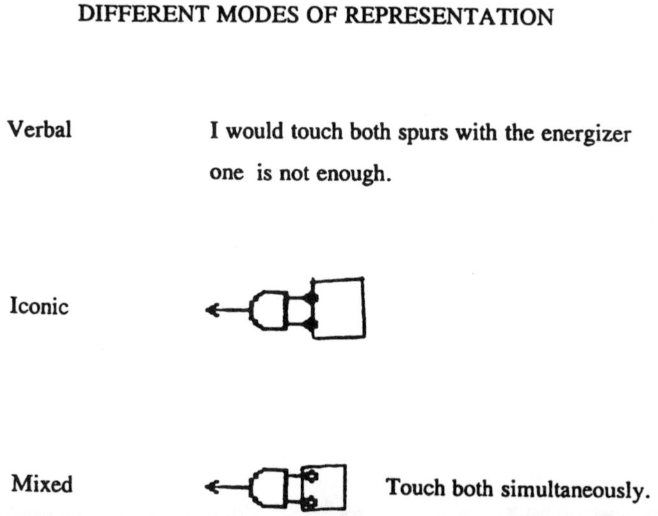

DIFFERENT MODES OF REPRESENTATION

Verbal — I would touch both spurs with the energizer one is not enough.

Iconic

Mixed — Touch both simultaneously.

Figure 11

Examples of different modes of representation: Verbal, iconic, and mixed. These examples are responses to the questions about an *and*-gate. From "Cognitive Socialization by Computer Games in Two Countries: Inductive Discovery or Mastery of an Iconic Code," by P. M. Greenfield, L. Camaioni, P. Ercolani, L. Weiss, B. A. Lauber, and P. Perucchini, 1994, *Journal of Applied Developmental Psychology, 15*, p. 73. Copyright 1994 by Ablex. Reprinted with permission. (See also Greenfield & Cocking, 1996.)

A

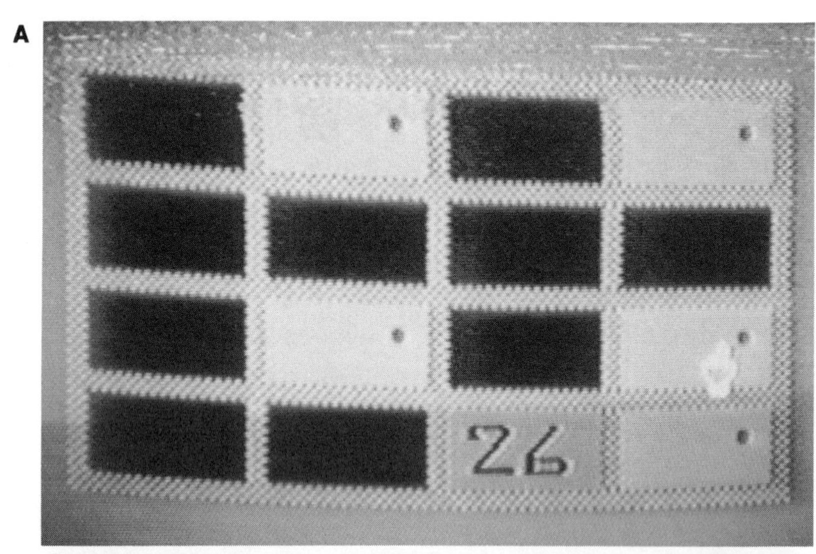

B

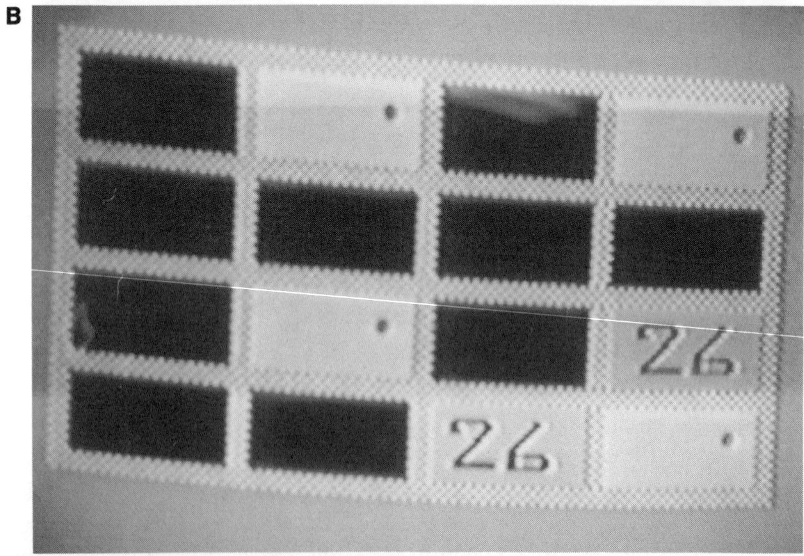

Figure 12

Computer version of memory game Concentration. (A) Player has found one *26* and placed the cursor to open a second "virtual door." (B) The door has opened, revealing a successful search for the matching *26*.

Figure 13

Mechanical version of Concentration memory game. Each door has a knob that lifts up. A pair of 66s has been found.

whereas students in Los Angeles were predominantly iconic; this difference was statistically significant. Its cause, we hypothesized, lay in the greater diffusion in the United States, compared with Italy (in the late 1980s), of all the electronic media that feature iconic imagery. Whereas the experimental effect described earlier implicates a causal mechanism for enhanced use of iconic imagery, this cross-cultural population difference implicates the role of the everyday environment in producing the real-world effect. The two findings complement each other; together, they suggest that the use of iconic imagery, not only in an experimental situation but also in the everyday environment, can have an impact on representational habits. It is therefore likely and logical that part of this real-world effect occurs on nonverbal intelligence tests.

The Raven Progressive Matrices and Cultural Bias

One might ask why the Raven is so culturally sensitive. The Raven is the test most often used to demonstrate large historical gains in performance IQ (Flynn 1984, 1987). It was designed to be "culture free." More recent terminology has included "culture fair" and "culture reduced." However, all these terms are misnomers. Instead, the Raven constitutes a conventionalized cultural genre.

The matrix is a culture-specific form of visual representation. To solve matrix problems, one must understand the complex representational framework in which they are presented. Figure 14 presents an example of a simple item from the Raven. To solve this or any other matrix item, one needs to know, for example, that a matrix is organized in rows and columns (Figure 14). Matrices presuppose much more than "acquaintance with certain simple shapes" (Flynn, 1994, p. 620).

The next sample (Figure 15) is a more complex item, developed by Carpenter, Just, and Shell (1990) to be analogous to items on the Raven. It illustrates additional conventionalized knowledge required by the test. It requires understanding of an ordinal relationship among the columns and among the rows as well as specific knowledge concerning what mental operations are relevant to perform on the test matrix. This item involves the figure addition rule (Carpenter et al., 1990): The operation of adding the black part of the figures from each row from left to right yields the figures in the far right column; the operation of adding the black part of the figures in a column from top to bottom yields the figures in the bottom row. In fact, research has shown that there are five rules that can generate all of the answers to matrix items (Carpenter et al., 1990). The understanding of these rules is culture specific. There is nothing in the matrix figures themselves that tells what operations to perform on them.

Indeed, I would like to make the case that the comprehension of matrices is socialized and taught in a particular cultural environment, the school. My evidence comes from research in Nabenchauk, a Zinacante Maya community in Chiapas, Mexico. There, the first matrices, cross-stitch patterns laid out on graph paper, were introduced into the

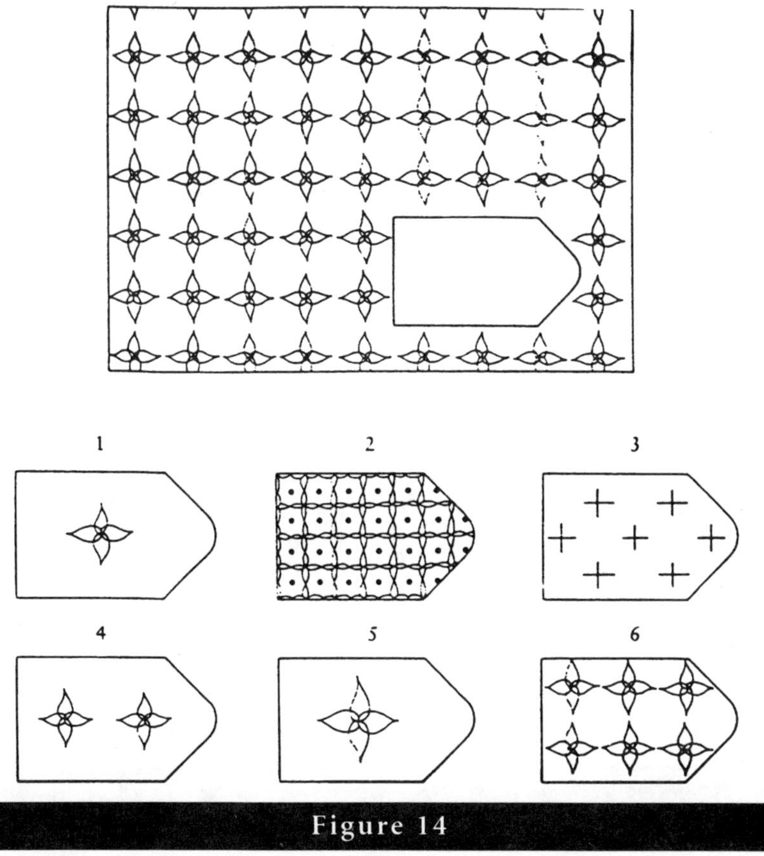

Figure 14

An example of a simple item from the Raven Standard Progressive Matrices. From the six inserts at the bottom of the figure, the participant must select the one that logically fits in the matrix above. Figure A5 of the Raven Standard Progressive Matrices, by J. C. Raven. Copyright 1938, 1976 by J. C. Raven Ltd. Reprinted with permission.

then agrarian community by school teachers. A more recent example is shown in Figure 16.

Currently, our research shows a statistically significant association between use of these patterns and some (vs. no) schooling (Greenfield & Maynard, 1997). That is, women who have had a few years of schooling are significantly more likely to use these patterns for embroidery or weaving than women who have never been to school. However, note

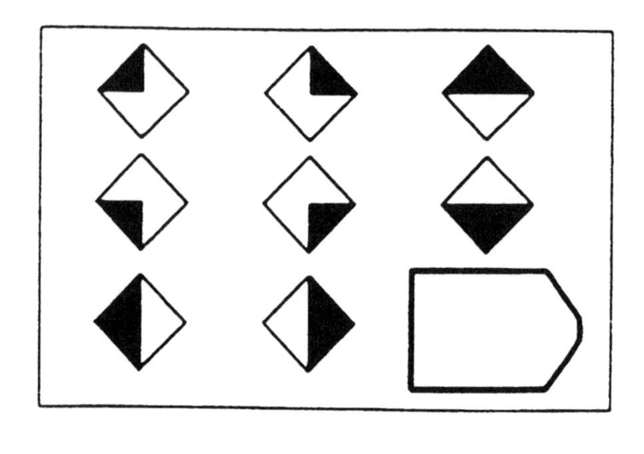

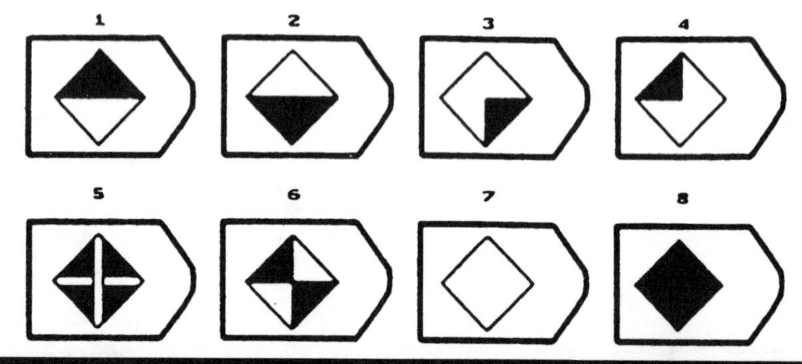

Figure 15

An example of an additive matrix item. The correct answer is *8*. From "What One Intelligence Test Measures: A Theoretical Account of the Processing in the Raven Progressive Matrices Test," by P. A. Carpenter, M. A. Just, and P. Shell, 1990, *Psychological Review, 97*, p. 409. Copyright 1990 by the American Psychological Association. Reprinted with permission.

that this, the only matrix in the Zinacantec cultural environment, does not involve any ordinal relationship between columns and rows. The matrix is used simply to indicate relative positions that can be used in transferring the pattern from paper to textile, either embroidered or woven. This is the simplest form of matrix; however, experience with it could form a cognitive foundation for ordered matrices, which could be introduced at a later point.

Figure 16

Cross-stitch pattern laid out on graph paper. The first type of matrix in use by the Zinacantec Maya in Chiapas, Mexico.

The association of schooling and use of cross-stitch patterns among the Zinacantec Maya people suggests that historically, in U.S. society, the advancement of basic literacy (which has continued even into the current century) would have improved performance on the Raven. In addition, higher levels of schooling seem required for the complex matrices on the Raven that combine multiple rules. Ombredane (1956, cited in Wober, 1975) reported a study in Africa that found divergence on the Raven between schooled and unschooled persons beginning at age 12–13 years. This finding could be extremely relevant to the Flynn effect for particular periods in different countries when postprimary education was greatly expanded. For example, there was a 10-fold increase in rates of secondary education in the United States in the first third of the 20th century (*Encyclopaedia Britannica,* 1972).

In addition, the use of matrices has become increasingly culturally diffused in recent years in the United States and other technologically advanced countries because of computers; a good example is the use of popular spreadsheet programs such as Microsoft Excel. These programs are simply blank matrices, organized by columns and rows, to be filled in by the user. Clearly, such a program requires users to represent their data mentally in matrix form, while providing practice in the use of this representational format.

From Film to Video Games: Shifting Visual Perspective

Another visual skill required by some performance tests is the ability to shift visual perspective. There are two major points concerning this skill. First, such tests are based on culture-specific conventions for representing spatial information and require culture-specific knowledge. Figure 17 provides an example of a spatial ability item that requires a shift in imagined perspective. This item from the Guilford–Zimmerman Aptitude Survey requires one to identify the upside-down alarm clock on the left, mentally rotate it a quarter turn to the right, as indicated by the ball with an arrow on it; and finally match the resulting visual perspective with one of the five choices on the right.

This item is extremely culture specific. Note that it requires knowl-

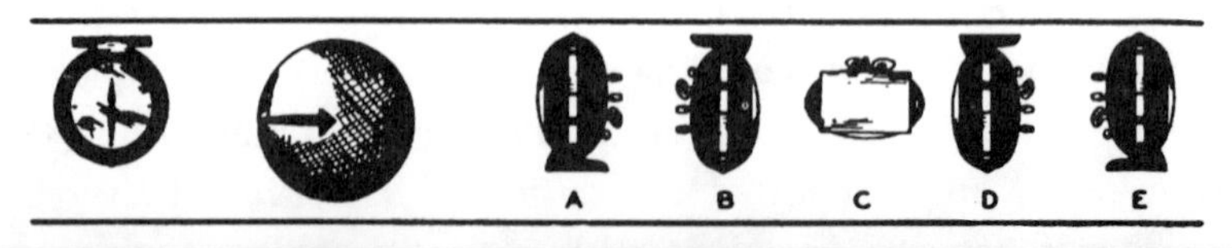

Figure 17

An item from the Guilford–Zimmerman Aptitude Survey: IV. Spatial visualization, Form B, 1953. From "Further Evidence of Sex-Linked Major-Gene Influence on Human Spatial Visualizing Ability," by R. D. Bock and D. Kolakowski, 1973, *American Journal of Human Genetics, 25*, p. 2. Copyright 1973 by the University of Chicago Press. Reprinted with permission.

edge of what the back of an alarm clock looks like. It also requires knowledge of the arrow as a visual symbol as well as even more specific knowledge that the arrow as portrayed does not symbolize a horizontal direction on a flat plane but rather represents rotation in the third dimension.

The second major point is that this culturally based skill in mentally shifting visual perspective has been used and developed, in turn, by the cultural technologies and genres of film and video games. Compare the test item shown in Figure 17 with the item shown in Figure 18. The latter is drawn from an analogous test of shifting visual perspective, one that has been shown to be related to the understanding of film editing techniques (Salomon, 1979). Such techniques can be used to show different perspectives on a scene; an example is the classic reaction shot, in which views of two faces interacting with each other are intercut in alternation.

Video games require and develop even more than film in the arena of visual perspective-taking skills. First, video games provide much more incentive to understand shifting perspectives (or any other visual convention) than does film; to fail to understand is to fail at the game. The player, moreover, must not only understand shifts of visual perspective but also actively coordinate them in playing the game. Figure 19 provides an example of a video game from the 1980s that requires the player to coordinate mentally two visual perspectives. This game, called Tranquility Base, involves coordinating a long shot and a close-up of terrain, as players land their spacecraft. Such coordination of perspectives has become ever more complex as the games have evolved and become more realistic; the advent of CD-ROM games such as Earthsiege, with its multiple three-dimensional perspectives, has increased the complexity even more.

But in the domain of computer technology, the representation of shifting perspective is not limited to video games. Complex perspective shifts and coordination are also intrinsic to using computer-assisted design (CAD) software for the home and popular programs like Adobe Photoshop. Of course, certain professions like architecture and photography here always required the ability to represent and comprehend

Imagine that you are the girl sitting on the window sill. How would you see the painter?

Figure 18

Test item from the Test of Changing Points of View. From *Interaction of Media, Cognition and Learning* (p. 164), by G. Salomon, 1979, San Francisco: Jossey-Bass. Copyright 1994 by Lawrence Erlbaum and Associates, Inc. Reprinted with permission.

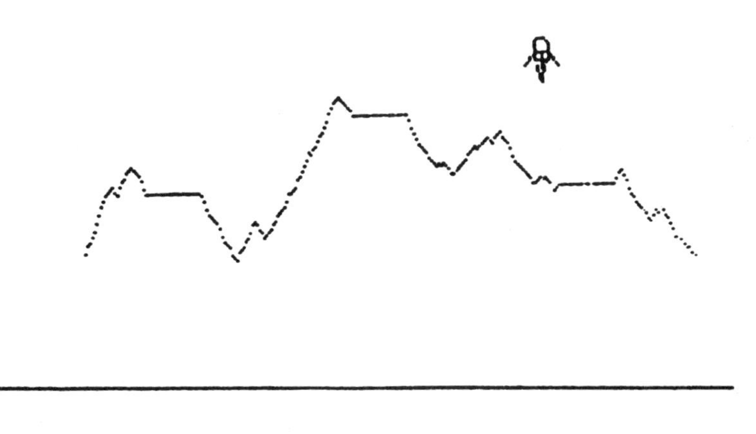

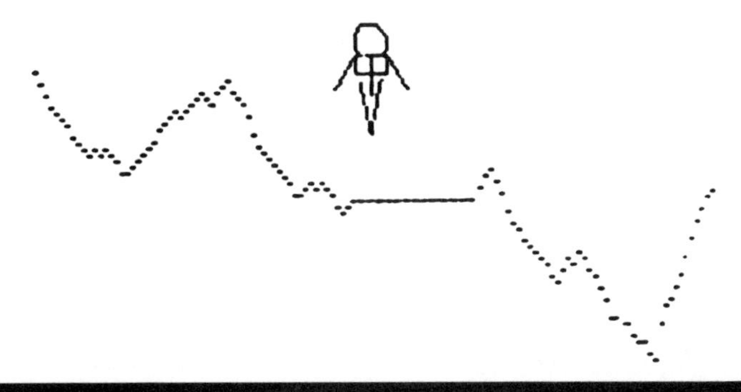

Figure 19

A long shot (top) and a close-up (bottom) of terrain, as a player lands a spacecraft in Tranquility Base, a computer game from the early 1980s. From *Mind and Media: The Effects of Television. Video Games, and Computers* (p. 116), by P. M. Greenfield, 1984, Cambridge, MA: Harvard University Press.

perspective shifts using both computer and other, older technologies (e.g., blueprints). Currently, however, this skill has a new, all-pervasive functionality. This can perhaps be best illustrated by the variety of medical imaging techniques such as ultrasound and positron-emission tomography (PET) for which the scans must be read by physicians and scientists but which are also seen en masse by laypeople in the United States and other technologically advanced countries (e.g., ultrasound during pregnancy).

The Modest Rise in Verbal IQ Scores and the Considerable Decline in Verbal SAT Scores

My analysis of mechanisms turns now to the modest historical rise in verbal IQ, combined with a simultaneous decline in verbal SAT scores. I begin by considering the role of vocabulary in creating this apparent paradox. Vocabulary is one of the most important motors for both verbal IQ and verbal SAT scores. My first hypothesis is that television has driven basic vocabulary up a bit in the population as a whole (verbal IQ) but that the decline of reading for pleasure (Beentjes & Van der Voort, 1988; Duchein, 1993) has driven the literary vocabulary required by the verbal SAT down. My second hypothesis is that the context-dependent grammatical structures favored by television and telephone differ from the context-independent structures that are so crucial for the verbal SATs. My third hypothesis is that the very nature of literature is being transformed by the electronic media. Each of these hypotheses will be developed more fully.

Vocabulary Development

There is evidence that vocabulary is learned from TV (Ball & Bogatz, 1970; Bogatz & Ball, 1971; Rice, Huston, Truglio, & Wright, 1990), which is a mass medium. However, this product of TV is a basic vocabulary, because the vocabulary of television is quite a limited one. On the basis of talking to teachers, Healy reported that "unless students read a lot on their own, their vocabulary growth slows down somewhere near the fourth grade level—approximately the level of media language" (1990, p. 100).

Similarly, the vocabulary used on IQ tests such as the Wechsler is quite basic. This, however, is not true of the vocabulary used on the verbal SAT, which is a more literary vocabulary. Compare the vocabulary list that must be defined for the vocabulary subtest of the WAIS-R (Wechsler, 1981) with a sample vocabulary list from the antonym section of the verbal SAT (Brownstein, Weiner, & Green, 1989). The range of difficulty of the two lists is quite different. The WAIS-R vocabulary subtest ranges from *bed* and *ship* (practice items) to more difficult items like *vehement* or *fallacious* (simulation items). In contrast, the range of the practice SAT antonym test is much harder; it runs from *fertile* to *stolid* and *expatiate.* One can imagine *vehement* being used on a TV talk show. However, it is difficult to imagine items such as *expatiate* ever being used in such a context.

This analysis is confirmed by research. Holding education constant, Glenn (1994) found a decline in vocabulary between 1974 and 1990 (cf. the work of Huang & Hauser, chapter 12, this volume). This decline was associated with a decline in the reading of newspapers and other print sources. One of the sources of the reading decline was the rise of television viewing.

Grammar

The verbal SAT requires analysis of passages from literature and science. TV ill prepares the viewers for the complex syntax that such passages use. TV shows for children use significantly more simple sentences and significantly fewer complex sentences than do children's picture books (Fasick, 1973). Content analysis of TV shows for children has also shown that the most common sentence structure in this genre is the incomplete sentence. Children's picture books, in contrast, use significantly more complete sentences (Fasick, 1973). Clearly, comparable comparisons of adult books and TV shows would reveal the same pattern.

The relative frequency of the incomplete sentence on TV has important significance for the verbal SAT, which tests the use of *decontextualized language.* Reading-comprehension passages on the SAT are decontextualized in the sense that all the information needed to answer a question must be found in the passage. If students use any outside knowledge to answer the questions, they are in trouble. TV language, in sharp contrast, is not decontextualized. As in face-to-face conversations, the propositions or ideas in those incomplete sentences are completed by the visual image.

One way in which the *contextualized* communication of television as a visual medium is manifest is in the high frequency of vague reference, that is, the use of pronouns or general terms of reference without antecedents that specify the referent. In television, the visual context is often relied on to provide this specification. In purely verbal media such as print or radio, the linguistic message must be self-contained. For example, Greenfield (1984) found that the announcer in a televized broadcast of a baseball game often took the identity of the batter for granted in announcing what the batter did. Faced with the same situation, a radio announcer always identified the batter. Compared with his colleague on TV, his verbal messages were more context independent.

There is also evidence that this phenomenon has a cognitive effect on the viewer. My colleagues and I did an experimental study comparing recall of radio and TV in elementary schoolchildren (Greenfield & Beagles-Roos, 1988). Recall of a TV story elicited more vague reference (e.g., using "he" or "she" or "it" without prior specification of the referent) than did recall of the same radio story by a given child. The same mechanism was at work: The participant was letting memory of an image tell part of the story (even though participants had been told that the listener had not viewed the story). The hypothesis is that this mode of processing and communicating information is a detriment to performance on the comprehension part of the verbal SAT. These passages must be comprehended without reference to any extralinguistic images.

Another area in which electronic media have promoted more contextualized language use than their predecessors is in the replacement of letter writing by telephone calls and now electronic mail. In telephone conversations particularly, the sentences of one person are conventionally completed by the conversational partner. An example is in the following interchange: "How are you?" "Fine." The second person does not make the complete sentence "I am fine." That would be considered impolitely redundant; instead, the second person uses the informational context provided by the first person to specify to whom "fine" refers. The same information would be conveyed in a letter by a complete sentence (e.g., "I am fine"). Similarly in electronic mail, replies often include the instigating message and depend informationally on its context. For example, I find myself often replying to an electronically mailed suggestion for action by simply attaching the single word *okay* to the message sent to me. Hence, the ascendance of these new forms of communication, along with the concomitant decline of letter writing, provides more practice in contextualized forms of verbal communication and less practice in decontextualized forms.

The Transformation of Literature by Electronic Media

Popular literature is more and more derived from television and film (e.g., the popular book about the TV show *The Real, Real World* and the numerous bestsellers resulting from the televised trial of O. J. Simpson). Popular literature has also become increasingly visual and interactive (e.g., the electronically assisted push-button book starring the outrageous animated TV characters Beavis and Butthead; Doyle, 1995). Literature becomes increasingly like TV. Because of this transformation of print by the electronic media, it follows that even the written vocabulary and grammar to which people are exposed is becoming increasingly basic.

In summary, TV slightly raises the average level of vocabulary on a mass scale; however, with the decline of reading purely literary works for pleasure and the increase of popular literature derived from television, the literary vocabulary required on the verbal SAT does not get developed. In addition, TV, telephone, and electronic mail develop contextualized language use, whereas reading literary works and writing letters develop the decontextualized language (Greenfield, 1972) required by the verbal SAT. From a historical perspective, TV has been in ascendance since the 1940s in the United States; the telephone's ascendance as a mass medium began at the end of the 19th century (Schwartz, 1994). The relative importance of print, both reading and writing, has been in a corresponding decline, particularly the reading of purely literary works for pleasure and the writing of letters.

CONCLUSION

There has been a change in the balance of print and visual media with the development of new modes of technology and visual communication; the balance has shifted toward images and diagrams (chapter 5, this volume). This change has produced the pattern of the Flynn effect, involving greater historical gains on visual and spatial than on verbal intelligence tests. For the United States, the focus on these technologies in the present chapter is consistent with Flynn's recent finding of an IQ spurt in this country between 1972 and 1989 (chapter 2, this volume); in other words, the Flynn effect accelerated in this period. Such a spurt is exactly what my model would predict. Because of its interactivity, I would expect modern computer technology, which became a mass medium in this period, to accelerate the Flynn effect on visual tests. Unlike film, computers provide constant feedback that should accelerate the development of the visual skills they require. Finally, I believe that the visual skills required by current games—such as Earthsiege, where, for example, the game is played from the visual perspective of both robots and the planes that oppose them (*Interaction*, 1996, p. 17)—have far outpaced the difficulty level of current nonverbal IQ tests. (A similar point is made by Williams in chapter 5, this volume.) It would not surprise me if ceiling effects had begun to show up on performance IQ tests.

The analysis presented in this chapter leads to the conclusion that nonverbal IQ tests are, in fact, more, not less, culture sensitive than verbal tests. Indeed, this fact has been known since the 1950s, mainly because of research in Africa (see Wober, 1975, for an excellent summary of research on ability testing in Africa). Why is this the case? One important part of the answer is that verbal tests get translated, whereas nonverbal ones do not, yet nonverbal, or performance, tests do rely on their own language of visual conventions. As I have shown, even the term *culture reduced* is a misnomer for visual tests such as the Raven Progressive Matrices.

If IQ Is Going Up, Are People Becoming More Intelligent?

This is an important question raised by Flynn (1984, 1987, 1994). My answer is *yes*, but in very specific ways. These ways include everyday examples such as provided by the legion of computer-literate hackers. They also include creative leaps. The recent discovery of a new muscle in the jaw by a dentist, Dr. Gary Hack, is just such an example. The report on National Public Radio news is instructive for this purpose.

> Robert Siegel: Now, the question that your discovery provoked immediately was how can somebody find anything new in a field so long gone over as anatomy and human anatomy of the head?
>
> Dr. Gary Hack: The anatomy textbooks teach a very precise way of dissecting this area, and that is from the side of the head. You go in from the side to this area behind the eye. If you do that, you cannot appreciate this structure. You must go from the front of the head, which is a novel and unique approach. But if you keep an open mind that there are new structures to be found and that to truly understand three-dimensional relationships, you must dissect from unique and novel approaches, you see things that others have not appreciated. (*All things considered*, 1996)

The preceding dialogue indicates that this creative discovery of a new muscle in the jaw was a result of the development of perspective-taking abilities, the same ones used in film and TV (cf. Figure 18) and in video games and software applications such as CAD software and Adobe Photoshop. Informal educational experiences with video games and other perspective-switching software would, I hypothesize, provide a certain type of cognitive socialization. In the course of this socialization, these external software tools would be appropriated and become internal cognitive tools (Salomon, Perkins, & Globerson, 1991; Saxe, 1991). As internal tools, they can be used for discoveries, such as that made by Dr. Hack.

The Cultural Evolution of IQ

To conclude, the Flynn effect is an example of the historical evolution of culturally phenotypic intelligence, as depicted in Figure 1. It is not the evolution of "general intelligence." However, the fact is that general intelligence must always be instantiated in a specific cultural form. Culture takes general intelligence and makes it specific.

REFERENCES

All things considered. (1996, February 14). *NPR Fax Service, No. 025,* pp. 2–3.

Arkana, M. L., & Snow, M. (1978). *Psychological tests and social work practice.* Springfield, IL: Charles C Thomas.

Ball, S., & Bogatz, G. (1970). *The first year of Sesame St.* Princeton, NJ: Educational Testing Service.

Beentjes, J., & Van der Voort, T. (1988). Television's impact on children's reading skills: A review of research. *Reading Research Quarterly, 23,* 389–413.

Bock, R. D., & Kolakowski, D. (1973). Further evidence of sex-linked major-gene influence on human spatial visualizing ability. *American Journal of Human Genetics, 25,* 1–14.

Bogatz, G., & Ball, S. (1971). *The second year of Sesame St.* Princeton, NJ: Educational Testing Service.

Brownstein, S. C., Weiner, M., & Green, S. W. (1989). *How to prepare for the Scholastic Aptitude Test* (SAT., 15th ed.) New York: Barron.

Bruner, J. S., Olver, R. R., & Greenfield, P. M. (1966). *Studies in cognitive growth.* New York: Wiley.

Cahan, S., & Cohen, N. (1989). Age versus schooling effects on intelligence development. *Child Development, 60,* 1239–1249.

Carpenter, P. A., Just, M. A., & Shell, P. (1990). What one intelligence test measures: A theoretical account of the processing in the Raven Progressive Matrices test. *Psychological Review, 97,* 404–431.

Cole, M., & Cole, S. R. (1993). *The development of children.* New York: Scientific American Books.

Dasen, P. (1984). The cross-cultural study of intelligence: Piaget and the Baolé. In P. S. Fry (Ed.), *Changing conceptions of intelligence and intellectual functioning: Current theory and research* (pp. 107–134). Amsterdam: North Holland.

Doyle, L. (1995). *This sucks, change it!* New York: Pocket Books.

Duchein, M. A. (1993). Remembrance of books past ... long past: Glimpses into aliteracy. *Reading Research and Instruction, 33,* 13–28.

Duranti, A., & Ochs, E. (1986). Literacy instruction in a Samoan village. In B. B. Schieffelin & P. Gilmore (Eds.), *Acquisition of literacy: Ethnographic perspectives* (pp. 213–232). Norwood, NJ: Ablex.

Encyclopaedia Britannica. (1938). United States. In *1938 Britannica book of the year* (pp. 677–688). Chicago: Author.

Encyclopaedia Britannica. (1972). United States (Vol. 22, pp. 578–742). Chicago: Author.

Encyclopaedia Britannica. (1990). United States. In *1990 Britannica book of the year* (pp. 740–743). Chicago: Author.

Encyclopaedia Britannica. (1994). United States (Vol. 29, pp. 149–457). Chicago: Author.

Fasick, A. M. (1973, February). Television language and book language. *Elementary School English,* pp. 125–131.

Flynn, J. R. (1984). The mean IQ of Americans: Massive gains 1932–1978. *Psychological Bulletin, 95,* 29–51.

Flynn, J. R. (1987). Massive IQ gains in 14 nations: What IQ tests really measure. *Psychological Bulletin, 101,* 171–191.

Flynn, J. R. (1994). IQ gains over time. In R. J. Sternberg (Ed.), *Encyclopedia of human intelligence* (pp. 617–623). New York: Macmillan.

Glenn, N. D. (1994). Television watching, newspaper reading, and cohort differences in verbal ability. *Sociology of Education, 67,* 216–230.

Greenfield, P. M. (1972). Oral or written language: The consequences for cognitive development in Africa, the United States and England. *Language and Speech, 15,* 168–178.

Greenfield, P. M. (1984). *Mind and media: The effects of television, video games, and computers.* Cambridge, MA: Harvard University Press.

Greenfield, P. M., & Beagles-Roos, J. (1988). Radio vs. television: Their cognitive impact on children of different socioeconomic and ethnic groups. *Journal of Communication, 38*(2), 71–92.

Greenfield, P. M., Brannon, C., & Lohr, D. (1994). Two-dimensional representation of movement through three-dimensional space. *Journal of Applied Developmental Psychology, 15,* 87–103. Reprinted in P. M. Greenfield & R. R. Cocking (Eds.), 1996. *Interacting with video* (pp. 169–185). Greenwich, CT: Ablex.

Greenfield, P. M., Camaioni, L., Ercolani, P., Weiss, L., Lauber, B. A., & Perucchini, P. (1994). Cognitive socialization by computer games in two countries: Inductive discovery or mastery of iconic code? *Journal of Applied Developmental Psychology, 15,* 59–85. Reprinted in P. M. Greenfield & R. R. Cocking (Eds.). (1996). *Interacting with video* (pp. 141–167). Greenwich, CT: Ablex.

Greenfield, P., & Maynard, A. (1997). *Women, girls, apprenticeship, and school-*

ing: A longitudinal study of historical change among the Zinacantecan Maya. Washington, DC: American Anthropological Association.

Healy, J. M. (1990). *Endangered minds: Why children don't think and what we can do about it.* New York: Simon & Schuster.

Interaction. (1996, Spring). Bellevue, WA: Sierra On-Line.

Klineberg, O. (1935). *Race differences.* New York: Harper & Row.

Laosa, L. (1978). Maternal teaching strategies in Chicano families of varied educational and socioeconomic levels. *Child Development, 49,* 1129–1135.

McFie, J. (1961). The effect of education on African performance on a group of intellectual tests. *British Journal of Educational Psychology, 31,* 232–240.

Mundy-Castle, A. C. (1974). Social and technological intelligence in Western and non-Western cultures. *Universititas, 4,* 46–52.

Neisser, U. (1976). General, academic, and artificial intelligence. In L. Resnick (Ed.), *The nature of intelligence* (pp. 135–144). Hillsdale, NJ: Erlbaum.

Okagaki, L., & Frensch, P. A. (1994). Effects of video game playing on measures of spatial performance: Gender effects in late adolescence. *Journal of Applied Developmental Psychology, 15,* 33–58. Reprinted in Greenfield, P. M., & Cocking, R. R. (Eds.). (1996). *Interacting with video* (pp. 115–140). Greenwich, CT: Ablex.

Ombredane, A. (1956). Etude psychologique des Noirs Asalampasu. I. Le comportement intellectuel dans l'épreuve du Matrix-Couleur. *Mémoires de l'Academie Royale des Sciences Coloniales, 1re Classe, 6,* fasc. 3.

Ombredane, A., Robaye, F., & Robaye, E. (1957). Etude psychotechnique des Baluba. Application experimentale du test d'intelligence. Matrix 38 à 485 noirs Baluba. *Mémoires de l'Academie Royale des Sciences Coloniales, 1re Classe, 6,* fasc. 5.

Ramey, C. (1996, April). Mother–child interaction and intelligence. In *Intelligence on the rise? Secular changes in I.Q. and related measures.* Paper presented at a conference sponsored by the Emory Cognition Project and the American Psychological Association, Atlanta, Georgia.

Rice, M. L., Huston, A. C., Truglio, R., & Wright, J. C. (1990). Words from Sesame St.: Learning vocabulary while viewing. *Developmental Psychology, 26,* 421–428.

Robinett, W. (1982). *Rocky's boot.* Portola Valley, CA: Learning Company.

Salomon, G. (1979). *Interactions of media, cognition and learning.* San Francisco: Jossey-Bass.

Salomon, G., Perkins, D. N., & Globerson, T. (1991). Partners in cognition:

Extending human intelligence with intelligent technologies. *Educational Research, 20*(3), 2–9.

Saxe, G. (1991). *Culture and cognitive development.* Hillsdale, NJ: Erlbaum.

Scheibel, A. (1996, March). *Introduction to Symposium on the Evolution of Intelligence.* Center for the Study of Evolution and the Origin of Life, University of California, Los Angeles.

Schmidt, W. H. O. (1960). School and intelligence. *International Review of Education, 6,* 416–430.

Schwartz, M. (Ed.). (1994). Telecommunication systems. In *Encyclopaedia Britannica* (Vol. 28, pp. 473–504). Chicago: Encyclopaedia Britannica.

Scriven, M. (1987). Taking games seriously. *Education Research and Perspectives, 14*(1), 82–135.

Serpell, R. (1993). *The significance of schooling: Life journeys in an African society.* Cambridge, England: Cambridge University Press.

Thorndike, R. L., Hagen, E. P., & Sattler, J. M. (1986). *Stanford–Binet Intelligence Scale, Item Book 4.* Chicago: Riverside.

Uribe, F. M. T., LeVine, R. A., & LeVine, S. E. (1994). Maternal behavior in a Mexican community: The changing environments of children. In P. M. Greenfield & R. R. Cocking (Eds.), *Cross-cultural roots of minority child development* (pp. 41–54). Hillsdale, NJ: Erlbaum.

Wechsler, D. (1981). *WAIS-R manual.* Cleveland, OH: The Psychological Corporation.

Wechsler, D. (1991). *WISC-III manual.* Cleveland, OH: The Psychological Corporation.

Wheeler, L. R. (1970). A trans-decade comparison of the IQ's of Tennessee mountain children. In I. Al-Issa & W. Dennis (Eds.), *Cross-cultural studies of behavior* (pp. 120–133). New York: Holt, Rinehart & Winston.

Wober, M. (1974). Towards an understanding of the Kinganda concept of intelligence. In J. W. Berry & P. R. Dasen (Eds.), *Culture and cognition* (pp. 261–280). London: Methuen.

Wober, M. (1975). *Psychology in Africa.* London: International African Institute.

5

Are We Raising Smarter Children Today? School- and Home-Related Influences on IQ

Wendy M. Williams

I took a good deal o' pains with his eddication,
sir; let him run in the streets when he was
very young, and shift for hisself.
It's the only way to make a boy sharp, sir.

Charles Dickens, *The Pickwick Papers,* 1836

If children grew up according to early indications,
we should have nothing but geniuses.

Goethe

Too often we give children answers to remember
rather than problems to solve.

Roger Lewin

When I was a kid my parents moved a lot. . .
but I always found them.

Rodney Dangerfield

How can one explain the worldwide increase in IQ test scores over the past 60 years, often called the *Flynn effect*? Today's test takers are scoring on average a whole standard deviation higher than in 1932 (e.g., Flynn, 1987, 1994; see also chapter 2, this volume). My discussion focuses primarily on two groups of causal influences that may have contributed to this effect: changes in schooling and changes in the home environment. I discuss changes that might reasonably be expected to have increased test scores, as well as those that may have decreased them. My perspective is that there is a rich tapestry of interwoven factors affecting test scores, some positive and others negative. The net result is the Flynn effect.

I begin with a review of the magnitude of the effect. Looking at intelligence test score changes over time, Flynn found the following gains per generation of 30 years: 20 points for culturally reduced visual tests of "fluid intelligence" like the Raven Progressive Matrices, 10 points for fluid tests that are not culturally reduced (like verbal analogies or number series), 10–15 points for the Wechsler intelligence tests in general, 9 points for the Wechsler verbal and other verbal tests of "crystallized intelligence," and very small gains for vocabulary subtests and academic achievement tests (also of crystallized intelligence).[1] The crystallized gains are quite substantial and look small only when compared to the fluid gains on the Raven matrices.

The terms *fluid* and *crystallized intelligence* were originally coined to refer to on-the-spot reasoning and problem-solving (fluid) versus accumulated knowledge such as vocabulary and factual knowledge (crystallized; for a review, see Horn, 1994). In tests of crystallized intelligence, the test taker is asked questions such as these: What is the boiling point of water? Who wrote Hamlet? and Why do we need license plates? In tests of fluid intelligence, the test taker must perform tasks such as detecting regularities in patterns, rotating objects in space, reconstructing arrays, and completing matrices with missing parts. In a recent analysis, Flynn (1994) showed that a person born in 1877 who

[1]The term *culturally reduced* refers to tests for which cultural differences in performance have been minimized. Such tests are designed to reduce the test taker's need to rely on familiar cultural artifacts and references.

scored at what was then the 90th percentile on the Raven matrices (the best measure of fluid reasoning skills) would, with the same number of correct answers, score at the 5th percentile of the cohort born in 1967. These population increases in test scores are so enormous that it behooves everyone who uses test scores in research or in work with children or adults to think about what is causing them.

SCHOOL-RELATED FACTORS

School-related factors that may have influenced test scores include increases in school attendance, changes in children's visual world, declines in reading, changes in the content of instruction, declines in textbook level and instructional time, teaching to the test, and increases in educational spending.

Increases in Schooling and the Implications for IQ Test Performance

In the 1930s, mean educational attainment in the United States was 8–9 years, which meant that non-high school graduates were taking IQ tests. In the 1990s, mean educational attainment is 14 years (Bronfenbrenner, McClelland, Wethington, Moen, & Ceci, 1996). Today's test takers, compared to those of a half century ago, have had much more overall exposure to school. What is the significance of this school exposure to IQ test performance?

It is clear that neither fluid nor crystallized intelligence, as measured by contemporary IQ tests, are pure measures of innate endowment. For one thing, both types of intelligence are influenced by schooling. It is perhaps not surprising that crystallized intelligence would increase with exposure to schooling, because school directly teaches the vocabulary and facts tapped by crystallized measures. But although psychologists have generally supposed that fluid measures would be much less influenced by schooling, the data show that school experiences affect this type of performance as well (for a review of multiple studies, see Ceci, 1991).

Cahan and Cohen (1989), for example, used cohort-sequential analysis to show that exposure to schooling increases both crystallized and fluid measures of intelligence. They looked at over 11,000 Israeli fourth, fifth, and sixth graders, comparing children who differed in age by only a couple of weeks but were in different grades because of the birthday cutoff for entering school in a given year. For instance, the study compared 10-year-olds with 4 versus 5 years of schooling. The effects of schooling were seen in both fluid and crystallized measures: For 9 of 12 tests administered, the effect of 1 year of schooling was larger than the effect of 1 year of age. Thus, in general, more school exposure results in higher IQ scores across the board.

Changes in Children's Games, Toys, and Visual World

This topic relates both to children's school and to their home experiences. I discuss it here because the concept of children's changing visual world is important in understanding school-related influences on IQ. Possibly contributing to gains in fluid intelligence are the diverse types of games, toys, and equipment children grow up with today compared to 40 or 50 years ago: computers, computer games, video games, and things that need to be assembled. Playing with these games and toys often requires mental manipulation and rotation, and the resulting new developmental experiences could shape visual–spatial abilities. Most children today have extensive experience with these games and toys. Schools in general, and even economically disadvantaged schools, now have computers for student use. Computers provide children with exposure to graphic designs, rotational movement, and images that may help train fluid reasoning skills. Raven matrices resemble the displays used in certain computer games (and, e.g., the Wallpaper screensaver program), and children often play these games for hours.

Children today also encounter forms of complexity while watching television that were unknown in the 1930s. Life as a whole is more complex for children today; they grow up inundated with stimuli from every corner. Many of these stimuli did not exist a half century ago (e.g., computers, computer games, sophisticated graphics, and other types of images in movies and on television). This information-rich stimulation may affect children's information-processing rates and capabilities. Furthermore, the world of today's children is littered (literally and figuratively) with cereal boxes and place mats at children's restaurants covered with mazes, games, and "find the 25 hidden or embedded animals" activities; McDonald's Happy Meal bags with block design problems; and similar packaging.[2] Many of these tasks are directly comparable to, and sometimes virtually identical to, questions on tests of fluid intelligence.

Declines in Reading

My research on why children do not read and how teachers and parents can encourage children to read (as well as numerous other studies) has shown that children spend between 5 and 8 hr a day watching television and playing computer games on television screens and in video arcades (e.g., Bronfenbrenner et al., 1996; Elley, 1994). Most children read as little as possible, often only 5 min a day (Williams, 1996; Williams, Blythe, White, Li, Sternberg, & Gardner, 1996). This means that today's children (compared to children in the past) have many fewer hours of exposure to the printed word; less familiarity with reading; and relatively more familiarity with geometric figures, patterns, and pictorial representations. Perhaps this lack of reading (and the associated decline in exposure to vocabulary words) has kept crystallized intelligence scores from increasing as much as fluid scores, whereas the increased exposure to pictures, figures, and shapes has contributed to fluid gains.

One test of this potential relationship would consist of comparing visual–spatial abilities (e.g., performance on Raven matrices) for children living in rural versus urban areas of the United States (with appropriate controls and measures of the home environment). One might also compare rural children who have computers and computer games in the home with rural children who do not. Again, my thesis is that one should think of the Flynn effect as the net gain in a tapestry of differentially weighted factors.

[2] Although a few magazines, newspapers, and comic books from the 1930s contained examples of maze problems and embedded figures tasks, the exposure of the average child to such stimuli was extremely limited in the 1930s compared with the 1990s.

The Matching of Instruction to the Demands of Different Types of IQ Tests

Broadly considered, trends in educational practice can either foster or inhibit the development of students' spatial abilities as measured on specific types of ability tests. In fact, the content of schooling has changed dramatically over the past half century. On the one hand, there is less emphasis on crystallized intelligence: Children today generally get less practice in and less learning time for fact memorization. Students' actual academic learning time, defined as engaged time spent on meaningful and appropriate tasks, has declined to 90 min per school day (Weinstein & Mignano, 1993). On the other hand, fluid thinking skills may be practiced more often today: A fourth-grade class may spend the entire day making Native American dyes and food colorings and creating bold geometric patterns in cloth instead of practicing the three Rs.

Most schools today allow children to work with math manipulatives—blocks, designs, and other three-dimensional representations—in place of more traditional paper-and-pencil math exercises. Therefore, today's children spend more time on math activities than on traditional types of math instruction. This "activities" focus and trend, which began around the late 1960s, exists across the curriculum. This focus has switched the emphasis of instruction onto interactive, two-way, child–child and child–teacher school experiences and away from rote solitary learning and memorization. So what, exactly, children learn is different today: They learn less declarative knowledge ("knowing that") and more procedural knowledge ("knowing how") and strategies. Perhaps this increase in procedural knowledge and knowledge of strategies, coupled with the practice with activities and manipulatives, translates into gains on measures of fluid intelligence.[3]

In addition, the emphasis on taxonomic, categorical reasoning, more often taught in the classrooms of the 1990s than the 1930s, may

[3]In general, schools today spend more time on hands-on activities and less time on traditional instruction. Rote, drill-based, and book-based solitary learning have been replaced throughout the U.S. school system by activities-based, exploratory, discovery-learning approaches, as influenced by the work of Piaget (e.g., 1970; Ginsburg & Opper, 1980).

increase performance on those fluid analogy problems that draw on well-known and shared crystallized knowledge, assumed to be constant across children. For example, "Apple and orange are alike because both are __ [fruits]" is a question involving taxonomic reasoning that loads on fluid intelligence. Series problems like "2, 4, 6, 8, __, 12" also require the use of well-known, crystallized knowledge to reason in a fluid way. In sum, for fluid intelligence tasks involving reconstructing patterns, rotating objects, and solving mazes, and for similar questions measuring abstract-reasoning ability, children today may benefit from many types of contemporary school activities. However, for factual questions that tap crystallized intelligence, such as "What's the boiling point of water?" children of the 1930s may have had an advantage.

Declines in Textbook Level and Instructional Time

There are also other important changes in school experiences. Donald Hayes and his colleagues (Hayes, Wolfer, & Wolfe, 1996) analyzed the complexity of words, sentences, and concepts appearing in textbooks used in different grades throughout the century. They found that textbooks have been "dumbed down" two or more grade levels in the last 30 years. Dumbing down of texts means that children learn less factual information; this would tend to decrease crystallized intelligence. Perhaps the associated decline in actual instruction time (and, specifically, traditional instruction time) has left children with more time for activities that build fluid reasoning skills, thus contributing to the Flynn effect.

School time spent on actual instruction has decreased sharply since the 1940s. The over 1,000 hr per year of instruction mandated by most states often translates into only about 300 hr of quality academic learning time (defined as time spent engaged in meaningful and appropriate tasks; Weinstein & Mignano, 1993). Three hundred hours divided by 187 school days equals about 90 min of actual quality learning time per day. Almost every study examining time and opportunity to learn has found that the amount of content covered predicts the amount learned (Berliner, 1988). Thus, learning of facts, information, and vocabulary has almost certainly decreased. This trend has probably exerted downward pressure on crystallized intelligence scores. The fact that crystallized scores have not declined suggests the existence of offsetting or compensatory activities and experiences that exert an upward pressure. These may include extracurricular activities, educational television, exposure to Happy Meal bags and cereal boxes with vocabulary words, and so on.

Teaching to the Test

A major focus of education since the 1960s and 1970s is referred to by teachers and administrators as "teaching to the test." Teachers are highly pressured by parents and administrators to keep students' test scores high (Hanushek, 1986). Teachers' jobs, principals' raises, and even the value of real estate in school districts are affected by test scores (Woolfolk, 1995). Because tests are best at measuring knowledge of facts and basic skills, some teachers may focus on these facts and skills to the exclusion of other aspects of the curriculum. Many teachers also spend a great deal of time on direct training of test-taking skills. There is a general acknowledgment by teachers, administrators, and parents that teaching to the test has become more prevalent and emphasized because of the U.S. society's increased emphasis on test performance.

Smith (1991) found that preparing for, giving, and recovering from standardized tests took an average of 100 hr of instructional time per year. One hundred hours is a significant proportion of the 1,000 hr of mandated school time for most states, especially when one considers that students receive only about 300 hours of actual direct instruction per year (Weinstein & Mignano, 1993). Therefore, one might expect gains in both fluid and crystallized intelligence; better test-taking skills in general would lead to overall higher scores, and drilling on facts and vocabulary would lead to increased crystallized scores.

U.S. society virtually guarantees that only children who do well on standardized tests have access to the best schools, programs, and resources. Because a child's future depends in large part on her or his test performance today, teachers spend significant classroom time teaching students how to do well on upcoming ability and achievement tests, which are given each year in many schools. When new curricula are announced, teachers immediately want to know if these curricula will help children score higher on tests. Parents and school administrators are also heavily invested in the test-score issue (Hanushek, 1986).

Much of this teaching to the test involves the direct teaching of strategies for guessing and test taking (such as how to "outsmart" the test developer and figure out the right answer). In addition, teachers sometimes drill students more on basic facts as a method of teaching to the test, but this takes far more time than does basic strategy training. Almost all teachers stress how to score higher by being testwise, and this strategy training would be likely to increase IQ scores across the board. Extra fact teaching—if teachers choose to prepare children by stressing the learning of extra facts, vocabulary, and so on—should exert upward pressure on crystallized scores.

My research on practical intelligence for school, conducted with colleagues at Yale and Harvard Universities, resulted in a book written to help teachers, administrators, and parents increase students' practical intelligence so that these students could profit more from school (Williams et al., 1996). Although the book contains sections designed to help students read critically, write better prose, and do better on homework, what teachers and administrators focused on was the section called "Preparing for Tests." Students were highly motivated to absorb this part of the curriculum: Many had been told since they were 4 years old that their test scores would either provide or deny them access to everything they wanted in life.

The study by my colleagues and I of the importance of students' practical intelligence for school showed that students who learned specific test-taking strategies did better as the year progressed on all types of tests, compared to matched control students in the same schools not exposed to the training. The point is that test-taking skills can be trained and can increase students' scores measurably. The success of school-based and commercial programs to increase students' scores on tests of admission to college, graduate school, and professional school provides additional testimony to just how open to training effects test performance is. Parents in New York City even train their 4-year-olds in how to do well on screening tests to gain admission to exclusive kindergartens (Gardner, 1996)!

Short high school-based training programs result in average gains on the Scholastic Aptitude Test (SAT) of 10 points in verbal scores and 15 points in math scores, whereas longer, commercial programs cause gains of 50 to as many as 200 points for some test takers (Owen, 1985). In a meta-analysis of 40 studies of aptitude and achievement test training, Kulik, Kulik, and Bangert (1984) found that more substantial gains occurred when students practiced on a parallel form of the test for brief periods. Therefore, how coaching programs are designed makes a difference in the score gains realized by students. Similarly, direct practice on IQ-test-like items, which is common in many schools, may have a significant effect on students' test performance.

Test-familiarity-oriented training and strategy training are both important to students' scores. Familiarity with the procedures of the test (which may lead to greater self-confidence and feelings of assurance), practice with answer sheets, and familiarity with specific question types all help students perform better (Anastasi, 1988). In addition, students are helped by instruction in general cognitive strategies such as general problem-solving skills, defining the question or problem correctly, knowing what is relevant to solving the problem and what is not, considering and analyzing alternative solution paths, choosing the best alternative, monitoring progress toward a solution, and checking work. Training in these skills helps students do better on many types of tasks (Anastasi, 1988; Williams et al., 1996). As I discussed earlier, the Flynn effect is largest for tests, such as the Raven matrices, which have been thought to be "culturally reduced" and, consequently, unresponsive to training in how to be testwise. However, Schliemann and Simoes (1989) found that Brazilian children's performance on the Raven matrices varied as a function of exposure to schooling. Thus, even performance on the Raven matrices may be influenced by test-strategy training and exposure to test taking. In general, there is substantial evidence that "testwise" training increases test performance, and today's students are likely to receive more of this type of training than did the students of a half century ago.

Increases in Educational Spending

Additional significant changes in students' school experiences may have resulted from increases in educational spending. Has education been improving? Are children learning more today in school? Much more money is spent on education today than in the past (Grissmer, Kirby, Berends, & Williamson, 1994; Hanushek, 1986): Public spending on elementary and secondary education was $18 billion in 1960, $45.7 billion in 1970, $108.6 billion in 1980, and $132.9 billion in 1983. In real inflation-controlled dollars, this represents a 250% increase. However, expressed as a percentage of the gross national product, the increase is more modest: 3.6% in 1960 compared with 4.0% in 1983. In addition, the population was increasing in size, so that expressing spending for education on a "per pupil" basis may be more appropriate. In 1960, spending was $1,711 per pupil; in 1990, spending was $4,775 per pupil (U.S. Bureau of the Census, 1992, p. 138). Consider a quote by Rothstein and Miles (1995): "With appropriate inflation adjustment, it appears that total real education spending per pupil increased by 61% from 1967 to 1991. Admittedly, this is a substantial increase, but it is much less than the 200% spending increase commonly assumed to have occurred" (p. 2).

However one examines the trend, its direction has clearly been up. But where has this money gone—on what, exactly, has it been spent? It turns out that much of this increase in expenditure has gone toward paying teachers higher salaries (Grissmer et al., 1994; Hanushek, 1986). Also, The Education for All Handicapped Children Act (Public Law 94-142), Title I, and Title IX have all resulted in greater expenditures on previously inadequately served children in the attempt to provide all children with equal access to the educational system. Therefore, it is difficult to say whether the quality of education for the average child has improved, or rather, if the nature of education and the populations served have simply changed. Still, it seems plausible that money spent on education might contribute to the Flynn effect through many potential mechanisms. It is possible that these funds have enhanced the quality of education for all students, thereby improving average intelligence test performance (for more detailed discussions, see chapters 10, 11, and 12, this volume).

HOME-RELATED FACTORS

Home-related factors that may have influenced test scores include the trend toward smaller families, greater educational attainment of parents, urbanization, changes in parental style, changes in stress levels of pregnant women, and improvements in health and nutrition.

Trend Toward Smaller Families

Fewer children per family, which has been the trend for nonwelfare families, generally means more resources per child (Hernandez, 1997). Nonwelfare-family income per child has increased over the time frame of the Flynn effect (Ceci, 1996; Grissmer et al., 1994). Real income has not grown, but dollars per child have. Clearly, more dollars per child can mean more resources per child, and the potential IQ-enhancing value of a family's financial resources is obvious (e.g., providing computers, tutors, summer enrichment camps, and so on). Increases in IQ may thus reflect the increase in IQ-enhancing experiences provided by today's parents compared to parents during the Depression and World War 2, for example. (Flynn, 1987, noted that with increased affluence comes an increase in IQ: He found an average 2.5-point IQ gain per 30-year generation due to a general increase in socioeconomic level.)

Greater Educational Attainment of Parents

As parents in successive cohorts become more educated, their children grow up in households that are more likely to stress education. Better educated parents are also able to teach their children the types of information relevant to IQ test performance. Thus, parents' educational level is an important potential contributor to the Flynn effect. In addition, even if the parents themselves are not well educated, research has demonstrated the significance of children's interacting with one educated adult in their lives. One educated adult who can provide mentoring and role modeling can help inner-city children escape the cycle of poverty (Blau, 1981). Blau found that African American innercity children with "better outcomes" had at least one adult with an educational orientation in their lives (this adult could be a parent, other relative, friend, or other interested person); interacting with one person with an educational focus is essential to the child's outcome. Therefore, the broad societal trends toward greater educational attainment (Ceci, 1991) affect not only the children of more educated people, but also other children with whom these people come into contact.

What trends have characterized the educational level of parents? Educational attainment of nonminority-group parents has increased by 70% and that of minority-group parents by 350% from 1973 to 1990 (Bronfenbrenner et al., 1996; Grissmer et al., 1994; also see chapter 11, this volume). Data from Hernandez (1995) indicate that in 1920, 18% of mothers and 16% of fathers had attained an educational level of 4 years of high school or more. In 1983, 81% of mothers and 86% of fathers had attained that level. The increase has been enormous by any standard. The trend basically hit asymptote (i.e., leveled off) in about the early 1970s. These increases in the educational attainment of parents may have contributed to gains in children's IQ scores. Direct evidence for the effects of educational attainment on IQ test performance can be found in reviews by Ceci (1991) and Ceci and Williams (1997) as well as in chapter 11 (this volume).

Trend Toward Urbanization: The Decline of Rural, Agricultural Childhood

Over the past half century, the population of many countries has become increasingly urbanized. The rural agricultural lifestyle has declined (Hernandez, 1995, 1997); today's children are more likely than ever to grow up in urban and suburban communities. Part of the information saturation that marks childhood today is associated with this move to the cities. Life in rural communities more often connects day-to-day activities to basic survival (e.g., concerns about weather; maintenance of buildings, vehicles, and machinery; or protection of water and food supplies). In urban and suburban areas (with the exception of high-crime city regions), the problems of living are of different types.

Whereas city dwellers may think about bus schedules, locations of businesses and stores, finding and keeping jobs, and avoiding troubled sections of town, rural dwellers may think about maintaining well-water quality during periods of drought, caring for livestock, and stocking sufficient food to last through winter snowstorms.

Rural and urban dwellers both face the need to survive, but the problems they face require different types of thinking. The thinking associated with city life may more closely approximate the types of questions on many IQ tests. For example, rural children's knowledge of farm crop rotation and livestock maintenance is unlikely to be as relevant to IQ test performance as is urban children's knowledge of complex bus schedules and current events. It is not only the immersion in certain types of information that could affect IQ test performance; children growing up in rural communities often help around the farm from an early age and may have less spare time for playing computer games, watching television, mall hopping, and the like.

Ceci (1991) and Flynn (1987) have demonstrated that the IQs of urban children are higher than those of rural children. This difference may reflect the greater cognitive stimulation and test-relevant exposure children experience in urban environments. Regardless of the precise causes, the fact that more children now live in urban as opposed to rural areas than was the case a half century ago means that IQs would be expected to have increased because of this demographic transition. Flynn (1987) also showed mild increases in IQ due to the effects of urbanization in France around the period of World War 2. In summary, it is possible that the population's move to the cities has contributed to the Flynn effect.

Changes in Parental Style

Are today's parents raising their children differently from the parents of a half century ago? Could prevailing styles of parenting be contributing to the Flynn effect? To answer this question, I first consider the evidence regarding the effects of parental style on children's IQ (for a review, see Williams, 1994). Next, I consider the evidence that parental styles have changed. A caveat regarding the behavior-genetics literature is also discussed. However, given the fact that environmental influences must be responsible for the Flynn effect, this discussion emphasizes environmental as opposed to genetic factors and issues.

Parental style has been defined as "a constellation of attitudes toward the child that are communicated to the child and that, taken together, create an emotional climate in which the parent's behaviors are expressed" (Darling & Steinberg, 1993, p. 488). Baumrind (e.g., 1991) described parental behavior as fitting within three general types or patterns: *authoritarian*, *permissive*, and *authoritative*.

Parents with an authoritarian style try to shape and control their children's behaviors, which they evaluate against a set of rigid standards. Authoritarian parents emphasize obedience, respect for authority, hard work, and traditional values, and they discourage authentic communication in favor of the "listen and obey" mode. Baumrind (1971, 1973) described authoritarian parents as being high in demandingness and low in responsiveness toward their children. The permissive parental style describes parents who give their children considerable freedom. Permissive parents have a tolerant and accepting attitude toward their children, rarely punish them, and make few demands and place few restrictions on them. The authoritative parental style refers to parents who set clear standards and expect their children to meet them, treat their children maturely, and use discipline when appropriate to ensure that rules are followed. These parents encourage their children to develop independence and individuality, and consequently they practice open communication in which children's points of view and opinions are considered.

Baumrind's research explored the interrelationship of parental style and children's cognitive and social competence. She began by studying preschool children to learn what effects parental style had on the children's intelligence and personality. Later, Baumrind and other researchers studied children in middle school and high school, children of different races and ethnic groups, and children of different socioeconomic backgrounds. Representative studies are reviewed later to provide an overview of the evidence for the effects of parental style on children's cognitive competence.

Bee and her associates investigated the mother–child relationship as a predictor of later IQ and language development in children (Bee et al., 1982). They found that infant physical status was a poor predictor of 4-year IQ or language, that child performance was a poor predictor before 24 months of age and good thereafter, and that family ecology (i.e., the composition of the household) predicted child IQ and language within a low-education subsample but not among mothers with more than a high school education. However, the quality of mother–infant interaction was one of the best predictors at every age tested; it was as good as actual child performance in predicting IQ and language development.

Estrada, Arsenio, Hess, and Holloway (1987) found that the affective quality of the mother–child relationship when the child was 4 years old was associated with mental ability at age 4, IQ at age 6, and school achievement at age 12. The associations remained significant even after the effects of mother's IQ, socioeconomic status, and children's mental ability at age 4 were taken into account. The authors suggested that affective relationships influence cognitive development through the parent's willingness to help children solve problems, through the development of children's social competence, and through the encouragement of children's exploratory tendencies. Hess, Holloway, Dickson, and Price (1984) found that maternal measures taken during preschool years predicted school readiness at age 5 and achievement test performance in Grade 6. However, the prediction was stronger at age 5 than age 12, meaning that the mother's influence on school achievement was stronger during preschool years.

Dornbusch and his colleagues (Dornbusch, Ritter, Leiderman, Roberts, & Fraleigh, 1987) examined the relation of parenting style to adolescent school performance in a sample of 7,836 high school students. The authors found that both authoritarian and permissive styles were associated with lower grades, whereas authoritative parenting was associated with higher grades. In a similar study investigating parenting practices and adolescent achievement in 6,400 high school students, Steinberg and his colleagues focused on the impact of authoritative parenting, parental involvement in schooling, and parental encouragement to succeed on adolescent school achievement (Steinberg, Lamborn, Dornbusch, & Darling, 1992). The authors found that authoritative parenting led to better school performance and stronger school engagement. They also found that parental involvement with schooling was a positive force in adolescents' lives when the parents had an authoritative style but less so when the parents had other styles.

The processes through which parental behavior affects a child's development were studied by Rogoff and Gardner (1984), who watched 32 middle-class mothers preparing their 6- to 9-year-old children for a memory test. The mothers guided the children in transferring relevant concepts from more familiar settings to the relatively novel laboratory task, thus assisting the children in mastering the task and in developing methods for completing similar future tasks. Formal attempts to measure the processes through which parental style influences child development in the context of more typical parent–child interactions have often focused on the home environment (e.g., Bradley & Caldwell, 1984). Researchers have evaluated various features of parenting behavior in the home, such as maternal responsivity, maternal acceptance of the child, maternal involvement, language stimulation, and encouragement of social maturity, through the use of the Home Observation for Measurement of the Environment (HOME) Inventory.

In a study by Elardo, Bradley, and Caldwell (1977), various aspects of the early home environment were related to language development at age 3. These aspects were the emotional and verbal responsibility of the mother, the provision of appropriate play materials, and maternal involvement with the child. Bradley and Caldwell (1984) found that HOME scores from age 2 predicted intelligence test scores at ages 3 and 4½ as well as first-grade achievement test scores. Later, Bradley, Caldwell, and Rock (1988) examined children as infants and at age 10, finding significant correlations between home environments measured at both 2 and 10 years and children's achievement test scores and classroom behavior.

In summary, there is considerable evidence for a strong association between parental style and children's cognitive competence. The key word here is *association*: Little evidence has demonstrated clear cause-

and-effect relationships between parental style and children's intelligence. Scarr (e.g., 1985, 1992) noted that mothers with higher IQs tend to have children with higher IQs and that these mothers also tend to have more effective parental styles. Therefore, the findings showing a link between parental style and children's cognitive competence were interpreted by Scarr as being due to the effect of shared genetically transmitted intelligence between mother and child. Scarr noted that parents' behaviors are correlated with their children's ultimately because of shared genes and that what is observed in the world of parenting and child development is explicable even if parents have no effects on their children or vice versa. (Not all developmentalists share this view; see the October 1993 issue of *Child Development* for opposing viewpoints.)

Researchers in behavior genetics have noted that family influences on children's IQs decrease dramatically throughout development, resulting in a significant reduction in these environmental effects on intelligence by adulthood (Plomin & Daniels, 1987). Studies of adoption have shown that parental style of adoptive parents has little or no effect on IQ (Scarr, 1997). Scarr concluded that

> although estimates of shared nongenetic effects on individual differences in IQ from twin and adoption studies vary from 0% to 10%, one can say that they appear to be rather small in adult populations in Western societies.... Socialization Theory, focused on parental rearing practices, which presumably vary mostly between families, accounts for no more than 10% of the variation in intelligence in these populations. (1997, pp. 27–28)

Of course, the behavior-genetics findings in general must still be reconciled with the ample evidence for environmental effects on intelligence, including studies such as Cahan and Cohen's (1989) cited earlier, which documents significant increases in IQ, and particularly in fluid intelligence, directly attributable to the influences of schooling.

Is there any hard evidence for a *causal* relationship between parental style and children's IQ? Two studies are relevant, one experimental and one correlational. In 1978, Riksen-Walraven conducted a study of 100 Dutch mothers' interactions with 9-month-olds. This research looked at parental responsiveness to infants and, especially, at the role of parental stimulation of infants in the infants' development. Mothers were randomly assigned to four groups characterized by different types of interaction; the amounts of interaction were constant across the groups. These types of interaction differed in quality and timing.

One group of mothers was instructed not to direct the child's activities too much, but to give the child the opportunity to find things out for her- or himself, to praise the child for these efforts, and to respond to the child's initiations of interactions. Basically, these mothers were taught to be responsive to their child (who was in control) and to support the child's initiatives, a parental style that creates a type of mediated-learning experience. Another group of mothers was told to speak often to their infants and to initiate interactions frequently; these mothers controlled the interactions instead of responding to the child. A third group was instructed to do a mixture of what the first two groups were doing. A fourth, control group, was given no instructions.

Three months later, the researcher observed and tested all infants. The mothers' behaviors differed significantly from group to group in accord with the instructions they had received. Infants of mothers who had been encouraged to be responsive showed higher levels of exploratory behavior than any other group and preferred novel to familiar objects. These babies also learned more quickly in a contingency task; that is, infants randomly assigned to a condition of greater maternal responsiveness showed enhanced cognitive functioning. The conclusion is that different styles of parenting may cause differential cognitive development in children.

In correlational research supporting the same conclusion, Bettes (1988) investigated the effects of maternal depression on mothers' speech to their infants and ultimately on the infants' cognitive and emotional development. The participants were 36 mothers and their 3- to 4-month-old infants. Ten of the mothers were rated as depressed. The mothers wore tape recorders to capture the vocalizations between them and their infants. When the babies cooed, the nondepressed mothers were responsive and cooed back quickly. Depressed mothers had a greater latency before responding to their babies; depressed mothers' vocal patterns were *not* tied to their children's vocal output. This latency meant that babies of depressed mothers had less predictability in controlling their environment. There was no difference in the vocalization patterns of infants of nondepressed versus depressed mothers at the start of the experiment (when the infants were 3–4 months old), but after 6–9 months the children of depressed mothers had shut down linguistically.

Bettes thus showed in a mini-longitudinal study that caregiver behavior matters. Depressed mothers failed to modify their behavior according to the behavioral development of their infants; these mothers were slower to respond to vocalizations, and eventually their infants started to vocalize less than infants of nondepressed mothers. Bettes stated that her study provided evidence for how maternal depression increases the infant's risk for psychopathology. For the present purposes, Bettes's study provides additional evidence that *how parents interact with their children affects the children's cognitive development.*

Given the evidence of the effects of parental style on cognitive development, the question left to consider is whether styles of interacting with children have changed for parents over the last half century or so, the time frame of the Flynn effect. Bronfenbrenner (1985, 1989) and Bronfenbrenner and Ceci (1994) have looked at changes over time in the nature and type of parental attention. In particular, the authors considered maternal responsiveness versus maternal directedness. They found that mothers' perceptions and ideas about how to mother, from the 1940s to the present, have changed for middle-class mothers. "How-to-parent" books and magazine articles have proliferated, and simultaneously, the content of the advice given to parents has changed dramatically, with associated changes in parents' beliefs and attitudes. The old literature stressed feeding on a strict schedule and using ample discipline. Since the 1940s, however, middle-class mothers have moved toward feeding on demand and responding more to their child. Early in the century, lower class parents were permissive and middle-class parents were rigid and controlling. In about the 1940s, this pattern shifted: Lower class parents became more rigid and authoritarian, and middle-class parents became more permissive (Bronfenbrenner, 1985; Bronfenbrenner & Ceci, 1994).

The point is that parenting norms have changed; middle-class mothers have been reading how-to books and have been influenced by these materials about recommended child-rearing practices. Thus, one may conclude that maternal responsive behavior may be more common in the 1990s than it was in the 1930s, although during the last 20 years this trend may have reversed somewhat (Bronfenbrenner et al., 1996). There are fewer children per mother today, which allows for more responsiveness. It is possible that these changes in responsiveness have contributed to the Flynn effect.

However, it is also true that, today, more mothers work outside of the home. To the extent that a mother is not available to interact with her child, this might argue for the opposite direction of developmental effects (if the quality of child care is not high and if one can assume lower responsiveness by care providers). The trend for more and more mothers to work outside the home has existed since World War 2: In 1949, 10% of mothers worked outside the home, compared to over 60% in 1994 (Bronfenbrenner et al., 1996). Once again, the picture is not straightforward when it comes to understanding the causes of the Flynn effect.

Changes in Stress Levels Affecting Pregnant Women and Their Babies

Another factor that might contribute to the Flynn effect is a change over the past half century in the level of stress experienced by pregnant women and the effects of this stress on the developing fetus. Huttunen and Niskanen (1978), two Scandinavian pediatricians and demographers, studied women whose husbands died *either* during a pregnancy (n = 168) or in the 9 months following the birth of a child (n = 167). The authors compared the outcomes for the children (born between 1925 and 1957) in these two groups to examine the effects of stress while the baby is in utero compared to the same stress once the baby is born. Babies whose fathers died when they were in utero were at an increased risk for later mental disorders and had an increased frequency of psychiatric disorders. But the effects were not mediated by birth complications: There were no more birth complications in the experimental than in the control group. It is reasonable to infer that the psychosocial risk factors that resulted in the decrements noted by Huttunen and Niskanen would be associated with cognitive decrements as well (Rutter, 1989).

Huttunen and Niskanen expressed the belief that the causes of the negative effects in the experimental group were related to the direct biological stresses on the mothers while the babies were in utero. (Remember that all the children grew up fatherless; the issue was simply whether the father died before or after the baby's birth.) The causal mechanisms may involve hormones and stress on the part of the mother; stress hormones may damage a child's system in utero. The authors did note, however, that the emotional state of mothers widowed a few months earlier as opposed to later may have created a different bonding and early-life experience for the babies in the experimental group (whose fathers died during the mothers' pregnancy) compared with the babies in the control group (whose fathers died after the baby's birth).

Are stresses greater or lesser on pregnant women today compared with a half century ago? In the 1930s, mothers contended with war, famine, disease, dangerous childbirth conditions and high childbirth risks, economic scarcity, poverty, and exhaustion due to caring for multiple children. In the 1990s, mothers have better health care, better access to food and good nutrition, a far easier time with childbirth because of medical advances, and fewer babies. But today there are other types of stresses associated with modern life; for example, women often work at demanding careers, have only short maternity leaves, and experience "drive-through births," sometimes with hospital stays of only 24 hr. There are also the stresses of crime, violence, and urban decay for city dwellers (Bronfenbrenner et al., 1996).[4]

[4] More generally, it is worth noting that children born during periods of economic scarcity generally have lower IQs. The heritability of IQ goes down in periods of scarcity because of ceiling effects (Bronfenbrenner & Ceci, 1994). In other words, children with the genetic capacity to score well can have their scores depressed because of poor developmental conditions. Ceci interpreted data from a study by Sundet, Tambs, Magnus, and Berg (1988) of secular trends in the heritability of intelligence test scores in Norwegian twins; he stated that these data reveal that coefficients of

Improvements in Health and Nutrition

Improvements in general health and nutrition of developing children and pregnant women must also be considered as potential contributors to the Flynn effect (see chapters 6 and 7, this volume, for detailed discussions). It seems reasonable that health and nutrition might influence cognitive development and performance. However, there is confusion regarding the enduring cognitive effects of early malnutrition (Strupp & Levitsky, 1995; for a review, see Ricciuti, 1993). Animals exposed to early malnutrition show lasting changes in emotionality, motivation, and anxiety. These factors affect all aspects of behavioral functioning, including cognition. Animals rehabilitated after the nutritional deprivation show lasting changes in cognitive flexibility and, possibly, susceptibility to proactive interference. However, the effects of nutritional deprivation have not even been studied for many cognitive processes; therefore, the data are inconclusive.

On the one hand, Stein, Susser, Saenger, and Marolla (1972) found that nutritional deprivation did not affect cognitive performance. Their study of the Nazi siege of The Netherlands evaluated the cognitive effects of a severe wartime famine in the winter of 1944–1945 that affected many pregnant women. Despite the severity of the famine, it yielded no cognitive effects when the women's children, now at 19 years of age, were tested before entry into the Dutch military. The 100,000 men born during the height of the famine at the end of World War 2 did not differ in IQ from men not exposed to a famine while in utero.[5]

On the other hand, Donald Bundy and colleagues (e.g., Bundy &

---

heritability of IQ tend to be lower during periods of war and economic depression. Thus, the performance of children on IQ tests is affected by prevailing economic conditions, and the poorer economic conditions at the beginning of the period under consideration (1930s) may have contributed to the Flynn effect. As economic conditions improved, one might expect IQ scores to have improved, as well. As described earlier, Flynn (1987) found that one generation of affluence (improvement in socioeconomic level) led to a 2.5-point gain in IQ.

[5]Note that the data from Sundet et al. (1988) just described involved heritabilities, which are range dependent, inasmuch as they were based on intraclass correlations. The data from Stein et al. reflected absolute levels of performance (mean IQs). Changes in means can proceed independently of changes in population variances (see Bronfenbrenner & Ceci, 1994). In other words, it is theoretically possible for heritabilities to go down while in the same population mean IQ goes up or vice versa. In practice, however, heritabilities tend to go down when the IQs of people with the genetic capacity to score high are depressed, perhaps owing to environmental assaults.

Cooper, 1989; Nokes & Bundy, 1994), reporting on World Health Organization–sponsored studies of the role of parasite infection in cognitive performance, argued that giving infected schoolchildren one 15¢ pill improved school and test performance dramatically. Students who were repeating grades and achieving extremely poorly became more alert and started doing well. These studies also investigated the effects of micronutrient administration along with the anthelminthic drugs. The main finding is that children with parasite infections are lethargic and unable to concentrate, and they do poorly in school. Proper nutrients and better medical care can result in major improvements in these children's cognitive functioning.

Members of the Child-to-Child Trust in London (Bailey, Hawes, & Bonati, 1994; Lansdown, 1995) have studied and disseminated literature on the topic of mothers' understanding of the role of good nutrition and good-health habits in countries all over the world. The data show that when mothers understand the mechanisms of promoting health, including how to provide proper nutrition, children are more likely to attend school for more years, perform better on tests, and succeed within the school environment. Therefore, the mother–child relationship can potentially lead to increases in children's IQ scores if the mothers have more knowledge of health and nutrition, behave in accordance with this knowledge, and train their children in positive health and nutrition habits.

Due to widespread health-education campaigns, such as Women, Infant, and Children (WIC) food programs and social-science-based food education programs, mothers' understanding of how to promote children's health has increased in the United States and other developed countries over the past 60 years. Nutrition (including micronutrient supplementation) and medial care (including immunizations) have improved. It is possible that these changes have contributed to the Flynn effect.

CONCLUSION

I have attempted in this chapter to scan the topography of world changes during the second half of this century, focusing on those that might be contributing to the Flynn effect. My thesis is that a tapestry of forces can be identified, some of which exert upward pressure on test scores and some of which exert downward pressure. I have suggested some highly probable causes of upward pressures on test scores (e.g., increased educational attainment of parents, decreased family size and increased family resources, and increased school attainment of children) and also some possible causes (e.g., shifts in parental style). I have also presented a list of factors that might cause downward trends (e.g., decreased emphasis on fact learning in school). The logical next step would be to link each factor with specific changes in fluid and crystallized intelligence scores and to examine these trajectories in various subgroups. Only by understanding how specific causal effects combine can one truly come to understand why IQ scores are escalating.

The Flynn effect describes a general trend, but patterns in the Flynn-effect data may help provide a more detailed picture pointing toward additional causal factors. For example, researchers should consider the specific types of items on IQ tests, look at what types of items and even what specific items have shown the largest changes over time, and note what types of items have not shown substantial changes. If patterns of increases and decreases in IQ scores correlate with social or economic trends, such as relative rates of children living in poverty, additional insight into potential mechanisms may be gained. To understand developmental factors that affect scores over time, one must look not only at global population trends, but also at specific, local changes and correlates of overall patterns of changes in IQ test performance on specific types of items as well as on specific items. Researchers must consider societal trends, parenting trends, and schooling trends from decade to decade and try to link these patterns of changes to patterns of changes in test scores. It seems wise to set future sights on this agenda.

REFERENCES

Anastasi, A. (1988). *Psychological testing.* New York: Macmillan.

Bailey, D., Hawes, H., & Bonati, G. (Eds.). (1994). *Child-to-child.* London: The Child-to-Child Trust.

Baumrind, D. (1971). Current patterns of parental authority. *Developmental Psychology Monograph, 4,* 1–103.

Baumrind, D. (1973). The development of instrumental competence through socialization. In A. D. Pick (Ed.), *Minnesota Symposium on Child Psychology* (Vol. 7, pp. 3–46). Minneapolis: University of Minnesota Press.

Baumrind, D. (1991). Parenting styles and adolescent development. In J. Brooks-Gunn, R. Lerner, & A. C. Peterson (Eds.), *The encyclopedia of adolescence* (pp. 746–758). New York: Garland Press.

Bee, H. L., Barnard, K. E., Eyres, S. J., Gray, C. A., Hammond, M. A., Speitz, A. L., Snyder, C., & Clark, B. (1982). Prediction of IQ and language skill from perinatal status, child performance, family characteristics, and mother-infant interaction. *Child Development, 53,* 1134–1156.

Berliner, D. (1988). Simple views of effective teaching and a simple theory of classroom instruction. In D. Berliner & B. Rosenshine (Eds.), *Talks to teachers* (pp. 93–110). New York: Random House.

Bettes, B. (1988). Maternal depression and motherese: Temporal and intonational features. *Child Development, 59,* 1089–1096.

Blau, Z. S. (1981). *Black children/White children: Competence, socialization, and social structure.* New York: Free Press.

Bradley, R. H., & Caldwell, B. M. (1984). The relation of infants' home environments to achievement test performance in first grade: A follow-up study. *Child Development, 55,* 803–809.

Bradley, R. H., Caldwell, B. M., & Rock, S. L. (1988). Home environment and school performance: A ten-year follow-up and examination of three models of environmental action. *Child Development, 59,* 852–867.

Bronfenbrenner, U. (1985). Freedom and discipline across the decades. In G. Becker & L. Huber (Eds.), *Ordnung und Unordnung [Order and disorder] Hartmut von Hentig* (pp. 326–339). Weinheim, Federal Republic of Germany: Beltz Verlag.

Bronfenbrenner, U. (1989). Ecological systems theory. In R. Vasta (Ed.), *Annals of Child Development Research, 6,* 185–246.

Bronfenbrenner, U., & Ceci, S. J. (1994). Nature–nurture reconceptualized in developmental perspective: A bioecological model. *Psychological Review, 101,* 568–586.

Bronfenbrenner, U., McClelland, P., Wethington, E., Moen, P., & Ceci, S. J.

(1996). *The state of Americans: This generation and the next.* New York: Free Press.

Bundy, D., & Cooper, E. (1989). Trichuris and trichuriasis in humans. *Advances in Parasitology, 28,* 107–173.

Cahan, S., & Cohen, N. (1989). Age versus schooling effects on intelligence development. *Child Development, 60,* 1239–1249.

Ceci, S. J. (1991). How much does schooling influence general intelligence and its cognitive components? A reassessment of the evidence. *Developmental Psychology, 27,* 703–722.

Ceci, S. J. (1996). *On intelligence.* Cambridge, MA: Harvard University Press.

Ceci, S. J., & Williams, W. M. (1997). Schooling and intelligence. *American Psychologist, 52*(10), 1051–1058.

Darling, N., & Steinberg, L. (1993). Parenting style as context: An integrative model. *Psychological Bulletin, 113,* 487–496.

Dornbusch, S. M., Ritter, P. L., Leiderman, P. H., Roberts, D. F., & Fraleigh, M. J. (1987). The relation of parenting style to adolescent school performance. *Child Development, 58,* 1244–1257.

Elardo, R., Bradley, R., & Caldwell, B. M. (1977). A longitudinal study of the relation of infants' home environments to language development at age three. *Child Development, 48,* 595–603.

Elley, W. B. (Ed.). (1994). *The IEA study of reading literacy.* Elmsford, NY: Pergamon Press.

Estrada, P., Arsenio, W. F., Hess, R. D., & Holloway, S. D. (1987). Affective quality of the mother–child relationship: Longitudinal consequences for children's school-relevant cognitive functioning. *Developmental Psychology, 23,* 210–215.

Flynn, J. R. (1987). Massive IQ gains in 14 nations: What IQ tests really measure. *Psychological Bulletin, 101,* 171–191.

Flynn, J. R. (1994). IQ gains over time. In R. J. Sternberg (Ed.), *The encyclopedia of human intelligence* (pp. 617–623). New York: Macmillan.

Gardner, R. (1996, March 4). Poor little smart kids. *New York Magazine,* pp. 51–55.

Ginsburg, H., & Opper, S. (1980). *Piaget's theory of intellectual development.* Englewood Cliffs, NJ: Prentice-Hall.

Grissmer, D. W., Kirby, S. N., Berends, M., & Williamson, S. (1994). *Student achievement and the changing American family.* Santa Monica, CA: Rand Institute on Education and Training.

Hanushek, E. A. (1986). The economics of schooling: Production and efficiency in public schools. *Journal of Economic Literature, 24,* 1141–1177.

Hayes, D. P., Wolfer, L. T., & Wolfe, M. F. (1996). Schoolbook simplification and its relation to the decline in SAT-verbal scores. *American Educational Research Journal, 33,* 1–18.

Hernandez, D. J. (1995). Changing demographics: Past and future demands for early childhood programs. *The Future of Children: Long-Term Outcomes of Early Childhood Programs, 5*(3), 145–160.

Hernandez, D. J. (1997). Child development and social demography of childhood. *Child Development, 68*(1), 149–169.

Hess, R. D., Holloway, S. D., Dickson, W. P., & Price, G. G. (1984). Maternal variables as predictors of children's school readiness and later achievement in vocabulary and mathematics in sixth grade. *Child Development, 55,* 1902–1912.

Horn, J. L. (1994). Theory of fluid and crystallized intelligence. In R. J. Sternberg (Ed.), *The encyclopedia of human intelligence* (pp. 443–457). New York: Macmillan.

Huttunen, M. O., & Niskanen, P. (1978). Prenatal loss of father and psychiatric disorders. *Archives of General Psychiatry, 35,* 429–431.

Kulik, J. A., Kulik, C. C., & Bangert, R. L. (1984). Effects of practice on aptitude and achievement test scores. *American Educational Research Journal, 21,* 435–447.

Lansdown, R. (1995, June). *Context and assessment.* Presentation and comments at Wellcome Trust conference, A healthy body and a healthy mind? Worcestershire, England.

Nokes, C., & Bundy, D. (1994). Does helminth infection affect mental processing and educational achievement? *Parasitology Today, 10*(1), 4–8.

Owen, L. (1985). *None of the above: Behind the myth of scholastic aptitude.* Boston: Houghton Mifflin.

Piaget, J. (1970). *Science of education and the psychology of the child.* New York: Orion Press.

Plomin, R., & Daniels, D. (1987). Why are children in the same family so different from one another? *Behavioral and Brain Sciences, 10,* 1–60.

Ricciuti, H. N. (1993). Nutrition and mental development. *Current Directions in Psychological Science, 2*(2), 43–46.

Riksen-Walraven, J. M. (1978). Effects of caregiver behavior on habituation rate and self-efficacy in humans. *International Journal of Behavioral Development, 1,* 105–130.

Rogoff, B., & Gardner, W. (1984). Adult guidance of cognitive development. In B. Rogoff & J. Lave (Eds.), *Everyday cognition: Its development in social context* (pp. 95–116). Cambridge, England: Cambridge University Press.

Rothstein, R., & Miles, K. H. (1995). *Where's the money gone? Changes in the level and composition of education spending.* Washington, DC: Economic Policy Institute.

Rutter, M. (1989). Isle of Wight revisited: Twenty-five years of child psychiatric epidemiology. *Journal of the American Academy of Child and Adolescent Psychiatry, 28,* 633–653.

Scarr, S. (1985). Constructing psychology: Making facts and fables for our times. *American Psychologist, 40,* 499–512.

Scarr, S. (1992). Developmental theories for the 1990s: Development and individual differences. *Child Development, 63,* 1–19.

Scarr, S. (1997). Behavior-genetic and socialization theories of intelligence: Truce and reconciliation. In R. J. Sternberg & E. Grigorenko (Eds.), *Intelligence, heredity, and environment* (pp. 3–41). Cambridge, England: Cambridge University Press.

Schliemann, A. D., & Simoes, P. U. (1989). O que estamos avaliando com os testes de inteligencia? [What are we evaluating with intelligence tests?] In T. Carraher, A. Schliemann, & E. L. Buarque (Eds.), *Anais do Simposio Latino Americano de Psicologia do Desenvolvimento* (pp. 27–31). Recife, ISSBD/Editora Universitaria da UFPE.

Smith, M. L. (1991). Put to the test: The effects of external testing on teachers. *Educational Researcher, 20*(5), 8–11.

Stein, Z., Susser, M., Saenger, G., & Marolla, F. (1972). Nutrition and mental performance. *Science, 178,* 708–713.

Steinberg, L., Lamborn, S. D., Dornbusch, S. M., & Darling, N. (1992). Impact of parenting practices on adolescent achievement: Authoritative parenting, school involvement, and encouragement to succeed. *Child Development, 63,* 1266–1281.

Strupp, B. J., & Levitsky, D. A. (1995). Enduring cognitive effects of early malnutrition: A theoretical reappraisal. *Journal of Nutrition, 125,* 2221S–2232S.

Sundet, J., Tambs, M., Magnus, P., & Berg, K. (1988). On the question of secular trends in the heritability of intelligence test scores: A study of Norwegian twins. *Intelligence, 12,* 47–59.

U.S. Bureau of the Census. (1992). *Statistical abstract of the United States.* Washington, DC: U.S. Government Printing Office.

Weinstein, C. S., & Mignano, A. (1993). *Elementary classroom management: Lessons from research and practice.* New York: McGraw-Hill.

Williams, W. M. (1994). Parenting and intelligence. In R. J. Sternberg (Ed.), *The encyclopedia of human intelligence* (Vol. 2, pp. 787–791). New York: Macmillan.

Williams, W. M. (1996). *The reluctant reader.* New York: Warner Books.

Williams, W., Blythe, T., White, N., Li, J., Sternberg, R., & Gardner, H. (1996). *Practical intelligence for school.* New York: HarperCollins.

Woolfolk, A. E. (1995). *Educational psychology.* Boston: Allyn & Bacon.

6

The Role of Nutrition in the Development of Intelligence

Marian Sigman and Shannon E. Whaley

The documented rise in IQ scores in industrialized countries discussed in this volume occurred over the 40- to 50-year period between the 1930s or 1940s and the early 1980s. Gains have been of about 15 IQ points, appeared in many countries in Europe and North America as well as Japan, and have been strongest in the kinds of abilities tapped by performance scales and the Raven Progressive Matrices (see chapter 2, this volume). The position to be taken in this chapter is that improvements in nutrition in these countries are likely to account, at least in part, for the rise in IQ scores. This conclusion is based on the research literature, reviewed in this chapter, indicating that nutrition and intelligence are linked. The purpose of this review is to present these findings and, at the same time, to outline the limitations in this research literature in the hope of encouraging more investigations of this important topic.

The research literature on the association between nutrition and cognition is more limited than one might imagine given the practical importance of the question. Most investigators have focused on young children because of the evidence of a sensitive period both in growth and in the development of cognitive skills. Moreover, many authors have investigated the effects of moderate-to-severe forms of malnutrition that occur more frequently in developing countries than in industrialized nations. Although crucially important, this focus on moderate-to-severe malnutrition ignores the consequences on cognitive skills of the more prevalent, milder forms of nutritional deficiencies.

The effort to identify links between nutrition and cognition is daunting. Research into the effects of nutrition on cognition has to overcome two problems. The first is that the measurement of the adequacy of nutrition is difficult. The most frequently used assessments of body size, commonly referred to as *anthropometry* in the nutrition literature, have serious limitations in that physical size is affected by many factors other than nutrition, such as genetic endowment and illness. In addition, anthropometry does not reveal much about the quality of nutrition in that specification of particular nutritional deficiencies is not obvious from measures of body size. Information on nutritional intakes is important, along with anthropometry, in assessing nutritional adequacy. However, nutritional adequacy is still difficult to assess because absorption of vitamins and minerals depends on interactions among nutrients as well as interactions between the characteristics of the eater and the food ingested. Authors of other chapters of this volume describe the difficulties of assessing intelligence, but the measurement of nutrition is also not easy.

The second problem in identifying the effects of nutrition on intelligence is that, in human populations, nutritional adequacy varies with all the other determinants of intelligence. Children who are less well fed are also less well supervised, stimulated, and educated than children who are more adequately nourished (Wachs, 1995). They are conceived by more poorly endowed and nourished parents, have more stressful fetal and birth experiences, and suffer more illnesses because of poor nutrition and associated environmental conditions. Researchers have attempted to circumvent the confounding of nutrition with other factors in a variety of ways that are discussed in the following review.

Besides the difficulties in linking nutrition to cognition, the attribution of historical changes in IQ to improvements in nutrition in industrialized countries is problematic for other reasons. First, it is not clear how to document changes in nutrition over time. One method would be to use surveys of food consumption. However, as mentioned earlier, food intake data have some limitations. Moreover, little monitoring of food consumption has been carried out on the national level until relatively recently. In the United States, the first progress report on the National Nutrition Monitoring system was published in 1986. The U.S. Department of Agriculture conducted six national household consumption surveys from the 1930s through the 1970s, but information on individual consumption was not collected until 1965. Other than the data in the first progress report, no historical comparisons of these data have been published, at least to our knowledge and that of the nutritionists we have consulted. Therefore, improvements in dietary intakes are not documented, except in a few countries such as Japan (Takahashi, 1986). Increases in heights of the populations of industrialized countries provide the clearest evidence that nutrition has improved (see chapter 7, this volume). Even if parallel improvements in nutrition and intelligence test scores could be demonstrated in the same countries and at the same time periods, it would be impossible to know whether nutritional improvements were crucial, because other environmental factors, such as health and education, may have improved along with nutrition. The literature using changes in height as a marker for nutritional improvement is reviewed in chapter 7, this volume, by Martorell, who concludes, on the basis of his review, that improvements in nutrition cannot account for the rise in IQ, an argument that is debated at the conclusion of this chapter.

There are four areas of research that are relevant to the question of whether nutrition and intelligence are associated: studies of children hospitalized for severe malnutrition, intervention studies of moderately malnourished children, correlational studies of food intake and cognitive performance, and vitamin and mineral supplementation studies carried out in both developing and industrialized countries. These four areas of research are reviewed, with particular focus on the mechanisms underlying associations between nutrition and intelligence.

EFFECTS OF SEVERE MALNUTRITION ON CHILDREN

The investigation of the effects of malnutrition on children began with studies of severely malnourished children. These studies compared children who had been hospitalized for kwashiorkor, marasmus, or edema with children who had not been hospitalized and did not have these conditions (Birch, Pineiro, Alcalde, & Cravioto, 1971; Evans, Moodie, & Hansen, 1971; Galler, Ramsey, Solimano, & Lowell, 1983; Hertzig, Birch, Richardson, & Tizard, 1972). Although it is often assumed that severe bouts of malnutrition doom a child to cognitive delays, the links between severe malnutrition and cognitive deficits may not be so direct. In reality, most children who suffer from short episodes of severe malnutrition are also chronically undernourished, which makes any effects of the severe episodes of malnutrition difficult to isolate. Furthermore, the strength of effects of severe malnutrition on cognitive development is not as clear as one might expect. In studies in which children were assessed during and shortly after the period of severe malnutrition, the children showed dramatic acute effects in many areas of functioning. However, the effects did not appear to be specific to malnutrition in that children hospitalized for this reason did not differ from children hospitalized for other conditions. The children appeared to recover from these acute episodes, although there was some indication of slower recovery for young infants.

The most frequently used research design identified children at school who had been hospitalized from malnutrition earlier in life and compared them to schoolchildren who had never been hospitalized. Overall, the results of these studies showed that the schoolchildren who were previously malnourished had lower IQs, poorer cognitive abilities and school achievement, and greater behavioral problems than controls matched on family background.

A variant of this research design is the comparison of the cognitive abilities of the hospitalized children with those of their nonhospitalized siblings. These studies had less consistent results than the comparison of hospitalized children and nonrelated controls. However, the evidence does not rule out the effects of chronic poor nutrition, which is likely to be shared by the siblings in the same family. The effects of severe malnutrition appear to be reversible in some cases in that some children who suffer from short periods of malnutrition owing to medical conditions (Berglund & Rabo, 1973; Klein, Forbes, & Nader, 1975; Lloyd-Still, Hurwitz, Wolff, & Shwachman, 1974) or traumatic events (Winick, Meyer, & Harris, 1975) recover completely, and malnourished children who are adopted into more affluent families have only moderate aftereffects of their early malnutrition.

An obvious difficulty in interpreting the studies of children hospitalized for malnutrition is that many characteristics of the home environment of hospitalized children are likely to be inferior in comparison to the home environments of nonhospitalized children. The use of sibling controls remedies this confound to some degree, but there may be genetic and nonshared environmental factors that differentially affect siblings. Therefore, the lower intelligence level of the hospitalized children may be due to their poorer genetic endowment, early interaction experience, health care, or family security as well as their poorer nutrition.

A related issue is whether the factor that is primarily affected by acute malnutrition is intelligence. Some interesting studies have shown that children who have suffered from malnutrition also behave differently in a variety of ways (see, e.g., Galler et al., 1983). First, their motor abilities are often reported to be compromised. In addition, teachers have reported that they are less attentive, more distractible, and less emotionally controlled than their peers and that they have poor relationships with others. Observational studies of younger children have supported these findings; formerly severely malnourished children were reported to play less with toys and engage less in task-focused behavior (Galler et al., 1983). Thus, their limited cognitive skills may result from deficits in attention or motivation that compromise their learning and ability to perform on intelligence tests.

The hypothesis that the link between malnutrition and intelligence test scores is mediated by motivational and emotional responsiveness is supported by findings from the literature on malnourished rats. Whereas authors of early studies of malnourished rats proposed that malnutrition caused irreversible brain dysfunction resulting in cognitive deficits, the recent consensus has been that nutritional deficiencies are associated with changes in emotional reactivity and motivation (Levitsky & Strupp, 1995; Strupp & Levitsky, 1995). Specifically, malnourished rats were reported to show increased motivation to find food in maze tasks; increased emotionality and anxiety in such tasks, as evidenced by increased spilling of food; and increased sensitivity to aversive reinforcement. Such emotional and motivational changes are now posited to be determinants of cognitive functioning, suggesting that further study of such reactivity should be carried out to establish definitive mediational links between malnutrition and cognitive skills.

Overall, the findings from research studies of children hospitalized for severe malnutrition have supported the hypothesis that severe malnutrition, particularly when it is followed by chronic undernutrition, limits intellectual development and functions. Although the severity of malnourishment in the children in these studies exceeded the level found regularly in most populations, particularly those in industrialized countries that have shown a rise in intelligence, this may have been less true earlier in this century.

EFFECTS OF FOOD SUPPLEMENTATION ON THE COGNITIVE DEVELOPMENT OF YOUNG CHILDREN

In an impressive body of research, an experimental approach has been used to investigate the links between nutrition and cognition. The majority of these studies were carried out in developing countries (Guatemala, Colombia, Indonesia, Jamaica, Mexico, and Taiwan), although one was carried out in New York City. In most cases, the populations in the areas studied suffered from moderate, rather than severe, malnutrition. The studies, most of which were conducted in the late 1960s and early 1970s, involved interventions directed at the gestating mother and very young child (Chavez & Martinez, 1975; Grantham-McGregor, Powell, Walker, & Hines, 1991; Husani, Karyadi, Husani, Sandaja Karlyadi, & Pollitt, 1991; Joos, Pollitt, Mueller, & Albright, 1983; McKay,

Sinisterra, McKay, Gomez, & Lloreda, 1978; Pollitt, Gorman, Engle, Martorell, & Rivera, 1993; Rush, Stein, & Susser, 1980; Waber et al., 1981).

Because the studies were generally designed and conducted independently, there is little overlap in their questions or hypotheses, research designs, interventions, or assessment methods. For example, some researchers focused on identifying the critical age period for intervention; some compared nutritional supplementation with parenting interventions and with its use in combination with parenting interventions; and some aimed to determine the necessary duration of supplementation. In most of these studies, mothers were recruited for the intervention either in pregnancy or in the first 2 years of the child's life.

Research Design

In all the studies, calories, protein, and micronutrients were provided for the supplemented participants. The duration of supplementation varied widely across studies, lasting as briefly as pregnancy in the New York City study and as long as 7 years after birth in the Guatemala study. The samples were selected in most of the studies using available population data concerning anthropometry and illness. To our knowledge, none of the researchers collected baseline nutritional data, and only a few monitored food intake during the period of supplementation. The latter is important, because food provided is not always consumed, or, at least, not consumed by the person to whom it is provided. Young malnourished children are difficult to feed and may not eat the food that is available, or children and families may distribute the food among other family members or peers. One estimate is that only half the food provided to a family during a famine is consumed by the family; the rest is given away or sold. In addition, mothers may reduce the amount of food given to a child whom they know is being fed in a community center or school in order to feed other family members more adequately. In the studies in which intake was monitored, net increases from supplementation ranged from 100 to 200 kilocalories (referred to as *calories* in everyday language) a day. Growth was assessed in most of these studies; the children showed at least modest increases in height compared with controls, an observation confirming the improvements in nutrition.

As in any area of research, certain constraints influenced these research designs. In the case of supplementation studies, an important ethical problem exists concerning the way in which the control group is treated. This is particularly problematic in studies of long-term supplementation because investigators are necessarily uncomfortable about depriving a part of the group for extended periods. Moreover, the participation rate is likely to decline in samples who are not provided any benefit for their participation. For this reason, some supplementation is often provided to control participants, particularly in studies of some duration. In the studies carried out in Taiwan and Guatemala, the control groups were provided with extra calories and multivitamins, and researchers in the New York City study distributed multivitamins to pregnant control mothers. The effects of the intervention have to be large enough, therefore, to be discriminated from the effects of the control supplementation.

An additional reason for providing nutrients to the control group is that the provision of food always changes the environmental system in many ways other than the nutritional. Food is so central to the lives of human beings that the psychological implications of being provided with nourishment cannot be overlooked. Anyone who wants to increase attendance at a meeting, seminar, or party knows the value of providing free food even to people who are amply fed. The provision of food to people with limited access to food may have powerful psychological effects, changing their moods, perceptions, and aspirations. This is a factor that is frequently ignored not only in studies of malnutrition but in government actions that affect food policy.

The distribution of food may also change the lives of the participants in other ways. In the Guatemala project, food was distributed at central locations twice a day, and participants were allowed to consume as much as they wanted. This meant that participants were in close social contact at those times. Moreover, the type of food provided and the age of the participant affected participation. In two villages, a sweetened drink with micronutrients was provided, whereas in two other villages, a more highly fortified beverage was made available. Mothers brought their infants and young children to receive the fortified beverage more often than mothers brought their young children to receive the sweetened drink. Therefore, there was greater attendance in the experimental villages by mothers and young children, so that the experimental group not only received more nutrients but, possibly, received more social interaction as a consequence. The situation was reversed when the children matured; older children were more fond of the sweetened drink, and attendance rose with the children's age in the villages in which that was given. The investigators in the Guatemala project were wise enough to keep attendance records so that statistical adjustments could be made in computing the effects of nutritional supplementation.

Research Results

In most of the supplementation studies, an infant or preschool developmental scale was used as the major outcome measure. A recent meta-analysis showed that the most consistent effect of supplementation on young infants was on motor skills (Pollitt & Oh, 1994). Infants in the supplemented groups had better developed motor skills than control infants across the age range tested, from 8 to 24 months. This is an important effect in that infant motor skills are predictive of later cognitive abilities among children in developing countries (Pollitt & Oh, 1994; Whaley, Sigman, Espinosa, & Neumann, in press).

Supplementation improved mental abilities in older infants, aged 18 to 24 months, and in preschool children. However, the effects are inconsistent both within and across studies. As an example, in the Guatemala study, there was a significant effect on the same measure at ages 4 and 5 but not at ages 3 and 6.

In the New York City study, there was no effect of maternal supplementation during pregnancy on infant mental and motor skills, but supplemented infants habituated more rapidly than control infants and showed better representational play skills. Rate of habituation is measured by showing an infant a stimulus repeatedly and recording the amount of decline in attention or the number of trials needed to reach a fixed baseline. Habituation is a sensitive measure of information processing in infants; infants who habituate rapidly tend to be more intelligent later in life (Bornstein & Sigman, 1986). Infant rate of habituation is a more effective predictor of later cognitive abilities than is performance on the infant developmental scales. Moreover, habituation rate has been shown to be sensitive to malnutrition in a study carried out in India (Rose, 1994). Representational play skills reveal the level of symbolic understanding that preschool children possess. Studies have shown that level of play in high-risk samples predicts later cognitive and language skills. Therefore, improvements in both habituation and representational play in the New York sample suggest that prenatal supplementation had some important impact on the children's potential development.

Studies of American preterm infants, many of whom are born small even for their gestational ages, have shown that these infants process information more slowly than full-term infants whose size is appropriate for their gestational age (Sigman, 1976; Sigman & Parmelee, 1974). Moreover, children born preterm have somewhat lower intelligence as a group and suffer far more school learning problems than would be expected for their intelligence level (see Friedman & Sigman, 1992). In a recent study, infant attention, measured at expected date of birth, and quality of caregiving, measured 1 month later, interacted to predict intelligence at age 18 (Sigman, Cohen, & Beckwith, 1997). Children who processed information quickly and had verbal caregivers during infancy had IQs that were 20 points higher than those of children who processed slowly and had less stimulating caregivers. Because mothers who are poorly nourished have higher rates of preterm birth, these findings may be indirect evidence of constitutional limitations in infants as a result of poor nutrient intakes during the gestational period. At the same time, this study shows the potential influence of the rearing environment, as do the studies of malnourished children in developing countries.

An educational intervention was provided in addition to supplementation in three studies. In the most elegantly designed study (in

Jamaica), there was a comparison of the effects of nutritional supplementation alone, family intervention alone, and the combination of nutritional supplementation and family intervention. Families were enrolled in the study when their stunted children were between 9 and 24 months of age and remained in the study for 2 years. There were significant effects of supplementation and intervention alone on the locomotor, performance, and overall developmental quotients of the children as well as an additive effect of supplementation with intervention. Therefore, both nutritional and psychosocial interventions contributed to the child's level of functioning, and a combination of the two made a significantly greater contribution to level of functioning than either in isolation.

Five of the supplementation studies included follow-up assessments well after the interventions had terminated. In most of the studies, the supplemented groups had higher scores on some measures of intellectual ability or achievement than the nonsupplemented groups. For example, in the Mexican study, the supplemented boys scored higher on the Raven matrices than the nonsupplemented boys, but there was no difference in the scores of the girls. In the Cali (Colombia) study, where intervention included both supplementation and educational intervention, IQ increased with increasing length of intervention. Reading readiness but not arithmetic or basic knowledge scores were higher in the supplemented group in the Bogotá study.

The most ambitious study with the best organized follow-up was conducted in Guatemala. Effects of supplementation were found in terms of educational achievement in such areas as arithmetic skills, vocabulary, reading achievement, and overall knowledge. There were fewer group differences on measures of information processing. The effects of supplementation were greatest for the poorest children.

For school-aged children, the interactions of nutritional adequacy and the quality of schooling cannot be overlooked. In the Guatemala study, the effects of supplementation increased with increasing grade attained. In other studies, the quality of the educational experience moderated the effect of nutritional intervention. For example, in a study of schoolchildren in Jamaica, Chang, Walker, Himes, & Grantham-

McGregor (1996) found an interaction between dietary supplementation and quality of the classroom. Dietary supplementation was not effective in changing behavior or achievement for children in disorganized classrooms. These findings are of particular significance for understanding the rise in intelligence in industrialized countries. Improvements in nutrition co-occurred with increases in the number of children enrolled in the educational system in most countries. Good nutrition may enable children to learn more effectively, but educational opportunities must be available to provide this learning.

Consequences of Sporadic Food Shortages

Chronic undernutrition is often accompanied by sporadic shortages in food owing to crop failure or some form of social disorganization. Children whose diets are at the margins of adequacy may periodically have almost nothing to eat. This certainly happened in the United States during the early years of the century and may continue to happen to parts of the U.S. population even now.

The effects of food shortages have not been identified because these are obviously unplanned. In the first author's own research, a food shortage unexpectedly occurred when the rains failed during the course of a longitudinal study in rural Kenya. Because the researchers were involved in measuring food intake, anthropometry, and children's skills, the consequences of this famine could be documented (McDonald, Sigman, Espinosa, & Neumann, 1994). It became obvious that the schoolchildren suffered the most during this food shortage. The diet of children between 18 and 30 months of age was generally maintained so that these children did not lose weight, although they did not gain either. However, their mothers and the school-aged children lost weight as food became unavailable. The activity level of the schoolchildren on the playground declined as did their attentiveness to tasks in the classroom. Intelligence test scores did not change over this very short period of 3 months. One intriguing finding emerged from a follow-up several years later: Arithmetic skills, but not reading and writing skills, were lower in children who had lost more weight during the famine than in children who had lost less weight. Because arithmetic learning is sequential, it is possible that these children lost out on learning some basic skills during the period of food shortage in a way that was not remedied because it was not recognized.

The effects of the food shortage can be attributed only partly to nutritional deprivation because the shortage also resulted in changes in family practices. During the food shortage, mothers spent less time caring for their toddlers, and school-aged children increased their caregiving responsibilities. Therefore, part of the effect of the food shortage was to divert the energies of the young children away from school. Their listlessness on the playground may have resulted from increased hunger or from increased home responsibilities. Whatever the cause, the schoolchildren had less energy to invest in classroom learning. These findings suggest that improvements in nutrition in industrialized countries manifested by greater food availability might have effects not only in terms of increasing the average level of food intake but in protecting families from the disruptions caused by shortages.

COGNITIVE DEVELOPMENT IN YOUNG CHILDREN IN RELATION TO FOOD INTAKE

For many years, the major research approach to the investigation of nutrition and cognition was to assess the intelligence of children who differed in height and weight. Lynn (1990) pointed out that the earliest of these studies was carried out in the last century (Porter, 1893). Numerous studies have shown that children who are taller and heavier tend to be somewhat smarter than smaller, leaner children, particularly in environments where food intake is limited (see Allen, 1995; Barrett, 1986; Lozoff, 1989; Pollitt, 1988; Simeon & Grantham-McGregor, 1990; and Wachs, 1995, for reviews). Taller and heavier schoolchildren had higher cognitive performance in Brazil (Paine, Dorea, Pasquali, & Monteior, 1992), in an inner city of the United States (Karp, Martin, Sewell, Manni, & Heller, 1992), in mainland China (Jamison, 1986), and in rural Kenya (Sigman, Neumann, Jansen, & Bwibo, 1989). In large studies in Europe and the United States, with samples ranging from 5,000 to 75,000, correlations between height and IQ ranged from .13 to .25

(Douglas, Ross, & Simpson, 1965; Scottish Council on Research in Education, 1953; Wilson et al., 1986). As discussed in the introduction to this chapter, the difficulty with interpreting these findings is that physical size is affected by many factors other than food intake. As an example, medically ill children may be both smaller and less cognitively skillful, so that the association between physical size and cognitive score may not be causative.

To address this issue, researchers recently have broadened the measures of nutrition to include not only anthropometry but assessments of food intake and blood chemistry. The Nutrition Collaborative Research Support Program (Nutrition CRSP) was organized to investigate the effects of mild-to-moderate undernutrition on the development and functioning of children and their parents in three different countries: Mexico, Egypt, and Kenya (Allen, 1995; Sigman, 1995). In all three countries, three different groups were targeted: pregnant women and their infants followed at least until the infants were 6 months of age, 18- to 30-month-old toddlers, and schoolchildren ages 7 and 8. Families with women who were likely to become pregnant, with 18-month-old toddlers, or with 7-year-old schoolchildren were recruited into the study, and their food intake was measured for 2 days a month over the course of 1 year. Intelligence was assessed in the target individuals using modified infant developmental scales and child and adult intelligence tests.

The project collected extensive data on food intake. The monthly measurements were converted into daily caloric intakes as well as intakes of protein, fat, carbohydrates, and calories from nonanimal and animal source food. The target individuals were weighed and measured weekly, and the health of all family members was surveyed monthly. The socioeconomic status of the family was measured, and the parents' literacy was assessed. The behavior of the schoolchildren in the classroom and on the playground was also observed and coded. Bimonthly observations of the caregiving of the infants and toddlers were made in Egypt and Kenya for two reasons. First, the researchers wished to have measures of the home environment so that the effects of psychosocial stimulation could be contrasted with the effects of nutrition. Second, measurements of the home environment were necessary for an understanding of the processes by which nutritional factors might affect the children's development.

Diets of children and adults in Egypt and Mexico provided sufficient calories and protein to prevent protein–energy malnutrition in these countries. This was not true in Kenya, where food intake was more limited. Although there did not seem to be a shortage of protein in the Kenyan diet, the caloric intake of many of the children was insufficient, and physical size and cognitive function were associated with caloric intake (Sigman, Neumann, Jansen, & Bwibo, 1989). In all three countries, cognitive abilities of the children were related to the quality of the diet as measured by the percentage of food from animal sources (Sigman, 1995). Children who were able to eat at least some animal-source food had higher scores on intelligence tests than children who ate no or very little food from animal sources. The associations between intake of animal-source food and cognitive performance were independent of confounding factors such as family socioeconomic status, illness, parental literacy, and level of social stimulation in the home (Sigman, McDonald, Neumann, & Bwibo, 1991; Sigman, Neumann, Baksh, Bwibo, & McDonald, 1989).

Animal-source food seems to be so important because it provides micronutrients that are otherwise unavailable to the children. For example, in Mexico the main source of iron and zinc for the children was tortillas, a major dietary staple. However, these minerals are not bioavailable because they bind to phytate in the tortillas. Animal-source food is necessary, therefore, to provide iron and zinc to Mexican children. Animal products compensate for similar micronutrient deficiencies in other countries.

The most important finding from the Nutrition CRSP study was that diet has to be conceptualized not only in terms of energy and protein but also in terms of the quality of the diet. Other studies have suggested that adequacy of micronutrient availability is related to cognitive function in children. Therefore, the rise in intelligence test scores in industrialized countries may be partly attributable not only to increased quantity of food but also to improved quality of diet; these countries provided school lunch and breakfast programs to children and lower cost food to parents.

EFFECTS OF MINERAL AND VITAMIN SUPPLEMENTATION ON CHILDREN

Studies of Iron-Deficiency Anemia in Young Children

An extensive body of research has suggested that iron-deficiency anemia has serious consequences for children's development. Iron-deficient anemic infants were found to be developmentally less advanced than nonanemic infants in a majority of studies (Grindulis, Scott, Belton, & Wharton, 1986; Lozoff, Brittenham, Viteri, Wolf, & Urrutia, 1982; Lozoff, Klein, & Prabucki, 1986; Walter, Kovalskys, & Steckel, 1983), although not in all (Delinard, Gilbert, Dodds, & Egeland, 1981; Johnson & McGowan, 1983). Iron-deficiency anemia in infancy also was found to predict lower competence many years later. This was true among Costa Rican children followed from infancy to 5 years of age (Walter et al., 1983) and among Israeli infants followed from infancy to 5 years and then to second grade (Palti, Pevsner, & Adler, 1983). Moreover, in two studies of 3- to 6-year-olds and two studies of schoolchildren, iron-deficient anemic children were less skillful than nonanemic children (Pollitt & Metallinos-Katsaras, 1990). As in the other forms of malnutrition discussed earlier, the extent to which these effects can be traced specifically to anemia is unclear because iron-deficient anemia is associated with birth complications, greater general malnutrition, poor environmental rearing conditions, and poverty.

Micronutrient deficiency, particularly iron deficiency, may be responsible for some of the effects often attributed to protein–energy malnutrition. For example, in the Nutrition CRSP, areas of study were chosen because of growth stunting in the local children. This selection was made on the assumption that this growth stunting was attributable to protein–energy malnutrition, an incorrect assumption as it turned out because intakes of energy and protein were adequate in both Egypt and Mexico. Pollitt (1995) suggested that the growth stunting may be due to micronutrient insufficiencies and pointed to a study in which differences in intelligence between groups of Mexican schoolchildren differing in anthropometry were no longer observed when the effects of anemia were considered. Micronutrient deficiency does appear to be important for stunting in both Egypt and Mexico, although morbidity owing to poor sanitation is also a contributing factor in Egypt (Neumann & Harrison, 1994). Moreover, Pollitt proposed that the beneficial effects of supplementation in the Guatemalan study may be as plausibly related to increases in iron intake as to the increases in energy and protein that were also accomplished by that intervention.

Studies that have explicitly aimed to improve the cognitive and behavioral skills of iron-deficient anemic children have had mixed results. The only clear finding is that short-term treatment of 5–11 days does not bring about more improvement in behavior and cognition than is found with a placebo. Two longer term studies (one in Chile and one in Costa Rica) that did not have control groups revealed only slight improvement in children whose hematological status had normalized (Lozoff, Jimenez, & Wolf, 1991; Walter, 1993). The authors of one study in the United Kingdom reported greater improvement in treated than nontreated groups (Grindulis et al., 1986), whereas researchers in Java demonstrated clinically significant improvements in developmental quotients in treated groups as opposed to the nontreated group (Idjradinata & Pollitt, 1993). In the only long-term longitudinal study that we are aware of, anemic Israeli infants provided with iron-fortified food for 3 months continued to have lower skills when of school age than nonanemic children followed over the same time period. Therefore, the research findings are mixed with only one study showing clear gains from iron supplementation, although the lack of control groups in some studies limits the conclusions that can be drawn from these investigations.

Studies of Vitamin and Mineral Supplementation in British and American Schoolchildren

A series of studies have been carried out in Great Britain and the United States in which groups of children were provided with vitamins and minerals, and their performance was tested before and after this supplementation. In the first of these, changes in nonverbal and verbal intelligence were compared in a group of Welsh 13-year-olds. Sixty children were provided with vitamin and mineral supplements over the course of 8 months, and 30 children were given placebos (Benton & Roberts, 1988). The supplemented group showed an increase in scores on the Calvert Non-Verbal Test, whereas the placebo group showed no increase. There was no group difference in the increase of scores on the verbal tests, suggesting that vitamin and mineral supplements most strongly affect nonverbal intelligence. Two studies were designed to replicate these findings, and neither succeeded (Crombie et al., 1990; Nelson, Naismith, Burley, Gatenby, & Geddes, 1990). However, one study used supplementation for only 28 days (Nelson et al., 1990), so that the intervention was not comparable to that of the original investigation. In another study, in Belgium, investigators found similar effects but only with boys who reported poor dietary intakes (Benton & Buts, 1990).

Two studies were carried out in California: one small study with 26 adolescents institutionalized by the juvenile justice system and one larger study with 410 8th- and 10th-grade students (Schoenthaler, Amos, Dorz, Kelly, & Wakefield, 1991; Schoenthaler, Amos, Eysenck, Peritz, & Yudkin, 1991). In the first investigation, children whose diets were supplemented for 13 weeks showed a change in Wechsler Intelligence Scale for Children (WISC) Performance IQ from 101 to 107 and no change in verbal IQ. The placebo group declined from 100 to 98.6 on retesting. An improvement of about 9 points would be expected on retesting after 3–5 weeks according to the WISC-R manual. In the second study, there was a significant gain in WISC Performance IQ among the children in a group supplemented at 100% of the U.S. recommended daily allowance (USRDA) but not in groups supplemented at 50% or 200% of the USRDA. Children supplemented at 100% of the USRDA improved by 13 points, 4 points more than would be expected from retesting alone. However, the score increase was mostly due to changes in scores on Object Assembly and Picture Arrangement, two subscales that show the most practice effects with retesting in standardization samples. An important conceptualization of the authors is that nonverbal skills should rise only in children who were originally deficient in micronutrients. To test this hypothesis, blood samples were collected for all children for whom consent was obtained. However, the results of the blood tests do not seem to have been published, so that the validity of this claim remains untested.

Overall, the research does not seem to be of sufficient quality to provide evidence one way or the other as to the benefits accruing from these micronutrient–vitamin interventions. The authors rarely reported the extent of compliance with the interventions, and the data are presented in ways that make it difficult to adjudicate the claims. For example, most of the investigators did not describe the intelligence tests used, and means and standard deviations were frequently not given. In studies in which these statistics were furnished, the representativeness of the samples was unclear. For example, the Welsh sample had a much higher nonverbal than verbal IQ, and both were higher than those for the Scottish replication study. Resolution of the issue of whether vitamin and mineral supplementation improves the cognitive abilities of children in the United States and Europe awaits a large-scale investigation with much more careful monitoring of interventions and testing than has been carried out at this time.

PROCESSES MEDIATING THE ASSOCIATION BETWEEN NUTRITION AND INTELLIGENCE

The preceding evidence suggests that nutrition and intelligence are linked in that children provided with better diets, in terms of both quantity and quality of food, showed superior performance on developmental and cognitive tests than children who were less well fed. An important follow-up question concerns the nature of the link between nutrition and cognition. One hypothesis is that poor diet results in limitations in brain development that are reflected in inferior cognitive functioning. This hypothesis does not account very well for the recovery from short-term malnutrition owing to illness or war often seen in children whose diets and quality of life improve. On the other hand, because these children function fairly well does not mean that they do not suffer from milder impairments in cognitive abilities. My group's research with preterm infants in the United States has identified innate styles of attention that predict differences in intelligence within the normal range 18 years later. Thus, there may be subtle effects on brain development that influence the child's ability to learn.

Another, equally plausible hypothesis is that lack of food intake, if it is not too severe, does not cause brain impairment but causes children to be inactive, irritable, or inattentive. Caregivers and teachers then adapt their response style to the child's behavior, depending on the age of the child and the context. If malnutrition occurs early so that growth falters by the 3rd or 4th month of life, as it often does in developing countries, caregivers may continue to carry infants who are small, passive, and irritable. The treatment of the infant as immature may start a vicious cycle in which the child's experiences are restricted so that the child acquires skills more slowly, leading to less than optimal intellectual development. There is some evidence for this pattern of development in that toddlers who are carried more tend to develop less well. Moreover, the level of motor skills is strongly predictive of later development in children in underdeveloped countries (Whaley et al., in press). This pattern of development is not restricted to malnourished populations. Infants who are at risk either because of birth complications or because of maternal alcohol or drug abuse during pregnancy are also more irritable and tend to elicit less than optimal caregiving (O'Connor, Sigman, & Kasari, 1993).

The social and cultural context in which the child is developing can never be overlooked in an attempt to understand the impact of nutrition on intelligence. We have reviewed evidence that the quality of the home and the school moderates the influence of nutrition on the acquisition of cognitive skills. Children in disorganized classrooms did not benefit as much from improved nutrition as children in well-organized classrooms (Chang et al., 1996). The way in which the culture interprets behavior also influences the link between nutrition and development. In two of the countries that my colleagues and I studied, Egypt and Mexico, the more active behavior of better fed girls was seen as disruptive rather than as interactive by their teachers, an interpretation that portends difficulties for these girls' eventual achievement.

An important moderator of the link between nutrition and cognition is the nature of the demands that the culture places on the individual for successful adaptation. Earlier in this century, the level of nutrition in developing countries may have been sufficient for individuals to acquire the skills needed for successful maintenance of themselves and their families. However, the level of skill required in the modern world, even in places like rural Kenya or Mexico, is rising. Individuals who can think abstractly, a skill taught in schools, are better adapted even when the culture depends mostly on subsistence farming. One may require different levels of nutrition to acquire and perform abstract reasoning than to learn and carry out more routine tasks, because levels of concentration required may be higher in the former than in the latter. Therefore, the quantity and quality of nutrition required may exceed that needed at previous times or in less sophisticated societies.

NUTRITION AND THE RISE IN INTELLIGENCE

We have reviewed the evidence that nutrition affects cognitive development; we believe that part of the rise in intelligence noted may be due to the improvements that have been documented in nutrition over the last 50 years. In our view, the evidence that nutrition and cognition are linked is compelling despite the methodological difficulties in proving this hypothesis. On the basis of this evidence, it seem unlikely that improvements in nutrition would not have produced populations that were better able to think and learn, the abilities that are tapped by intelligence tests. At the same time, we are convinced that the rise in intelligence is not due entirely to nutrition. Nutrition rarely works alone in shaping intellectual skills; better fed individuals can learn and perform better only if they have access to experiences that shape their development appropriately for the demands of their culture. Furthermore, nutritional improvements may be responsible for the rise in IQ at some points in a country's history and not others, depending on historical changes in nutrient availability, demands for abstract thinking, and exposure to the kinds of skills required by intelligence tests.

As mentioned earlier, Martorell argues in chapter 7, this volume, that the improvement in nutrition is unlikely to be responsible for the rise in IQ scores principally because the increases in height that serve as a marker for nutritional improvements level off in some countries yet intelligence continues to rise. We disagree with Martorell's argument for several reasons. First, as mentioned earlier, height is influenced by many factors, not only nutrition. Height may be genetically constrained so that few populations can be much taller than the Dutch (one of the world's tallest people), but cortical circuitry may not be similarly constrained genetically. In other words, there may be physical limitations on height but not on other forms of development. In addition, nutrition tends to have effects over generations, so that the adequacy of diet of maternal grandmothers may be as important as maternal diet. Moreover, there may be improvements in nutrition that are not reflected in height. For example, there is some indication that intakes of iron have been steadily rising since 1955 (U.S. Department of Health and Human Services and U.S. Department of Agriculture, 1986). Furthermore, as mentioned earlier, nutrition could have effects at one point in time but not in another. Thus, improvements in nutrition could have accounted for the rise in IQ in Holland earlier in this century, but other factors might be responsible for the current rise.

One of the points that seems most important to acknowledge is that the processes underlying the ability to reason, learn, and respond on tests of intelligence may be extremely complex. The extent to which individuals are motivated or confident is likely to have an enormous impact on what they are able to learn, which inevitably influences what abilities and achievements are manifested. In this light, the adequacy of nutritional intakes may have psychological effects far outreaching the physical effects. Children who grow up hungry are likely to have less sense of security and well-being than children who do not have to deal with frequent episodes of hunger. This psychological sense of well-being is likely to affect children's ability to reason and learn.

In conclusion, the rise in intelligence over the last 50 years is likely to have been influenced by the improving standards of nutrition. The improvement in nutrition was partly shaped by government policies that made food more affordable to the majority of the population. These government programs provided people who could not afford food with funds and food stamps, provided food to pregnant women, and created school breakfast and lunch programs to ensure that children were not hungry. A diminution of federal food programs has the potential to cause a fall in intelligence analogous to the rise that has been documented up to this time.

REFERENCES

Allen, L. (1995). Malnutrition and human function: A comparison of conclusions from the INCAP and nutrition CRSP studies. *Journal of Nutrition, 125,* 119S–1126S.

Barrett, D. E. (1986). Nutrition and social behavior. In H. E. Fitzgerald, B. M. Lester, & M. W. Yogman (Eds.), *Theory and research in behavioral pediatrics* (pp. 147–198). New York: Plenum Press.

Benton, D., & Buts, J. P. (1990). [Letter]. *Lancet, 335,* 1160.

Benton, D., & Roberts, G. (1988, January 23). Effect of vitamin and mineral supplementation on intelligence of a sample of school children. *Lancet, 331,* pp. 140–143.

Berglund, G., & Rabo, E. (1973). A long-term follow-up investigation of patients with hypertrophic pyloric stenosis—With special reference to the physical and mental development. *Acta Paediatrica Scandinavia, 62,* 125–129.

Birch, H. G., Pineiro, C., Alcalde, E. T. T., & Cravioto, J. (1971). Relation of kwashiorkor in early childhood and intelligence at schoolage. *Pediatric Research, 5,* 579–585.

Bornstein, M., & Sigman, M. (1986). Continuity in mental development from infancy. *Child Development, 57,* 251–274.

Chang, S. M., Walker, S. P., Himes, J., & Grantham-McGregor, S. M. (1996). Effects of breakfast on classroom behavior in rural Jamaican schoolchildren. *Food and Nutrition Bulletin, 17*(3), 248–257.

Chavez, A., & Martinez, C. (1975). Nutrition and development of children from poor rural areas: V. Nutrition and behavioral development. *Nutrition Reports International, 11,* 477–489.

Crombie, I. K., Todman, J., McNeill, G., Florey, D. D., Menzies, I., & Kennedy, R. A. (1990, March 31). Effect of vitamin and mineral supplementation on verbal and non-verbal reasoning of schoolchildren. *Lancet, 335,* 744–747.

Delinard, A., Gilbert, A., Dodds, M., & Egeland, B. (1981). Iron deficiency and behavioral deficits. *Pediatrics, 68,* 828–833.

Douglas, J. W. B., Ross, J. M., & Simpson, H. R. (1965). The relationship between height and measured educational ability in schoolchildren of the same social class, family size, and stage of sexual development. *Human Biology, 37,* 178–192.

Evans, D., Moodie, A., & Hansen, J. (1971). Kwashiorkor and intellectual development. *South African Medical Journal, 45,* 1413–1462.

Friedman, S., & Sigman, M. (1992). *The psychological development of low birthweight children.* Norwood, NJ: Ablex.

Galler, J. R., Ramsey, F., Solimano, G., & Lowell, W. (1983). The influence of early malnutrition on subsequent behavioral development. II. Classroom behavior. *Journal of the American Academy of Child Psychiatry, 22,* 16–22.

Grantham-McGregor, S., Powell, C. A., Walker, S. P., & Hines, J. H. (1991). Nutritional supplementation, psychological stimulation, and mental development of stunted children: The Jamaican study. *Lancet, 338,* 1–5.

Grindulis, H., Scott, P. H., Belton, N. R., & Wharton, B. A. (1986). Combined deficiency of iron and vitamin D in Asian toddlers. *Archives of the Diseases of Children, 61,* 843–848.

Hertzig, M., Birch, H., Richardson, S., & Tizard, J. (1972). Intellectual levels of school children severely malnourished during the first two years of life. *Pediatrics, 49,* 814–824.

Husani, M. A., Karyadi, L., Husani, Y. K., Sandaja Karlyadi, D., & Pollitt, E. (1991). Developmental effects of short-term supplementary feeding in nutritionally-at-risk Indonesian infants. *American Journal of Clinical Nutrition, 51,* 799–804.

Idjradinata, P., & Pollitt, E. (1993). Reversal of developmental delays in iron-deficient anemic infants treated with iron. *Lancet, 341,* 1–4.

Jamison, D. T. (1986). Child malnutrition and school performance in China. *Journal of Developmental Economics, 20,* 299–309.

Johnson, D. L., & McGowan, R. J. (1983). Anemia and infant behavior. *Nutrition and Behavior, 1,* 185–192.

Joos, B. K., Pollitt, E., Mueller, W. H., & Albright, D. L. (1983). The Bacon Chow Study: Maternal nutritional supplementation and infant behavioral development. *Child Development, 54,* 669–676.

Karp, R., Martin, R., Sewell, T., Manni, J., & Heller, A. (1992). Growth and academic achievement in inner-city kindergarten children. *Clinical Pediatrics, 31,* 336–340.

Klein, P. S., Forbes, G. B., & Nader, P. R. (1975). Effects of starvation in infancy (pyloric stenosis) on subsequent learning abilities. *Pediatrics, 87,* 8–15.

Levitsky, D. A., & Strupp, B. J. (1995). Malnutrition and the brain: Changing concepts, changing concerns. *Journal of Nutrition, 125,* 2212S–2220S.

Lloyd-Still, J. D., Hurtwitz, I., Wolff, P. H., & Shwachman, H. (1974). Intellectual development after severe malnutrition in infancy. *Pediatrics, 54,* 306–311.

Lozoff, B. (1989). Nutrition and behavior. *American Psychologist, 44,* 231–236.

Lozoff, B., Brittenham, G. M., Viteri, R. E., Wolf, A. W., & Urrutia, J. J. (1982). The effects of short-term oral iron therapy on developmental deficits in iron deficient anemic infants. *Journal of Pediatrics, 100,* 351–357.

Lozoff, B., Jimenez, E., & Wolf, A. (1991). Long-term developmental outcome of infants with iron deficiency. *New England Journal of Medicine, 325,* 687–694.

Lozoff, B., Klein, N. K., & Prabucki, K. M. (1986). Iron-deficient anemic infants at play. *Journal of Developmental and Behavioral Pediatrics, 7,* 152–158.

Lynn, R. (1990). The role of nutrition in secular increases in intelligence. *Personality and Individual Differences, 11,* 273–285.

McDonald, M. A., Sigman, M., Espinosa, M. P., & Neumann, C. G. (1994). Impact of a temporary food shortage on children and their mothers. *Child Development, 65,* 404–415.

McKay, H., Sinisterra, L., McKay, A., Gomez, H., & Lloreda, P. (1978). Improving cognitive ability in chronically deprived children. *Science, 200,* 270–278.

Nelson, M., Naismith, D. J., Burley, V., Gatenby, S., & Geddes, N. (1990). Nutrient intakes, vitamin-mineral supplementation, and intelligence in British schoolchildren. *British Journal of Nutrition, 64,* 13–22.

Neumann, C. G., & Harrison, G. G. (1994). Onset and evolution of stunting in infants and children: Examples from the Human Nutrition Research Collaborative Research Support Program. Kenya and Egypt studies. *European Journal of Clinical Nutrition, 48,* S90–S102.

O'Connor, M. J., Sigman, M., & Kasari, C. (1993). International model for the association among maternal alcohol use, mother–infant interaction, and infant cognitive development. *Infant Behavior and Development, 26,* 177–192.

Paine, P., Dorea, J. G., Pasquali, J., & Monteior, A. M. (1992). Growth and cognition in Brazilian school children: A spontaneously occurring intervention study. *International Journal of Behavioral Development, 15,* 160–183.

Palti, H., Pevsner, B., & Adler, B. (1983). Does anemia in infancy affect achievement on developmental and intelligence tests? *Human Biology, 55,* 183–194.

Pollitt, E. (1988). A child survival and development revolution. *Society for Research in Child Development Newsletter, 4.*

Pollitt, E. (1995). Functional significance of the covariance between protein energy malnutrition and iron deficiency anemia. *Journal of Nutrition, 8,* 2272S–2278S.

Pollitt, E., Gorman, K. S., Engle, P., Martorell, R., & Rivera, J. (1993). Early supplementary feeding and cognition: Effect over two decades. *Monographs of the Society for Research in Child Development, 58*(7, Serial No. 235), V–99.

Pollitt, E., & Metallinos-Katsaras, E. (1990). Iron deficiency and behavior: Constructs, methods, and validity. In R. J. Wurtman & J. J. Wurtman (Eds.), *Nutrition and the brain.* New York: Raven Press.

Pollitt, E., & Oh, S. (1994). Early supplementary feeding, child development and health policy. *Food & Nutrition Bulletin, 15,* 208–214.

Porter, W. T. (1893). The physical bases of precocity and dullness. *Transactions of the Academy of Sciences, 6,* 161–175.

Rose, S. A. (1994). Relation between physical growth and information processing in infants born in India. *Child Development, 65,* 889–903.

Rush, D., Stein, Z., & Susser, M. (1980). Diet in pregnancy: A randomized controlled trial of nutritional supplements. *Birth Defects, 16*(3), 1–197.

Schoenthaler, S. J., Amos S. P., Dorz, W. E., Kelly, M. A., & Wakefield, J., Jr. (1991). Controlled trial of vitamin–mineral supplementation on intelligence and brain function. *Personality and Individual Differences, 12,* 343–350.

Schoenthaler, S. J., Amos, S. P., Eysenck, H. J., Peritz, E., & Yudkin, J. (1991). Controlled trial of vitamin–mineral supplementation: Effects on intelligence and performance. *Personality and Individual Differences, 12,* 351–362.

Scottish Council on Research in Education. (1953). *Social implications of the 1947 Scottish Mental Survey.* London: University of London Press.

Sigman, M. (1976). Early development of preterm and full-term infants: Exploratory behavior in eight-month-olds. *Child Development, 47,* 606–612.

Sigman, M. (1995). Nutrition and child development: More food for thought. *Current Directions in Psychological Science, 4,* 52–55.

Sigman, M., Cohen, S. E., & Beckwith, L. (1997). Why does infant attention predict adolescent intelligence? *Infant Behavior and Development, 20,* 133–140.

Sigman, M., McDonald, M. A., Neumann, C., & Bwibo, N. (1991). Prediction of cognitive competence in Kenyan children from toddler nutrition, family characteristics, and abilities. *Journal of Child Psychology and Psychiatry, 32,* 307–320.

Sigman, M., Neumann, C., Baksh, M., Bwibo, N., & McDonald, M. A. (1989). Relations between nutrition and development of Kenyan toddlers. *Journal of Pediatrics, 115,* 357–364.

Sigman, M., Neumann, C., Jansen, A. A. J., & Bwibo, N. (1989). Cognitive abilities of Kenyan children in relation to nutrition, family characteristics, and education. *Child Development, 60,* 1463–1474.

Sigman, M., & Parmelee, A. H. (1974). Visual preferences of four-month-old preterm and full-term infants. *Child Development, 45,* 959–965.

Simeon, D. T., & Grantham-McGregor, S. (1990). Nutritional deficiencies and children's behavior and mental development. *Nutrition Research Review, 3,* 1–24.

Strupp, B. J., & Levitsky, D. A. (1995). Enduring cognitive effects of early malnutrition: Theoretical reappraisal. *Journal of Nutrition, 125,* 2221S–2232S.

Takahashi, E. (1986). Secular trends of female body shape in Japan. *Human Biology, 58,* 229–301.

U.S. Department of Health and Human Services and U.S. Department of Agriculture. (1986). *Nutrition monitoring in the United States—A report from the Joint Nutrition Monitoring Evaluation Committee* (DHHS Publication No. PHS 86-1255). Washington, DC: U.S. Government Printing Office.

Waber, D. P., Vuori-Christiansen, L. P., Oritz, N., Clement, J. R., Christiansen, N. E., Mora, J. O., Reed, R. B., & Herrera, M. G. (1981). Nutritional supplementation, maternal education, and cognitive development of infants at risk of malnutrition. *American Journal of Clinical Nutrition, 34,* 807–813.

Wachs, T. D. (1995). Relation of mild-to-moderate malnutrition to human development: Correlational studies. *American Journal of Nutrition, 125,* 2245S-2254S.

Walter, T. (1993). Impact of iron deficiency on cognition in infancy and childhood. *European Journal of Clinical Nutrition, 47,* 307–316.

Walter, T., Kovalskys, J., & Steckel, A. (1983). Effect of mild iron deficiency on infant mental development scores. *Journal of Pediatrics, 102,* 519–522.

Whaley, S., Sigman, M., Espinosa, M., & Neumann, C. (in press). Infant predictors of cognitive development in an undernourished Kenyan population. *Journal of Developmental and Behavioral Pediatrics.*

Wilson, D. M., Hammer, L. D., Duncan, P. M., Dornbush, S. M., Retter, P. L., Hintz, R. L., Gross, R. T., & Rosengeld, R. G. (1986). Growth and intellectual development. *Pediatrics, 78,* 646–650.

Winick, M., Meyer, K. K., & Harris, R. C. (1975). Malnutrition and environmental enrichment by early adoption. *Science, 190,* 1173–1175.

7

Nutrition and the Worldwide Rise in IQ Scores

Reynaldo Martorell

Flynn (1984, 1987, 1994) has documented a worldwide rise in IQ scores; gains are largest for tests that emphasize on-the-spot problem solving (fluid intelligence) rather than acquired skills, knowledge, and vocabulary (crystallized intelligence). Time series data, particularly from North America and Europe, show that the rise in scores goes back to when tests began to be used widely, in the 1930s. For some countries, the gains are about 15 IQ points or 1 standard deviation (*SD*). Indirect evidence has suggested that the rise in scores began as far back as the latter part of the 19th century, and thus the total gains in IQ may be larger than recorded. Trends continued into the 1980s in Scandinavian countries (Emanuelsson & Svensson, 1990; Teasdale & Owen, 1989) and in the United States (Horgan, 1995). Reports about trends in the 1990s are needed to assess continuity to the present.

The subject of this chapter is whether nutritional improvements in industrialized countries can account for this worldwide rise in IQ scores, which has come to be known as the Flynn effect. The diets in these countries have improved markedly over this century. Also, environmental sanitation and hygiene and the availability of public health measures have improved, which has reduced the burden of infection.

These changes in diet and health are generally believed to be the causes of the increases in adult height in industrialized countries, a trend that began in the middle of the 19th century and ceased in Scandinavian countries and the United States only recently.

The analysis begins with a review of the nature of undernutrition, with special emphasis on the meaning of height as an indicator of nutrition. This is followed by an examination of secular trends in height in industrialized countries. Next, two types of evidence about nutrition and cognitive development are reviewed: (a) studies of intellectual performance in undernourished children from developing countries and (b) studies of the effects of multinutrient supplements on IQ in schoolchildren from Europe and North America. In the final section, an integrated analysis is carried out that addresses the question of whether nutrition offers a plausible explanation for the Flynn effect.

THE NATURE OF UNDERNUTRITION

Most of the evidence linking poor nutrition and cognition has come from recent studies performed in developing countries. Because nutritional and health conditions in these countries today are similar to those that existed in Europe and the United States a few generations ago (Wray, 1978), it is important to understand the nature, causes, and manifestations of undernutrition in poor countries.

The four major nutritional problems of poor countries are protein–energy undernutrition (Waterlow, 1992), Vitamin A deficiency (Sommer & West, 1996), iron-deficiency anemia (Yip & Dallman, 1996), and iodine deficiency (Hetzel, Dunn, & Stanbury, 1987). Other nutritional problems have either a more limited geographical distribution or are considered to be of lesser health importance. Nutritional deficiencies tend to occur together and are more common among the poorest of families. These aspects need to be taken into account in interpreting the literature on nutrition and cognition. For example, it is hard to attribute effects to isolated nutrient deficiencies or to nutritional problems apart from the context of poverty (Martorell, 1997).

The immediate causes of nutritional problems, through which

larger social and environmental causes operate, can be grouped into those related to food, health, and care (Jonsson, Ljungqvist, & Yambi, 1993). Factors considered under "food" include all of what is fed to children, such as breast milk, complementary foods, and home family meals; among the determinants of whether nutritional requirements are met are amounts of foods provided, appropriateness for consumption by young children, and nutrient characteristics of the foods used. "Health" refers primarily to infections, such as diarrheal diseases, which spread through contaminated foods and liquids and are nutritionally important because they depress appetites and impair nutrient utilization of consumed food. "Care" is the complex of behaviors on the part of caretakers aimed at meeting the biological and psychosocial needs of children; rapid and proper management of childhood illness is an example of good care.

About 24% of children are born with low birth weight (LBW; i.e., < 2500 g) in the least developed countries, whereas in industrialized countries the figure is about 6% (UNICEF, 1996). Bangladesh, one of the world's poorest countries, had a prevalence of LBW of 50% in 1990; the Nordic countries, at the other extreme, had a prevalence of about 4% (UNICEF, 1996). The principal cause of LBW in developed countries is prematurity; however, the greater LBW prevalences of developing countries are explained, not by greater prematurity rates, but largely by the greater frequency of intrauterine-growth-retarded infants (IUGR; Villar & Belizán, 1982). IUGR babies are full-term neonates suffering from chronic undernutrition in utero, a condition whose causes include small maternal body size, poor diet during pregnancy, and perhaps anemia (Kramer, 1987).

Nutritional problems are most common and severe in the first 2–3 years of life for several reasons (Martorell, 1995). Young children have greater relative (i.e., per unit of mass) nutritional needs, in part because they are growing faster than at any other point in life, including adolescence. At the same time, young children have relatively small gastric capacities and thus require frequent meals with rich nutrient concentrations; however, in many poor settings, children receive infrequent dilute or bulky diets of low nutritional value. Immature immunological systems and increased exposure to infectious agents account for the high frequency of diarrheal and respiratory infections. Another reason for the vulnerability of small children is their complete dependence on others for care. These factors, along with intrauterine growth retardation, account for the fact that many nutritional problems are at their peak of incidence and severity in early childhood.

Growth failure is the most generalized and visible sign of malnutrition (Martorell, 1995). In the face of dietary deficiencies and frequent infections, nutrients are allocated preferentially to maintaining vital functions. Under these conditions, linear growth may be limited or completely arrested, and if necessary, muscle and fat may be metabolized for its nutrient content. This causes stunting (severe linear growth retardation, defined as height more than 2 *SD*s below the median in the reference population) and wasting (low weight for length, defined as weight more than 2 *SD*s below the median weight for height in the reference population) in many children. In the least developed countries, about 50% of children under 5 years of age are stunted, and 10% are wasted (UNICEF, 1996). In Bangladesh about 63% of young children are stunted, and 17% are wasted (UNICEF, 1996). Stunting and wasting are associated with impaired immunocompetence and greater mortality risks (Pelletier, Frongillo, Schroeder, & Habicht, 1995) in developing countries about 17% of newborns die before reaching 5 years of age (UNICEF, 1996), and 56% of these deaths are associated with growth failure (Pelletier et al., 1995). Once the turbulent first 3 years are over, however, survivors improve in health; growth rates become normal relative to international references, and children remain short but do not get relatively shorter (Martorell, Khan, & Schroeder, 1994). There may be some catch-up in growth during adolescence, associated with delayed maturation and more time to grow, but this is generally of limited significance and not enough to compensate for growth failure in early childhood. The small body size of adults, which characterizes most people from developing countries, is largely the result of nutritional forces operating during the intrauterine period and the first 3 years of life and is an excellent summary indicator of overall nutritional history.

THE SECULAR TREND IN HEIGHT

Height has been increasing over time in the industrialized countries. The best explanation for this trend is that it reflects improved nutrition as a result of rising dietary quality and reductions in infectious diseases. For example, Takaishi (1995) documented remarkable changes in the Japanese diet, not so much in terms of total protein and energy but in its composition and quality; major increases in consumption have taken place in fat, animal protein, calcium, and many micronutrients. The burden of infection has fallen dramatically in industrialized countries. For example, the principal causes of infant mortality in New York City at the turn of the 20th century were diarrheal and respiratory infections (Wray, 1978), the same two primary causes in today's least developed countries.

Arguments were presented in the preceding section that justify the use of measures of physical growth, such as height, as indicators of nutrition. A point also emphasized is that impaired growth is generally limited, and to a remarkable degree, to early childhood. Whereas secular trends in stature have been documented in children, the best and most complete data come from measurements of young men at induction into the army, particularly in the case of Europe (Schmidt, Jørgensen, & Michaelsen, 1995). Because the increases in height in adults are primarily driven by improvements in their growth as young children, the secular trend in height is a lagged indicator; that is, it reflects changing nutritional conditions as they existed when the adults measured were young children—in the case of 18-year-old army recruits, about 15 years prior to induction.

An important consideration in relating secular trends in stature to the Flynn effect is that of timing. When did the secular trend in height begin, and is it still continuing? In 1988, Tanner wrote the following concerning Europe:

> It is not clear when the present trend started, though an astonishing series of Norwegian growth data stretching back to 1741 indicate that little gain in adult height took place from 1760 to 1830, a gain of about 0.3 cm per decade took place from 1830

> to 1875, . . . to the present day. Danish data stretching back to 1815 show a similar lack of gain till [*sic*] about 1845. (p. 392)

Studies of Georgian convicts, Union army soldiers, and West Point cadets have suggested that average heights actually decreased by as much as an inch during the first half of the 19th century in the United States (Komlos & Katzenberger, 1995). The secular trend in height, in the United States as well as in northern Europe, most likely began in the second half of the 19th century. This means that the beginning of the secular trend in adult stature preceded the widespread use of intelligence tests in the 1930s by about 80 years. Changes in some areas of the world were slowed or interrupted during World Wars 1 and 2 but reemerged strongly during the postwar periods (Takaishi, 1995; Van Wieringen, 1986). The continuity of the secular trend in height into the present is a subject of current research.

National surveys suggest that the secular trend in height has leveled off in the United States. There are no differences, for example, in the stature of persons 18–24 years measured in the first (1971–1975) and second (1976–1980) National Health and Nutrition Examination Surveys (NHANES). Mean heights were 177.0 cm in men and 163.3 cm in women from NHANES I (Abraham, Johnson, & Najjar, 1979) and 177.0 cm in men and 163.4 cm in women from NHANES II (Najjar & Rowland, 1987). Mean heights of persons 18–24 years measured in Phase 1 of NHANES III (1988–1991) were similar to results from previous surveys: 176.5 cm in men and 163.9 cm in women (Martorell, unpublished analysis using sample weights). Data from Phase 2 of NHANES III (1991–1994) have not yet been released.

Data from Japan indicate that the trend continued until 1990, the last year reported by Takaishi (1995). The Scandinavian countries and The Netherlands reached a plateau during the 1980s, but heights continued to increase until 1990, the end point studied, in middle and southern European countries (Schmidt et al., 1995). These countries had not yet reached the low levels of postneonatal mortality achieved by the Scandinavian countries and The Netherlands, suggesting that optimal nutrition and health conditions had not yet been realized (Schmidt et al., 1995).

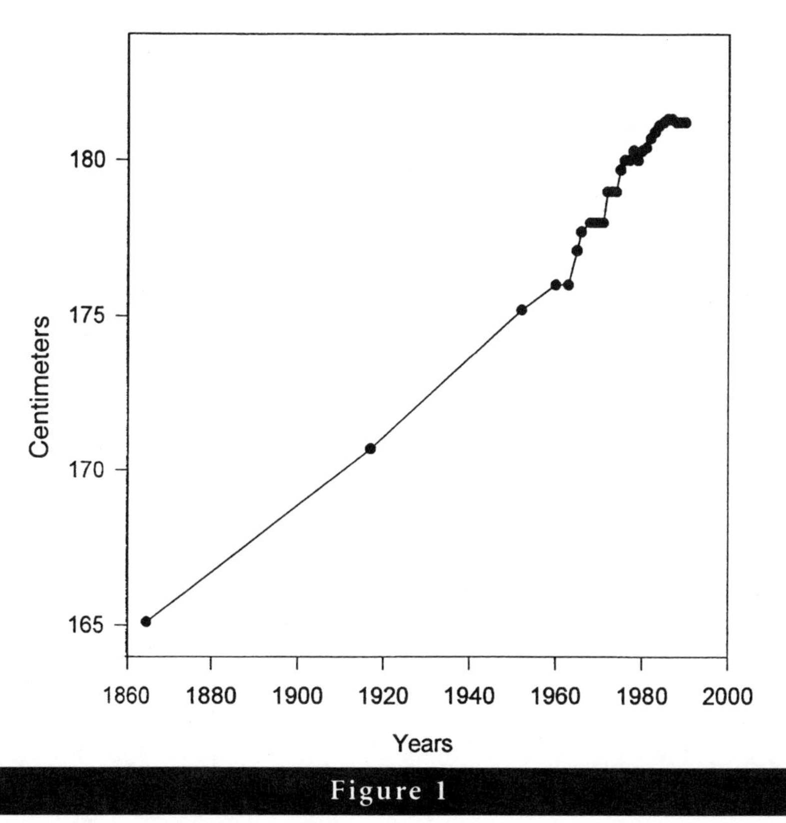

Figure 1

Height of Dutch recruits from 1865 to 1990.

Trends in the height of Dutch recruits are shown in Figure 1. The Dutch are the tallest population in the world, with a mean height for men of 181.2 cm (5 ft 11.3 in.) in 1990. Van Wieringen (1986) presented a more complete documentation of trends since 1860, showing these to have been approximately linear over a long period, although a plateau has been reached now. Differences in the stature of Dutch recruits measured over the years with respect to 1990 are shown in Figure 2. In 1865, Dutch recruits were 16.1 cm shorter than they are today. The difference was reduced to 10.5 cm by 1917 and to 5.2 cm by 1960. The difference with respect to 1990 became 1.0 cm or less for the first time in 1978. Since 1984, adult heights have differed by no more than 0.1 cm from the 1990 level. The secular trend has leveled off in Scandi-

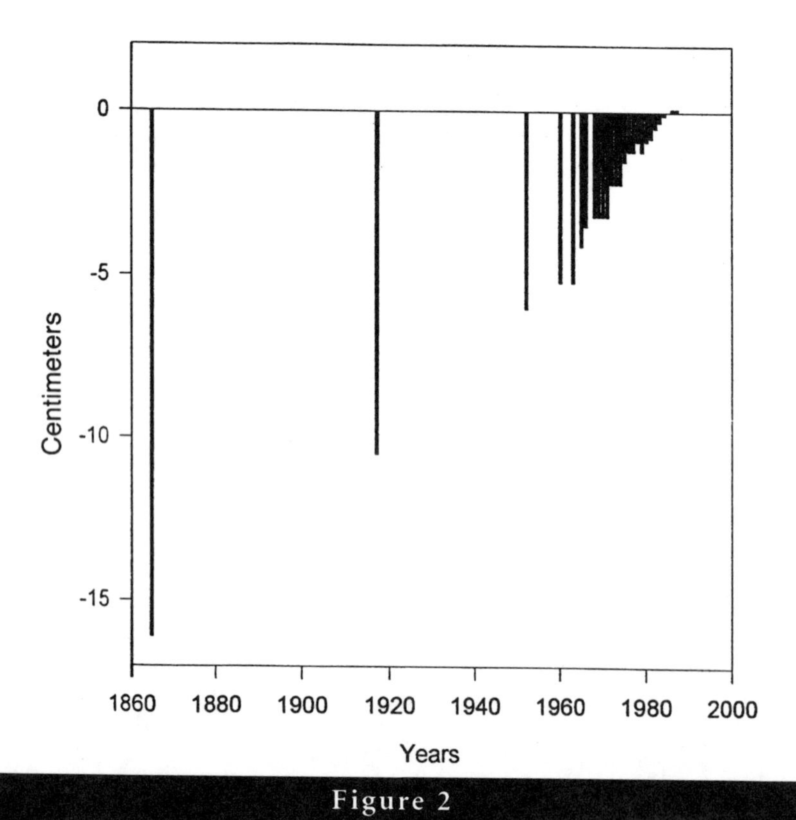

Figure 2

Difference in the stature of Dutch recruits with respect to the 1990 sample mean of 181.2 cm.

navian countries at slightly lower heights and at approximately the same time as in The Netherlands (Schmidt et al., 1995). For example, the height of Swedish recruits came within 1 cm or less of the 1990 mean for the first time in 1970 and ceased to increase altogether since 1980. Mean height was 179.4 cm in 1990 for Swedish recruits.

For the Scandinavian countries and The Netherlands, therefore, one would infer from the fact that adult heights reached current levels in about 1980 that optimal nutrition and health conditions during early childhood were largely achieved by the mid-1960s, and that since then, the growth of children has been unconstrained by undernutrition. This is consistent with postneonatal mortality rates, for which historically low levels were achieved in these countries in the mid-1960s (Schmidt et al., 1995). In the United States, conditions of optimal nutrition may have been achieved even earlier because adult heights reached current levels in the early 1970s.

UNDERNUTRITION AND COGNITION

Having discussed the nature of undernutrition and described secular trends in height, I turn now to the evidence linking poor nutrition during early life to impaired cognitive development. Rather than a review per se of this vast literature, I present the conclusions of a comprehensive review recently undertaken (Martorell, 1997).

Whereas it used to be thought that undernutrition affected cognition largely by damaging the brain during sensitive periods of early development, many other mechanisms, in addition to organic damage, are now recognized as important (Brown & Pollitt, 1996). For example, nutrition may affect cognitive development in children by limiting physical activity and reducing interaction with other people and with the environment. Timing is crucial for most of the mechanisms linking undernutrition to poor cognition proposed to date. Damage to the brain is more likely during the rapid phases of growth that occur during the prenatal period and infancy. Apathy and withdrawal, whether caused by diet, infection, or both, are more of a problem in young children.

Follow-up studies of infants of low birth weight have indicated that relative to controls of normal birth weight, performance at school age is reduced by about 6 IQ points (Aylward, Pfeiffer, Wright, & Verhulst, 1989; Paz et al., 1995; Teberg, Walther, & Peña, 1988). It is interesting that the prognosis is equally poor for premature and IUGR babies. However, most of the evidence is from developed countries, where the environment is likely to buffer against poor outcomes; it is possible that effects of low birth weight are greater in impoverished environments, such as exist in developing countries today or existed in Europe and North America a century ago.

Severe, clinical malnutrition (e.g., kwashiorkor, marasmus) and stunting are associated with a reduction of as many as 15 IQ points according to follow-up studies of cases and neighborhood or classroom controls (Grantham-McGregor, Powell, & Fletcher, 1989; Martorell, 1997; Wachs, 1995). Although differences remain after controlling for socioeconomic indicators, it is difficult to separate nutritional from social causes of poor performance; rather, these effects should be viewed as the product of a powerful interaction between poor nutrition and poverty. Experimental studies have shown that improvements in nutrition lead to long-lasting benefits and that these are greatest among those from poorer backgrounds (Martorell, 1977). In the follow-up study of Guatemalan participants exposed to nutrition interventions in early childhood, performance of adolescents and young adults was improved significantly on reading and vocabulary tests but not in the Raven Progressive Matrices (Pollitt, Gorman, Engle, Martorell, & Rivera, 1993). This remains the only study to have assessed the effects of a nutrition intervention in early childhood on later performance in youths and adults.

Certain micronutrient deficiencies are known to interfere with learning and cognition. Iodine deficiency, for example, has profound effects on development; severe prenatal deficiency leads to cretinism, a syndrome of developmental impairment and profound mental and motor aberration (Hetzel, 1989). Studies have now established that cretinism is the extreme of a continuum of functional impairment. Children in iodine-deficient areas are also affected. In a meta-analysis of experimental and case-control community studies, it was found that moderate iodine deficiency was associated with a reduction in IQ of 13.5 points (Bleichrodt & Born, 1994). Iron-deficiency anemia in young children is also associated with significant deficits in the Bayley Scales of Mental and Motor Development and other tests (Haas & Fairchild, 1989; Lozoff, 1990). As in stunting, it is difficult to isolate the effects of anemia from those of the setting of poverty in which anemia is generally found. Intervention studies in young children indicate significant effects in anemic participants, but effects on nonanemic, iron-deficient children are less certain.

A wide range of tests have been used in studies of undernutrition and cognition, ranging from those emphasizing problem solving (fluid intelligence) to those concerned with acquired skills, knowledge, and vocabulary (crystallized intelligence), with many combining both dimensions. The effects found have been as varied as the tests, and it would be incorrect to say that undernutrition affects only one dimension of intelligence. Certainly, there is no basis for claiming that effects are systematically greater for nonverbal than for verbal performance.

In summary, undernutrition in early childhood has large but general effects on intelligence. Stunting, the best overall indicator of nutritional deficiency, is associated with a deficit of about 1 *SD* in IQ score. Large effects have been found as well for anemia and for moderate iodine deficiency.

VITAMIN–MINERAL SUPPLEMENTATION AND NONVERBAL INTELLIGENCE

Over the last century, improvements in the standard of living in Europe and North America have resulted in better diets and reduced levels of infection, and through these mechanisms, there have been declining prevalences of undernutrition and improved growth in young children. Therefore, improved nutrition during early development is one possible explanation for the rise in test scores observed in these countries.

Nutrition may also affect test scores at older ages. It is possible that a percentage of schoolchildren in industrialized countries is deficient in one or more vitamins or minerals and that this deficient status affects intellectual performance adversely. It is less likely that nutrient intakes influence performance beyond levels of intake exceeding recommended dietary allowances (RDAs). In this regard, Whitehead (1991) noted that "no physiological explanation exists of how vitamin and mineral supplementation could affect brain function in a well-nourished subject" (p. 906).

Several studies have been carried out to test the effect of multiple nutrient supplements on performance (Table 1). These studies were carried out among schoolchildren in Europe and California, and all involved randomized, double-blind, placebo-controlled designs, the

Table 1

Studies of Vitamin–Mineral Supplementation and Nonverbal Intelligence

Study	Place	Age (years)	Duration	*N*	Results[a]
Benton & Roberts (1988)	Wrexham, Wales	12–13	8 months	90	Supplemented group increased by 9 IQ points, placebo group by 1 point, and "no tablet" group by 4.5 points. Similar effects in boys and girls.
Benton & Butts (1990)	Belgium	13	5 months	167	None overall. Significant increase (8 points) only in boys and only in those with poorer diets.
Benton & Cook (1991)	Wales	6	6–8 weeks	47	Significant increase (9.3 points). Effects larger in girls.
Nelson et al. (1990)	London	7–12	1 month	210	None.
Crombie et al. (1990)	Dundee, Scotland	11–13	7 months	86	None.
Schoenthaler et al. (1991)	California	12–16	3 months	615	Effects claimed for WISC-R nonverbal tests, one of several outcomes. Significant increase only in group receiving 100% RDA (3.6 points). Changes not significant in 50% RDA and 200% RDA groups.
Southon et al. (1994)	Norwich, England	13–14	14 months	51	None.

Note. All studies used randomized, double-blind, placebo-controlled designs. WISC-R = Wechsler Intelligence Scale for Children, Revised; RDA = recommended dietary allowance.
[a]None of the studies included effects on verbal IQ scores.

strongest possible type of design to test for such effects. The supplements included a large number of vitamins and minerals at levels equal to or greater than the RDAs. The RDA is typically set at 2 *SD*s above the estimated requirement in the population and should, in theory, satisfy the needs of all but about 2.5% of individuals. With one exception (Southon et al., 1994), these investigators did not characterize nutritional status of the participants using appropriate methods.

None of the authors of these studies reported effects on verbal intelligence, but some did report effects on tests of nonverbal function. The first of the series, by Benton and Roberts (1988), was a study among Welsh children, each of whom received a supplement (treated), a placebo (control), or neither (no tablet). The investigators found a large difference in performance on the Calvert Non-Verbal Test between treated and control children (8 IQ points) but less difference between treated and "no tablet" groups (3.5 IQ points). Thus, the effect ranged from 8 to 3.5 points depending on the reference group chosen. Effects were found to be similar in boys and girls. Two studies failed to replicate the findings (Crombie, 1990; Crombie et al., 1990; Nelson, Naismith, Burley, Gatenby, & Geddes, 1990); however, one of these studies had a duration of only 1 month, and its null findings may be questioned (Nelson et al., 1990). Crombie et al. (1990) measured dietary intakes using the 7-day weighed-record method and found that nutrient intakes were similar to those of the general population and to those of the children included in the previous study in Wales by Benton and Roberts (1988). Although Crombie found a small, nonsignificant difference in favor of the treatment group in performance on the Calvert Non-Verbal Test, the direction of effects was not consistent in three other tests of nonverbal performance. The investigators concluded that the findings of Benton and Roberts (1988) could not be confirmed.

In a subsequent experiment, Benton and Buts (1990) found that the nonverbal test performance of the treated group was not significantly different from that of the placebo group. However, a post hoc analysis revealed that among boys with poorer diets, defined on the basis of information from diaries kept by the children for 15 days, a significant improvement of 8 points was observed in the treated group. It is not clear what these results mean, because the validity of the dietary assessment method is questionable (Southon et al., 1994), and it is not clear why boys and not girls responded to treatment. Later, Benton and Cook (1991), in a study of Welsh children, found an increase of 9.3 points in subscales of the British Ability Scale. Contrary to the prior findings by Benton and Buts (1990), Benton and Cook (1991) found that the effect on performance was greater in girls than in boys. Also, Benton and Cook (1991) were not able to replicate convincingly previous findings of greater effects among those with poorer diets. Parents were asked to report their children's frequency of consumption of 24 foods. However, the only analyses reported involved information about sugar intake, presumably an indicator of diets poor in vitamins and minerals. The authors stated that the change in intelligence test scores among those supplemented was significantly correlated with sugar intakes.

The largest study to date is that of Schoenthaler, Amos, Eysenck, Peritz, and Yudkin (1991), who studied over 600 schoolchildren in California. A significant increase of 3.6 points in performance on the nonverbal Wechsler tests but not in other tests was observed, but only among the group receiving the RDA level. Those receiving half or twice the RDA had similar gains (1.0 and 1.4 points, respectively), which were not statistically significant. No information about nutrient intakes or micronutrient status was presented. The study by Schoenthaler et al. (1991) has been strongly criticized by many. Peto (1991) dismissed the findings because "so many *p* values could have been generated that even if vitamins are wholly ineffective it would not be particularly surprising for a few fairly extreme *p* values to be unearthed—indeed, it would be mildly surprising if they could not be" (p. 906). This point is raised as well by Blinkhorn (1991), who also comments on the inappropriateness of the statistical analyses and their presentation.

Southon et al. (1994) included careful measures of nutrient intakes using the 7-day weighed-record method and were the first to measure, as well, a range of biochemical indices of micronutrient status. The sample of 51 children from local schools in Norwich, England, were found to have diets that "could be considered as adequate," but some children had "low" biochemical values, particularly in the case of serum ferritin values in girls, among whom the prevalence of iron deficiency was found to be 21%. Dietary intake measures were found to correlate poorly with biochemical status, and on this basis, the authors cautioned against using diet measures alone to characterize nutritional status as done by Benton and Buts (1990). Finally, the researchers found that treatment with a vitamin and mineral supplement did not alter verbal or nonverbal performance and, furthermore, that this result occurred regardless of nutritional status.

Unfortunately, studies involving multiple-micronutrient supplements have not been carried out in developing countries, where deficiency is likely to be more common and severe than in Europe and North America. However, studies concerning iron supplementation and helminthic infections have provided useful information about the role of iron deficiency in intellectual performance.

Double-blind, placebo-controlled trials have been carried out in which treatment was given to anemic, iron-deficient (but nonanemic), and iron-replete (normal) schoolchildren (Pollitt, Hathirat, Kotchabhakdi, Misell, & Valyasevi, 1989; Soemantri, Pollitt, & Kim, 1985). Treatment was effective in correcting iron deficiency in both of these experiments. In Indonesia, there were no differences in IQ (using the Raven Progressive Matrices) between iron-status groups (9–11 years of age) prior to treatment (IQ was lower by 1.2 points in anemic relative to normal children, $p > .05$). Treatment for 3 months was found to improve school achievement test scores, but unfortunately, IQ was assessed only at baseline (Soemantri et al., 1985). In a similar study in Thailand (Pollitt et al., 1989) among 9- to 11-year-old children, on the other hand, no effect of treatment was found, over a 3-month period, on school achievement tests or on IQ measured by the Raven at baseline and posttreatment.

Studies have also been done to assess the effects of treating helminthic infections; these problems depress iron status and are also likely to have systemic, debilitating effects that can influence learning. No statistically significant effects on school achievement tests were found in treated Jamaican (Simeon, Grantham-McGregor, Callender, & Wong, 1995) or Guatemalan (Watkins, Cruz, & Pollitt, 1996) children; however, another Jamaican study revealed significant effects on auditory short-term memory and on scanning and retrieval of long-term memory (Nokes, Grantham-McGregor, Sawyer, Cooper, & Bundy, 1992). Finally, lack of breakfast prior to arrival at school has adverse effects on cognition, particularly on the speed of information retrieval in working memory; however, these effects may not apply to well-nourished children (Pollitt, 1995).

DISCUSSION

This chapter has emphasized two ways through which nutrition may affect intelligence. The first is through its impact on development during fetal life and early childhood, and the second is through effects of current nutritional status and dietary intake levels in the schoolchild.

Epidemiological studies, largely from developing countries, have revealed that nutritional deficiencies are most common and severe in early childhood. Malnutrition, identified through clinical signs of deficiency, intrauterine growth retardation, or postnatal growth failure, has profound but nonspecific effects on cognition. Although most of the evidence is from developing countries, present conditions in these countries are not unlike those that existed in Europe and North America several generations ago (Wray, 1978). In fact, because of public health advances such as vaccinations, the situation was probably worse in the presently industrialized countries. The infant mortality in New York City was 140 deaths per 1,000 live births in 1898 (Wray, 1978); in 1994, the corresponding statistic for the least developed countries was 108 deaths per 1,000 live births (UNICEF, 1996). The percentage of newborns of low birth weight in mid-19th-century Europe and North America may have been at about the level found today in the poorest countries, 24%; today, it is about 6%. Severe growth failure (stunting) probably affected half or more of children and now is an insignificant problem. Iodine deficiency was a major problem in the mountainous areas of Europe (e.g., the Alps) and in the Great Lakes areas of the

United States; salt iodization, which began as early as the 1920s, led to a remarkable decline in goiter and cretinism in Switzerland and other countries by the 1940s (Hetzel, 1989). Anemia (severe iron deficiency) was probably very common in children, and now it is relatively rare; the presence of low iron stores, on the other hand, probably still affects a minority of children, but previously the condition affected most of the population.

In short, nutritional problems, which are known to impair cognition significantly, vanished from northern Europe and the United States over the last century or so. This general improvement in nutritional status must have had an effect on both verbal and nonverbal performance. However, it is important to ask when these effects on performance were manifested. One may speculate that the effects were manifested primarily over the period when nutritional deficiencies were overcome, roughly from 1850 to 1960. It is difficult to conceive that Dutch and Scandinavian recruits tested since 1970 were malnourished in childhood to a degree that would have a significant effect on intelligence. The gains in adult height since 1970 have been very small (i.e., ~1 cm), and it is unlikely that these changes, although probably nutritional in nature, can account for the continued and pronounced rise in performance IQs. If nutrition were an important cause of the Flynn effect, a notable deceleration in the trend toward higher IQs would have been evident since the 1970s; in Scandinavian countries and The Netherlands, a plateau in performance should have been reached in the early 1980s, when adult stature reached current levels. Also, scores should not have risen at all in the United States since about 1970. The evidence concerning trends in test scores is limited. Several investigators have reported that trends continued unaltered in The Netherlands, Sweden, and the United States into the 1980s (Emanuelsson & Svensson, 1990; Horgan, 1995; Teasdale & Owen, 1989). Future investigators should pay close attention to the continuity of these trends and to the rate of change. Continued rises in tests scores during the 1990s and beyond in Scandinavian countries, The Netherlands, and the United States should exclude nutrition as the present driving force behind the changes.

Diets of schoolchildren have also improved over time, but the implications for cognition of these changes are less clear. A series of recent studies carried out in Europe and California suggest that multivitamin and mineral supplementation, including iron, does not improve verbal measures of intelligence among apparently healthy schoolchildren. The picture is mixed with regard to nonverbal measures of intelligence. On balance, there are considerable inconsistencies in the findings and no firm evidence of effects on nonverbal measures. Unfortunately, vitamin and mineral supplementation studies have not been carried out among children in developing countries, where poor status is more common and effects more likely to be found. In these areas, only experiments focusing on iron, directly through iron supplementation or indirectly by controlling helminthic infections, are available; results of these studies have suggested that improving iron status even in anemic children has little or no effect on IQ and achievement. Studies among schoolchildren concerning the effects of supplementation with vitamins and minerals are clearly needed in both developing and industrialized countries. These should be double-blind, placebo-controlled trials with sample sizes large enough to detect small changes (i.e., 3 IQ points) and with appropriate dietary and biochemical measures to characterize nutritional status. At present, it is concluded that the huge gains in performance registered in developed countries cannot be due to dietary improvements in schoolchildren.

Lynn (1990; see also chapter 8, this volume), in examining the same issue addressed in this chapter, has concluded that nutrition offers the best explanation for the Flynn effect. He has pointed out that the secular trends in height and intelligence are of about the same order of magnitude, 1 *SD*, and has stressed the persuasiveness of studies showing a link between poor nutrition and impaired cognition. In fact, the gain in height in The Netherlands and other countries is about 15 cm, equivalent to more than 2 *SD*s; at any rate, there is no reason to expect a one-to-one correspondence in the magnitude of the secular trends in height and intelligence. Teasdale and Owen (1989) discarded nutrition as an explanation, as I do, because there has been no deceleration of gains in intelligence test scores in recent data. Eysenck (1991) interpreted the literature on effects of vitamin and mineral supplementation on intelligence in schoolchildren as supportive of nutrition as an explanation for the Flynn effect; from the review of this literature presented in this chapter, I find the evidence inconclusive.

CONCLUSION

There is considerable evidence from today's poor countries that links poor nutrition to impaired cognitive development. Therefore, it is reasonable to suggest that the remarkable nutritional improvements that occurred in many countries during the second half of the 19th and first half of the 20th centuries probably resulted in improved intellectual performance. However, intelligence test scores continued to increase in northern Europe and the United States at an undiminished rate into the late 1980s, after the disappearance of nutritional deficiencies and poor physical growth during early childhood. Because information about trends is not available for the 1990s, the conclusion that nutrition cannot explain rises in intelligence since about 1970 must remain tentative. One key reason for rejecting nutrition as an explanation of recent trends in intelligence test scores is that cohorts of young adults in Scandinavian countries, The Netherlands, and the United States measured since 1970 are unlikely to have been undernourished during critical childhood phases of development, yet available evidence does not indicate a plateau in IQ scores, as would be expected. A second reason is that there is no conclusive evidence that greater intakes of vitamins and minerals affect cognitive performance among schoolchildren in developed countries.

REFERENCES

Abraham, S., Johnson, C. L., & Najjar, M. F. (1979). *Weight and height of adults 18–74 years of age: United States, 1971–74* (National Center for Health Statistics, Series 11, No. 211; DHEW Publication No. PHS 79-1659). Washington, DC: U.S. Government Printing Office.

Aylward, G. P., Pfeiffer, S. I., Wright, A., & Verhulst, S. J. (1989). Outcome studies of low birth weight infants published in the last decade: A metaanalysis. *Journal of Pediatrics, 115,* 515–520.

Benton, D., & Buts, J. P. (1990). Vitamin/mineral supplementation and intelligence. *Lancet, 335,* 1158–1160.

Benton, D., & Cook, R. (1991). Vitamin and mineral supplements improve the intelligence scores and concentration of six-year-old children. *Personality and Individual Differences, 12,* 1151–1158.

Benton, D., & Roberts, G. (1988). Effect of vitamin and mineral supplementation on intelligence of a sample of school children. *Lancet, 1,* 140–143.

Bleichrodt, N., & Born, M. P. (1994). A metaanalysis of research on iodine and its relationship to cognitive development. In J. B. Stanbury (Ed.), *The damaged brain of iodine deficiency: Neuromotor, cognitive, behavioral, and educative aspects* (pp. 195–200). New York: Cognizant Communication.

Blinkhorn, S. (1991). A dose of vitamins and a pinch of salt. *Nature, 350,* 13.

Brown, J. L., & Pollitt, E. (1996). Malnutrition, poverty and intellectual development. *Scientific American, 274,* 38–43.

Crombie, I. K. (1990). Vitamin/mineral supplementation and intelligence. *Lancet, 335,* 1159–1160.

Crombie, I. K., Todman, J., McNeill, G., Du, V., Florey, C., Menzies, I., & Kennedy, R. A. (1990). Effect of vitamin and mineral supplementation on verbal and non-verbal reasoning of school children. *Lancet, 335,* 744–747.

Emanuelsson, I., & Svensson, A. (1990). Changes in intelligence over a quarter of a century. *Scandinavian Journal of Educational Research, 34,* 171–187.

Eysenck, H. J. (1991). Raising I.Q. through vitamin and mineral supplementation: An introduction. *Personality and Individual Differences, 12,* 329–333.

Flynn, J. R. (1984). The mean IQ of Americans: Massive gains 1932 to 1978. *Psychological Bulletin, 95,* 29–51.

Flynn, J. R. (1987). Massive IQ gains in 14 nations: What IQ tests really measure. *Psychological Bulletin, 101,* 171–191.

Flynn, J. R. (1994). IQ gains over time. In R. J. Sternberg (Ed.), *Encyclopedia of human intelligence* (pp. 617–623). New York: Macmillan.

Grantham-McGregor, S., Powell, C., & Fletcher, P. (1989). Stunting, severe malnutrition and mental development in young children. *European Journal of Clinical Nutrition, 43,* 403–409.

Haas, J. D., & Fairchild, M. W. (1989). Summary and conclusions of the international conference on iron deficiency and behavioral development, October 10–12, 1988. *American Journal of Clinical Nutrition, 50,* 703–705.

Hetzel, B. S. (1989). *The story of iodine deficiency: An international challenge in nutrition.* New York: Oxford University Press.

Hetzel, B. S., Dunn, J. T., & Stanbury, J. B. (Eds.). (1987). *The prevention and control of iodine deficiency disorders.* Amsterdam: Elsevier.

Horgan, J. (1995). Get smart, take a test. *Scientific American, 273*(5), 12–14.

Jonsson, U., Ljungqvist, B., & Yambi, O. (1993). Mobilization for nutrition in Tanzania. In J. Rohde, M. Chatterje, & D. Morley (Eds.), *Reaching health for all* (pp. 185–211). New Delhi, India: Oxford University Press.

Komlos, J., & Katzenberger, P. (1995). The heights of Georgia convicts in the antebellum US, 1817–1885. In R. Hauspie, G. Lindgren, & F. Falkner (Eds.), *Essays on auxology* (pp. 285–290). Welwyn Garden City, Hertfordshire, England: Castlemead.

Kramer, M. S. (1987). Intrauterine growth and gestational duration determinants. *Pediatrics, 80,* 502–511.

Lozoff, B. (1990). Has iron deficiency been shown to cause altered behavior in infants? In J. Dobbing (Ed.), *Brain, behavior and iron in the infant diet* (pp. 107–131). London: Springer-Verlag.

Lynn, R. (1990). The role of nutrition in secular increases in intelligence. *Personality and Individual Differences, 11,* 273–285.

Martorell, R. (1995). Promoting healthy growth: Rationale and benefits. In P. Pinstrup-Anderson, D. Pelletier, & H. Alderman (Eds.), *Child growth and nutrition in developing countries* (pp. 15–31). Ithaca, NY: Cornell University Press.

Martorell, R. (1997). Early child development: Investing in our children's future. In M. E. Young (Ed.), *Proceedings of the World Bank Conference of Early Child Development: Investing in the Future, Atlanta, GA, April 8–9, 1996* (pp. 39–83). Amsterdam: Elsevier Science.

Martorell, R., Khan, L., & Schroeder, D. G. (1994). Reversibility of stunting: Epidemiological findings in children from developing countries. *European Journal of Clinical Nutrition, 48,* S45–S57.

Najjar, M. F., & Rowland, M. (1987). *Anthropometric reference data and prev-*

alence of overweight, United States, 1976–80 (National Center for Health Statistics, Vital and Health Statistics, Series 11, No. 238; DHHS Publication No. PHS 87-1688). Washington, DC: U.S. Government Printing Office.

Nelson, M., Naismith, D. J., Burley, V., Gatenby, S., & Geddes, N. (1990). Nutrient intakes, vitamin–mineral supplementation, and intelligence in British schoolchildren. *British Journal of Nutrition, 64,* 13–22.

Nokes, C., Grantham-McGregor, S. M., Sawyer, A. W., Cooper, E. S., & Bundy, D. A. P. (1992). Parasitic helminth infection and cognitive function in school children. *Proceedings of the Research Society of London Britannica, 247,* 77–81.

Paz, I., Gale, R., Laor, A., Danon, Y. L., Stevenson, D. K., & Seidman, D. S. (1995). The cognitive outcome of full-term small for gestational age infants at late adolescence. *Obstetrics and Gynecology, 85,* 452–456.

Pelletier, D. L., Frongillo, E. A., Jr., Schroeder, D. G., & Habicht, J.-P. (1995). The effects of malnutrition on child mortality in developing countries. *Bulletin of the World Health Organization, 73,* 443–448.

Peto, R. (1991). Vitamins and IQ. *British Medical Journal, 302,* 906.

Pollitt, E. (1995). Does breakfast make a difference in school? *Journal of American Dietetic Association, 95,* 1134–1139.

Pollitt, E. Gorman, K. S., Engle, P. L., Martorell, R., & Rivera, J. A. (1993). Early supplementary feeding and cognition. *Monographs of the Society for Research in Child Development, 58*(Serial No. 235), 1–22.

Pollitt, E., Hathirat, P., Kotchabhakdi, N. J., Misell, L., & Valyasevi, A. (1989). Iron deficiency and educational achievement in Thailand. *American Journal of Clinical Nutrition, 50,* 687–697.

Schmidt, I. M., Jørgensen, M. H., & Michaelsen, K. F. (1995). Height of conscripts in Europe: Is postneonatal mortality a predictor? *Annals of Human Biology, 22,* 57–67.

Schoenthaler, S. J., Amos, S. P., Eysenck, H. J., Peritz, E., & Yudkin, J. (1991). Controlled trial of vitamin–mineral supplementation: Effects on intelligence and performance. *Personality and Individual Differences, 12,* 351–362.

Simeon, D. T., Grantham-McGregor, S. M., Callender, J. E., & Wong, M. S. (1995). Treatment of *Trichuris trichiura* infections improves growth, spelling scores and school attendance in some children. *Journal of Nutrition, 125,* 1875–1883.

Soemantri, A. G., Pollitt, E., & Kim, I. (1985). Iron deficiency anemia and educational achievement. *American Journal of Clinical Nutrition, 42,* 1221–1228.

Sommer, A., & West, K. P., Jr. (1996). *Vitamin A deficiency: Health, survival and vision.* New York: Oxford University Press.

Southon, S., Wright, A. J. A., Finglas, P. M., Bailey, A. L., Loughridge, J. M., & Walker, A. D. (1994). Dietary intake and micronutrient status of adolescents: Effect of vitamin and trace element supplementation on indices of status and performance in tests of verbal and non-verbal intelligence. *British Journal of Nutrition, 71,* 897–918.

Takaishi, M. (1995). Growth standards for Japanese children: An overview with special reference to secular change in growth. In R. Hauspie, G. Lindgren, & F. Falkner (Eds.), *Essays on auxology* (pp. 302–311). Welwyn Garden City, Hertfordshire, England: Castlemead.

Tanner, J. M. (1988). Hormonal, genetic, and environmental factors controlling growth. In G. A. Harrison, J. M. Tanner, D. R. Pilbeam, & P. T. Baker (Eds.), *Human biology* (3rd ed., 390–393). New York: Oxford University Press.

Teasdale, T. W., & Owen, D. R. (1989). Continuing secular increases in intelligence and a stable prevalence of high intelligence levels. *Intelligence, 13,* 255–262.

Teberg, A. J., Walther, F. J., & Peña, I. C. (1988). Mortality, morbidity, and outcome of the small-for-gestational age infant. *Seminars in Perinatology, 12,* 84–94.

UNICEF. (1996). *The state of the world's children.* New York: Oxford University Press.

Van Wieringen, J. C. (1986). Secular growth changes. In F. Falkner & J. M. Tanner (Eds.), *Human growth: A comprehensive treatise* (Vol. 3, 2nd ed., pp. 307–332). New York: Plenum Press.

Villar, J., & Belizán, J. M. (1982). The relative contribution of prematurity and fetal growth retardation to low birth weight in developing and developed societies. *American Journal of Obstetrics and Gynecology, 143,* 793–798.

Wachs, T. D. (1995). Relation of mild-to-moderate malnutrition to human development: Correlational studies. *Journal of Nutrition, 125,* 2245S–2254S.

Waterlow, J. C. (1992). *Protein-energy malnutrition.* London: Edward Arnold.

Watkins, W. E., Cruz, J. R., & Pollitt, E. (1996). The effects of deworming on indicators of school performance in Guatemala. *Research Society of Tropical Medicine and Hygiene, 90,* 156–161.

Whitehead, R. G. (1991). Vitamins, minerals, schoolchildren, and IQ. *British Medical Journal, 302,* 906.

Wray, J. D. (1978). Child health in the Americas: A historical and global perspective. In S. J. Bosch & J. Arias (Eds.), *Evaluation of child health services: The interface between research and medical practice* (DHEW Publication No. NIH 78-1066, pp. 277–292). Washington, DC: U.S. Government Printing Office.

Yip, R., & Dallman, P. R. (1996). Iron. In E. E. Ziegler & L. J. Filer, Jr. (Eds.), *Present knowledge in nutrition* (7th ed., pp. 3–21). Washington, DC: ILSI Press.

8

In Support of the Nutrition Theory

Richard Lynn

I am alone among contributors to this volume in regarding improvements in nutrition as the sole factor responsible for the secular increase of intelligence. The possible effects of nutrition are discussed in this volume by Sigman and Whaley (chapter 6) and by Martorell (chapter 7). I first review their chapters and then summarize the arguments for my nutrition theory, which are set out fully in Lynn (1990).

SIGMAN AND WHALEY

Sigman and Whaley (chapter 6) devote much of their attention to the positive effects of giving nutritional supplements to children in third-world countries. This provided some support for the hypothesis that improvements in nutrition have contributed to the secular rise of intelligence in first-world countries, but it is only of a weak kind. Nutrition is generally poorer among contemporary third-world peoples than it was among first-world peoples in the 1930s (the baseline from which intelligence has increased), so the two situations are not comparable. In addition, Sigman and Whaley are critical of the studies reporting nutritional supplementation effects in the United States and Britain.

They argue that the evidence for a positive effect is slim and complain that the studies do not describe the intelligence tests used or present precise data on means and standard deviations. I think that these criticisms are unduly harsh. Nevertheless, Sigman and Whaley do conclude that "improvements in nutrition in these [industrialized] countries are likely to account, at least in part, for the rise in IQ scores" (p. 155). My objection to their chapter it not so much to what they say as to their failure to present the case adequately.

MARTORELL

Martorell (chapter 7) differs from Sigman and Whaley in that he rejects the possibility that nutrition has made any contribution to the secular increase of intelligence. He concludes his chapter by summarizing his two key reasons for rejecting the nutrition hypothesis, which I discuss in turn. First, he says that "cohorts of young adults . . . measured since 1970 are unlikely to have been undernourished during critical childhood phases of development" (p. 201). He argues that the presence of nutritional status can be inferred from height and that this had reached its maximum by the early 1960s in The Netherlands, the United States, and Scandinavia, yet intelligence test scores continued to rise in these recent cohorts for whom nutrition was optimal.

This argument does not stand up to examination. Concerning The Netherlands, Martorell is wrong when he says that height has not increased in young men born from the 1960s onward. His own Figure 1 shows that for the cohorts whose height was measured from 1978 onward, and who therefore were born from 1960 onward, height was increasing at approximately the same rate as over the preceding century. It was only in the last 5 years, 1985–1990, that heights have stabilized. To prove his point, Martorell would have to show that IQs continued to increase during these 5 years, yet he does not do this. Maybe intelligence test scores also stabilized in The Netherlands in these critical years.

The case based on American heights and IQs is equally unconvincing. The NHANES studies on the secular trend for height are conflicting. A comparison of the first and second studies for 1971–1975 and 1976–1980 shows no increase in height, but NHANES 3 (1988–1991) does show continuing increases in height for both males and females. Martorell does not offer any evidence on whether IQs have been rising in these recent American cohorts, but even if he had, it would not bear crucially on the nutrition theory because the evidence concerning height is inconclusive.

There is the same problem with the data from Sweden. Martorell says that heights have ceased to increase in Sweden since 1980, but so what? He would need to show that IQs in Sweden continued to increase since 1980 but does not do so. The only evidence providing some support for his view comes from Denmark. Citing Teasdale and Owen (1989), Martorell says that heights have stabilized among recent Danish military draft cohorts. This is not quite right. Teasdale and Owen showed only that the rate of increase slackened in the cohort born in 1967–1969 to about half that of earlier cohorts, although the secular increase of intelligence continued at the same magnitude. Even if the Danish evidence were confirmed, the result would not be fatal for the nutrition theory of the rise of IQ. It is possible that the neurological development of the brain and the physiological processes required for problem solving are more sensitive to suboptimal nutrition than is height. The result would be that as nutrition improved, the secular increase in height would slacken but that of intelligence would continue. A point might be reached at which there were no longer any secular increases in height but there were still secular increases in intelligence.

Martorell's second point is that "there is no conclusive evidence that greater intakes of vitamins and minerals affect cognitive performance among schoolchildren in developed countries" (p. 201). Yet of the seven studies summarized in his chapter, four showed that cognitive performance does improve after taking nutritional supplements; the other three found no effect. In a situation like this, positive results should be weighted more strongly than negative results, because the latter can arise through differing or faulty procedures. The three negative studies used different procedures, with different time periods for which the supplements were taken, different tests, and so on. Faults can easily occur in this kind of research. For instance, it is difficult to ensure that schoolchildren are taking the supplements as required, and if appreciable numbers fail to do so, the investigators will obtain a negative result. Positive results cannot arise through faulty investigative procedures of this sort. Therefore, the most reasonable reading of the seven studies is that nutritional supplementation probably does have a positive effect on the intelligence of some children.

Furthermore, even if Martorell were correct on this point, it would not disprove the nutrition theory of the secular rise of intelligence. The nutrition theory, at least in my formulation, places most weight on improvements in the nutrition received by the fetus from better nourished pregnant mothers and on improvements in nutrition of infants. Even if it turns our that administering nutritional supplements to adolescents for a few weeks fails to produce an improvement in their IQs, the nutrition theory of the secular rise of IQ would remain undamaged.

Martorell also argues that "there is no basis for claiming that [undernutritional] effects are systematically greater for nonverbal than for verbal performance" (p. 193). His chapter provides no evidence for this assertion. Indeed, his own work has shown the opposite; the nutritional supplements given to young children in his Guatemala study produced an increase in reading and vocabulary (crystallized IQ) but not in fluid IQ (the Raven Progressive Matrices). This result is, however, the reverse of what has typically been found in economically developed nations, namely, that fluid intelligence has increased more than crystallized intelligence. Nevertheless, all forms of intelligence have shown increases; the evidence can be found in Flynn (1984, 1987) and in Lynn and Hampson (1986).

THE NUTRITION THEORY

The nutrition theory of the secular rise of intelligence rests on five strands of evidence documented previously (Lynn, 1990). First, nutrition has undoubtedly improved in the economically developed nations from the 1930s up to 1978–1989, the latest years for which there are data showing IQ improvements, as shown in Lynn and Pagliari (1994). The improvement in nutrition is shown by secular increases in height and by surveys of nutritional intakes.

Second, the crucial point that Sigman and Whaley and Martorell fail to note is that head size and brain size have also increased during the last half century (Miller & Corsellis, 1977; Ounsted, Moar, & Scott, 1985). The significance of this fact is that brain size is a determinant of intelligence. What has occurred is that the improvements in nutrition have increased the growth of the brain, and probably also its neurological development, and this has increased intelligence.

Third, despite all the complexities of isolating effects of nutrition on intelligence from associated nonnutritional effects such as poverty, incompetent parenting, and so on, to which Sigman and Whaley draw attention, the impact of nutrition has been isolated in several studies of twins with different birth weights who have been found to have different IQs in adolescence (e.g., Churchill, 1965; Henrichsen, Skinhoj, & Anderson, 1986). The heavier of the two twins had the larger brain and the higher IQ. Twins with different birth weights arise because one fetal twin obtains better nutrition than the other, producing enhanced growth, increased weight, and a larger brain. These twins constitute perfect controls for all the usual confounding factors and provide the clearest evidence that the quality of prenatal nutrition does have permanent effects on intelligence through brain size and probably also through neurological development.

Fourth, the type of abilities that have shown the greater secular increases in intelligence are precisely those that are most affected by nutrition, namely the nonverbal reasoning abilities, sometimes called *fluid intelligence*, rather than the verbal-educational abilities, sometimes called *crystallized intelligence*. This difference shows up in the studies of twins with different birth weights cited above and also in nutritional supplement studies (e.g., Schoenthaler, Amos, Eysenck, Peritz, & Yudkin, 1991), both of which show that fluid intelligence is more sensitive to variations in nutrition than crystallized intelligence.

Fifth, there is strong circumstantial evidence for the nutrition theory of the secular increase in intelligence that makes it possible to rule out competing theories. None of the contributors to this volume have focused their attention on the fact that the full magnitude of the secular IQ increase, amounting to approximately 3 IQ points per decade, as assessed by the Wechsler tests, has occurred among 4- to 6-year-old children as well as among schoolchildren and adults (Flynn, 1984), and it has also been present among 1- to 2-year-olds assessed by infant intelligence and motor development tests (Hanson, Smith, & Hume, 1985). This can only mean that the factors responsible for the secular increases in intelligence are probably having their impact before the age of 2 years and certainly before the age of 4–6. This observation rules out the plethora of putative effects from cognitively stimulating innovations occurring later in childhood favored in this volume by Greenfield (chapter 4), Williams (chapter 5), and Ceci and colleagues (chapter 11), such as video games, radio and TV, children's puzzle books, better schooling, and the like. None of these things can have any effect because they do not operate on 2-year-olds and have minimal impact on 4- to 6-year-olds. In a field such as this one, in which there are too many hypotheses, it is important to weed a lot of them out, and this is effectively done by the evidence that 2- to 6-year-olds display the full magnitude of the IQ increase. In fact, this evidence weeds out all the competing hypotheses, so that only the nutrition theory remains.

GENUINE AND SPURIOUS IQ GAINS

None of the contributors to this volume have made what I regard as an important distinction between genuine and spurious increases in IQ. The IQ gains on the Wechsler tests, amounting to approximately 3 IQ points per decade, should be regarded as genuine increases of intelligence, but the larger gains obtained by 18-year-old military conscripts on the Raven Progressive Matrices, documented by Flynn (1987) in several European countries, are largely spurious. They are best interpreted as schooling effects. The Raven requires the application of the mathematical principles of addition, subtraction, progression, and the distribution of values, as shown in detail by Carpenter, Just, and Shell (1990). In the 3 decades (1950s–1980s) over which these increases in

Raven scores have occurred, increasing proportions of 15- to 18-year-olds have remained in schools, where they have learned math skills that they have applied to the solution of matrices problems.

The fact that 1980s cohorts have learned more math than 1950s cohorts and that this has enabled them to secure higher scores on the Raven should not be regarded as an increase of intelligence. What the later cohorts have gained is only an improvement of the specific problem-solving skills required for the Raven. It is a training effect, and it has been known for decades that scores on intelligence tests can be improved by training (e.g., Lloyd & Pidgeon, 1961).

Although these problem-solving skills have shown a secular improvement, it has been secured at the expense of other problem-solving skills that have declined. For instance, 15- to 18-year-olds who worked on farms in the 1950s would have known more about agriculture and would have performed better on tests of agricultural knowledge (e.g., How many kilograms of nitrogen a year should be used to fertilize 1 acre?) than later cohorts, who know less about agriculture because they are in schools.

Should one therefore simply shrug one's shoulders and conclude that all that is known is that some cognitive skills have improved over time, others have probably deteriorated, and maybe others have remained static? This is more or less the position adopted by Flynn (1984; see also chapter 2, this volume) when he argues that it is only intelligence test scores that have risen over time, whereas "real intelligence" has probably remained static. I disagree. Although the secular increase of standard matrices scores among European 18-year-olds does not reveal much of interest, the increase of scores on the Coloured Progressive Matrices of 2.7 IQ points per decade among British 5- to 11-year-olds over the 33-year-period 1949–1982 (Lynn & Hampson, 1986) is more significant because all the children involved were learning math in school in 1949 and in 1982, so that the school-learning factor is controlled. What this finding shows is a genuine rise of intelligence, not a rise of math skills transferred to solving similar intelligence test items. Notice that the magnitude of this secular rise of intelligence of 2.7 IQ points per decade is similar to the 3-IQ-point rise per decade found in a number of countries on the Wechsler. This should be regarded as a genuine rise of intelligence.

REFERENCES

Carpenter, P. A., Just M. A., & Shell, P. (1990). What one intelligence test measures: A theoretical account of the processing in the Raven Progressive Matrices test. *Psychological Review, 97,* 404–431.

Churchill, J. M. (1965). The relationship between intelligence and birth-weight in twins. *Neurology, 15,* 341–347.

Flynn, J. R. (1984). The mean IQ of Americans: Massive gains 1932 to 1978. *Psychological Bulletin, 95,* 29–51.

Flynn, J. R. (1987). Massive IQ gains in 14 nations: What IQ tests really measure. *Psychological Bulletin, 101,* 171–191.

Hanson, R., Smith, J. A., & Hume, W. (1985). Achievements of infants on items of the Griffiths scales: 1980 compared to 1950. *Child: Care, Health and Development, 11,* 91–104.

Henrichsen, L., Skinhoj, K., & Anderson, C. E. (1986). Delayed growth and reduced intelligence in 9–17 year old intrauterine growth retarded children compared with their monozygous co-twins. *Acta Paediatrica Scandinavia, 75,* 31–35.

Lloyd, F., & Pidgeon, D. A. (1961). An investigation into the effects of coaching on non-verbal test material with European, Indian and African children. *British Journal of Educational Psychology, 31,* 145–151.

Lynn, R. (1990). The role of nutrition in the secular increases of intelligence. *Personality and Individual Differences, 11,* 273–286.

Lynn, R., & Hampson, S. L. (1986). The rise of national intelligence: Evidence from Britain, Japan and the United States. *Personality and Individual Differences, 7,* 23–32.

Lynn, R., & Pagliari, C. (1994). The intelligence of American children is still rising. *Journal of Biosocial Science, 26,* 65–67.

Miller, A. K., & Corsellis, J. A. (1977). Evidence for a secular increase in brain weight during the past century. *Annals of Human Biology, 4,* 253–257.

Ounsted, M., Moar, V. A., & Scott, A. (1985). Head circumference charts updated. *Archives of Diseases of Childhood, 60,* 936–939.

Schoenthaler, S. J., Amos, S. P., Eysenck, H. J., Peritz, E., & Yudkin, J. (1991).

Controlled trials of vitamin–mineral supplementation: Effects on intelligence and performance. *Personality and Individual Differences, 12,* 351–362.

Teasdale, T. W., & Owen, D. R. (1989). Continuing secular increases in intelligence and a stable prevalence of high intelligence levels. *Intelligence, 13,* 255–262.

PART TWO

A Narrowing Gap in School Achievement

9

Trends in Black–White Test-Score Differentials: I. Uses and Misuses of NAEP/SAT Data

Robert M. Hauser

Until the mid-1980s, there was not much good news for those who believed that ability and achievement test differences between Blacks and Whites in the United States were the malleable products of environment or culture. On IQ tests and other similar tests, there were typical and persistent mean differences between Blacks and Whites on the order of 1 *SD*. If test scores were normally distributed, one would expect about 84% of Whites to exceed the mean score among Blacks and about 16% of Blacks to exceed the mean score among Whites. That is no longer the case on some cognitive tests; there has been a substantial convergence in the performance of Blacks and Whites. In this chapter, I review evidence about the timing of the changes in test scores and the types of tests in which they have occurred. Herrnstein and Murray (1994) reviewed some of this evidence in their recent best-seller *The Bell Curve*, but I demonstrate here that their account was far from adequate. Although the data are incomplete, on the basis of the standards of evidence routinely accepted in psychological research on test performance, I conclude that the available evidence of change is highly significant and incontrovertible.

TRENDS BEFORE 1970

Loehlin, Lindzey, and Spuhler (1975) offered two sources of data on trends in Black–White differences in IQ. First, they assembled data from Shuey's (1966) extensive review. Even though they acknowledged problems with the quality of Shuey's data, they argued that "gross changes over time should be detectable, even in the absence of single studies making well controlled comparisons over substantial time spans" (pp. 140–141). In a comparison of 259 studies of Black and White preschool children, elementary schoolchildren, and high school students, they found little evidence that test-score differentials had declined between the pre-1945 period and the time between 1945 and 1965. To be sure, the differential was unchanged (14 points) in individual tests administered to elementary schoolchildren, and it declined from 16 to 13 points in nonverbal group tests of elementary schoolchildren. At the same time, the gap increased from 9 to 16 points among preschool children, from 13 to 16 points in verbal tests among elementary schoolchildren, and from 11 to 19 points among high school students. Loehlin et al. (1975) concluded that the data "fail to suggest much change over time in black-white differences in the groups for which there is the most data—and probably the most representative data—the elementary school children" (p. 141).[1]

Second, Loehlin et al. (1975, pp. 142–144) compared the test-score distributions of Black and White military recruits in World War 1, World War 2, and the Vietnam War. These comparisons were hampered by changes in the tests (Army Alpha and Beta, Army General Classification Test, and Armed Forces Qualification Test, respectively), by changes in the intervals within which test scores were reported, and by changes in population coverage and definition. On the basis of the assumption that the score distributions were normal, the authors estimated mean IQ score differences. These differences were 17 points in World War I, 23 points in World War II, and 23 points in the Vietnam War. They again concluded that the test-score differences

[1]Those conclusions would not change had I also considered Osborne and McGurk's (1982) continuation of Shuey's review, which covered the period from 1966 to 1979.

were large but that no inference should be made with regard to trend: "Because there are a number of ways in which these samples are unrepresentative of the total U.S. male population, we are not willing to draw strongly the conclusion that the black–white gap in average measured ability has actually widened since the time of World War I" (p. 144). Of course, these data provided no information about Black or White women.

Relative to typical standards in the population sciences, all of these findings are of questionable validity. The population coverage is haphazard; scientific sampling is the exception; scores on various tests are rendered nominally comparable, merely by the assumption of normality in the trait distribution and by the expression of scores as deviations from the mean. In short, up to about 1980, data on trends in Black–White differences in IQ were probably of even worse quality than those available for global assessments of trends in test performance (Flynn, 1984, 1987). All the same, the Coleman–Campbell report of 1966, *Equality of Educational Opportunity,* must have resolved any doubts about Black–White differentials in the early 1960s with its finding that test-score differences among elementary and secondary students were roughly 1 *SD* in reading and verbal tests within every region of the United States (Coleman et al., 1966).

TRENDS SINCE 1970

The National Assessment of Educational Progress

In the 1980s, evidence of substantial aggregate change in Black–White test-score differences began to accumulate. The primary source of these new data has been the National Assessment of Educational Progress (NAEP), a large periodic national testing program with a complex sampling design (Zwick, 1992) that began in the early 1970s. Each participating student is asked to complete only part of each test, and scores for population groups are estimated from the incomplete data. Until recently, only a few social background characteristics were associated with each student observation, but the 1988 redesign is much richer in variables than its predecessor. Originally, the NAEP samples were not representative at the state level, but this has begun to change as NAEP has become the vehicle for measuring progress toward national educational standards.

The NAEP testing program includes both grade-level tests in Grades 4, 8, and 12, and age-specific tests at ages 9, 13, and 17. Although the national NAEP samples are relatively large (though decreasing in size) and well designed, there are some problematic issues in population coverage. Some schools refuse to participate. Students in special programs are not covered, and student absence and dropout create coverage problems by age 17. Moreover, there is some nonresponse among test takers to queries on racial or ethnic identification. At the same time, NAEP is plainly superior in design and coverage to previous mechanisms for monitoring children's academic performance at the national level and for specific age and population groups.

The age-specific NAEP tests, which cover youths in regular classrooms at every grade level, are designed to permit temporal comparisons of performance. NAEP tests are criterion-referenced, and they are administered in regular cycles of varying length, depending on the subject. Like most other investigators, I focus on the three tests that have been administered most frequently—reading, science, and mathematics—which have gradually been shifted from 4-year cycles to administration in every even-numbered year. The NAEP uses a repeated cross-section design. It is not a longitudinal study of individuals, so one cannot follow the development of individual performance across time. However, one particular advantage of the NAEP design is that the 2- or 4-year testing intervals are commensurable with the 4-year differences between age groups, so it is sometimes possible to follow the development of birth cohorts from age 9 to age 17 as well as to measure aggregate trends and differentials.

The Scholastic Assessment Test

A secondary source of data on trends in Black and White test-score performance is the Scholastic Assessment Test (SAT)—formerly, the Scholastic Aptitude Test—of the College Entrance Examination Board

(CEEB). Since the 1940s, the SAT has been administered regularly to college-bound seniors (and some juniors) in U.S. high schools. The SAT has two components, verbal (SAT-V) and quantitative (SAT-M), which are often used by colleges and universities in screening applicants for undergraduate admission. Perhaps because of the long decline in SAT-V scores that began in the early 1960s, there have been frequent well-publicized efforts to tie changes in SAT scores to school or youth policies. Trends in SAT performance hit the front page of *The New York Times* each year, and they are often used as key indicators of trends in how schools are performing, as well as in comparisons among groups of students. However, it is doubtful that changes in SAT scores demonstrate any true change in scholastic performance. The Preliminary Scholastic Aptitude Test (PSAT) has been administered since 1959 to a national sample of high school juniors (Solomon, 1983). The PSAT is a shorter version of the SAT, and the problems of self-selection in these samples are limited to those implied in reaching the junior year of high school. In the aggregate, there has been no trend in PSAT performance in the past 35 years (Berliner & Biddle, 1995, pp. 23–24).

The uses of SAT scores as social indicators are grossly disproportionate to their validity. Test takers are self-selected from among high school students who plan to attend colleges that require SAT scores. Selection is known to vary across time with respect to academic performance (rank in class), sex, minority status, socioeconomic background, and geographic origin. Presumably, these variations are in part a consequence of variations in the entrance requirements of colleges and universities and of changes in the demand for college education among American youths. Typically, SAT coverage is lower in the central states than on either coast because of competition from less expensive tests of the American College Testing Program (ACT). As Wainer (1987) stated,

> If we wish to draw inferences about all high school seniors, the possibly peculiar events that would impel someone to take the test or not makes these inferences difficult. These difficulties manifest themselves when we try to assess the significance of

> changes observed over time. Is the change due to more poorly trained individuals, to a broader cross-section taking the test or merely to a different cross-section of individuals deciding to take the test? (p. 2)

Problems in interpreting trends in the SAT have given rise to a minor industry of test-score adjustment and analysis. One major goal of the industry is to counter gross misinterpretations of trends and differentials in SAT scores, like meaningless state-to-state comparisons. For example, the highest scoring states are typically those, like Wisconsin, in which most students take the ACT, which is required by the University of Wisconsin System, and a small minority of elite students take the SAT (Wainer, 1985). A second major goal is to find out what can be learned from the SAT about trends and differentials in academic performance. This attempt has yielded a lot of clever and careful statistical work, beginning with efforts to explain the long-term decline in SAT-V scores, but it has provided few definitive answers about trends in academic performance (Alwin, 1991; College Entrance Examination Board, 1977; Flynn, 1984; Menard, 1988; Morgan, 1991; Murray & Herrnstein, 1992; Zajonc, 1976, 1986).

My favorite contribution to this literature is an elegant paper by Howard Wainer (1987). He showed that the uncertainty in SAT scores introduced by the average 12–14% nonresponse on the race–ethnicity question dwarfs the observed growth in minority SAT performance that occurred from 1980 to 1985. After observing the average verbal and math scores of White, minority-group, and nonresponding test takers, Wainer observed that if the scores of nonresponding test takers are the same as those of respondents of the same race or ethnic group, then it is possible to estimate the share of White and minority-group test takers among nonrespondents. Depending on whether one uses the verbal or math scores to make the estimates, this estimation procedure yields very different but rather high estimates of the share of minority-group students among nonrespondents. From 1980 to 1985, the estimated share of minority-group members among nonrespondents was never less than half and ranged as high as 70%, whereas the share of minority-group members was always estimated to be higher for mathematical than for verbal scores. The discrepant estimates invalidate the assumption that respondents and nonrespondents of the same ethnicity perform equally well, and the resulting uncertainty in test scores is larger than the observed changes in average test performance among minority test takers.

Black–White Differences in NAEP Scores

I think the uncertainties of the SAT data are far greater than those of NAEP, and for that reason I focus mainly on trends and differentials in performance on the NAEP. However, if one takes the scores at face value, there has also been a partial convergence in Black and White performance on the SAT. For the moment, I ignore the official reports of performance on the NAEP and offer a brief review of their presentation in secondary sources. I also postpone discussion of Herrnstein and Murray's (1994) treatment of the NAEP data to a later section.

Jones (1984, pp. 1209–1211) was one of the first to examine the Black–White convergence in test scores. NAEP tests in 1971, 1975, 1980, and 1982 showed declining Black–White differences in the percentage of correct responses on the NAEP reading and mathematics tests for children who were born after 1965. He also analyzed differentials in mathematics scores and suggested that the "difference between black and white students in algebra and geometry enrollment might be responsible for a large part of the white-black average difference in mathematics achievement scores" (1984, p. 1211).

As the evidence from the NAEP accumulated, others noted the trends. A 1986 report of the Congressional Budget Office (CBO), *Trends in Educational Achievement*, said that, with reference to the previous decline in academic achievement, "the average scores of black students declined less than those of non-minority students during the later years of the general decline; stopped declining, or began increasing again, earlier; and rose at a faster rate after the general upturn in achievement began" (Koretz, 1986, pp. 75–76). In reaching this conclusion, the CBO report relied mainly on trends in average proficiency scores during the first dozen years of the NAEP, but it also found corroborating evidence in the SAT, in nationally representative samples of high school seniors of 1971 and 1979, and in several state or local studies. Similarly, Humphreys (1988, pp. 240–241) reported substantial gains of Blacks relative to Whites at ages 9, 13, and 17 in the four NAEP reading assessments from 1971 to 1984.

The National Research Council's (NRC) 1989 report, *A Common Destiny: Blacks and American Society*, also noted trends in Black–White gaps in the NAEP assessments of reading, mathematics, and science through 1986 at ages 9, 13, and 17 (Jaynes & Williams, 1989, pp. 348–354).[2] Beyond finding signs of aggregate convergence, the NRC panel also disaggregated the Black–White differences by levels of proficiency and by region. For example, the panel found that "the broad pattern is one of improvement over time at each level of reading proficiency" (p. 349). Furthermore, they found that the same broad pattern of improvement occurred in each of four geographic regions. The NRC report also noted the large remaining cognitive gaps between Blacks and Whites, adding the evidence of a national literacy survey to that from the NAEP and the SAT.

In a fascinating preview of *The Bell Curve*, the late Richard Herrnstein (1990b) wrote a review of *A Common Destiny*, which appeared in *The Public Interest*. His main theme was that "*A Common Destiny* suffers . . . from one crucial failing: in assessing the gaps separating white and black Americans, it obstinately refuses to consider the evidence concerning racial differences at the individual level" (p. 4). Herrnstein claimed that the themes of the book were "rooted" in the discrimination model, that any Black–White differences were viewed as prima facie evidence of discrimination, whereas the book "ignores the alternative model, the 'distributional' model, which explains the overlapping of the populations and their differing averages by referring to characteristics of the populations themselves" (p. 6). Herrnstein mainly faulted *A Common Destiny* for failing to root its explanations in IQ differences between Blacks and Whites:

> *A Common Destiny* says almost nothing about differences be-

[2]I was a member of the National Academy of Sciences–National Research Council Committee on the Status of Black Americans and contributed to chapter 7, "The Schooling of Black Americans."

> tween blacks and whites on standardized tests of intelligence or cognitive aptitude; what little it says is mostly wrong. . . . Notwithstanding some vague hints in the book, there is no clear evidence that the gap between the races has been closing recently or that it shrank when the economic gap between the races was shrinking. (p. 7)

In the remainder of Herrnstein's review, he argued for the centrality of intelligence in accounting for racial differentials in the areas of economic status, crime, health, and housing. However, neither in his review nor in a subsequent exchange with the authors of the NRC report (Hauser, Jaynes, & Williams, 1990) did Herrnstein acknowledge the findings from the NAEP of decreasing differences in achievement test performance between Blacks and Whites. Rather, he reiterated, "the differences *are* intractable, for we do not know how to eliminate them" (Herrnstein, 1990a, p. 125).

After the hubbub about *A Common Destiny* subsided, other scholars continued to draw on the NAEP test series. Smith and O'Day (1991, pp. 72–77) reported declining test-score differences at ages 9, 13, and 17 in cohorts tested in reading, mathematics, and science from 1971 to 1988. With respect to reading scores, they observed that

> these are extraordinary data. By conservative estimate, they indicate a reduction in the gap between black and white students over the past twenty years of roughly 50 percent when the students are seventeen years old. Moreover, these reductions took place during the same time period as a striking decrease in dropout rates for black students. (p. 75)

Smith and O'Day further estimated that the reduction in the Black–White gap in mathematics was on the order of 25–40% and that in science it was roughly 15–25%. Finally, Grissmer, Kirby, Berends, and Williamson (1994, pp. 11–17) reported narrowing gaps between Blacks and Whites in reading and mathematics achievement using the NAEP data for the middle to late 1970s and for 1990 at ages 13 and 17. Miller (1995, pp. 45–59) offered a detailed review of the performance of

Blacks and Whites at ages 9 and 17 for each administration of the reading and mathematics assessments since 1971.

Although these reviews covered various years of the NAEP and differed, also, in their coverage of specific ages, tests, and functions of test performance, the reviews were unanimous in reporting an overall trend toward reduced Black–White performance differentials. The works cited were also unanimous in drawing attention to remaining large gaps in performance. There was relatively little attention given to the reasons for the gaps or for their partial closure. As noted earlier, Jones (1984) pointed to exposure to math courses as a remaining source of Black–White math score differences, and subsequently the NRC report also emphasized differential course taking as well as reduced segregation and increased compensatory education (Jaynes & Williams, 1989, pp. 350–352). Smith and O'Day (1991, pp. 79–84) offered no specific analyses of changes in test-score gaps but suggested that the gaps might be explained by improved social background and reduced poverty, increased access to preschool, reduced racial isolation—especially in the South—and changes in instruction and curriculum—especially increased emphasis on basic skills and minimum competencies. Grissmer et al. (1994, pp. xxv–xxxi) carried out detailed analyses of the effects of changes in family background on test scores. They found that changes in family background composition, especially improved maternal schooling and fewer siblings, accounted for about one third of the improvement in test scores among minority-group students from the 1970s to 1990.[3]

Achievement or Ability?

There is something schizophrenic in American opinion about cognitive ability and academic achievement. Americans think we value academic achievement and that it represents, to some degree, the kind of merit we want to see rewarded. We worry endlessly about trends and differentials in academic achievement. We spend a great deal of money to

[3]See Armor (1992) for similar findings.

create and improve it in the public schools, and we blame the schools because we think that they have not produced enough of it. We think that if academic achievement were higher, we would do better economically and socially, as individuals and local communities and in the world economy. Yet we grow rigid with apprehension when someone applies terms like *ability*, *intelligence*, or—worse yet—*IQ*, rather than academic achievement, to what are usually rather similar and highly correlated measures. We fret about the fairness of standardized tests, although lack of statistical bias is long established (Wigdor & Garner, 1982, p. 3), and we often disapprove—both personally and legally—of the mechanical use of achievement or ability test scores to make decisions about entry to jobs or to schools. Obversely, we have turned test preparation into a minor industry. Among college admission tests, we prefer the ACT to the SAT because it focuses relatively more on achievement than aptitude, and we applaud the revision of the latter for shifting in the same direction, yet the ACT and old SAT were highly correlated, as are the new and old versions of the SAT.

It is a serious question whether NAEP assessments—or the SAT for that matter—are truly tests of achievement, scholastic aptitude, ability, intelligence, or IQ. As a nonmember of the psychometric profession, I am inclined to join those who elide or ignore the distinction between achievement and ability (Jencks & Crouse, 1982). I do not believe that ability can be assessed without reference to past learning and opportunity to learn. Moreover, I think it is difficult to maintain sharp distinctions in test content between aptitude and achievement. Although I do not ignore the specific content of tests, I also think that any test performance partly indicates overall levels of realized ability.

For example, I think there is wide agreement that scores on the Armed Forces Qualification Test (AFQT) can justifiably be interpreted much like performance on an IQ test, and there is ample precedent for this, both in the historical development of the test and in its use by Loehlin et al. (1975) and others, including Herrnstein and Murray (1994). At the same time, there is a great deal of evidence that schooling raises scores on IQ tests (Ceci, 1991), and the best recent evidence is based on the AFQT, suggesting that each year of schooling raises IQ by about 2 to 3.5 points (Fischer et al., 1996; Korenman & Winship, 1995; Neal & Johnson, 1996).

Before its recent renaming as the Scholastic Assessment Test, the SAT was called the Scholastic Aptitude Test, although it was based on a general ability test—the original Army Alpha Test of World War 1 (Lemann, 1995). Because its purpose is to select among high school seniors, there are no age norms for the test. However, the eminent psychologist Julian Stanley and his associates have for years applied a set of age norms to SAT scores to select gifted younger students for special summer enrichment programs. For example, the gifted sixth, seventh, and eighth graders who took the SAT in the Midwest Talent Search in 1987 had combined scores of 793 (male students) and 656 (female students), compared with the combined scores of 1986 college-bound seniors of 938 (male students) and 877 (female students). Among the gifted younger students, average SAT scores increased regularly with age, from a combined score of 696 for those born in 1975 to 826 for those born in 1972 (Northwestern University, 1987).

If SAT scores rise regularly with age and exposure to schooling, do they not reflect achievement as well as aptitude or ability? Throughout *The Bell Curve*, Herrnstein and Murray (1994) played with the tensions and contradictions between Americans' images of ability and achievement, and they repeatedly shifted the line between the two to suit their rhetorical purposes. The SAT was portrayed at some times as a measure of "achievement," whose downward trend shows neglect of education among the cognitively gifted, and at other times as a measure of "intelligence," whose use in college entry demonstrates both the establishment of a national cognitive elite and the defects of affirmative action.

Herrnstein and Murray on Black–White Test-Score Trends

Although the NAEP tests appear to be heavily loaded on the achievement end of the spectrum of test content, there is also some precedent for treating them as tests of ability. For this reason, I consider in detail

Herrnstein and Murray's treatment of the NAEP findings in *The Bell Curve*. In my judgment, the changing test-score differentials were given minimal attention, and much of what was said about them was wrong. In all, about six pages of the main text of *The Bell Curve* (pp. 289–295) were devoted to the question "Is the difference in Black and White test scores diminishing?" (p. 289) within the 46-page chapter "Ethnic Differences in Cognitive Ability." Most of the data on trends in test-score differences were put into one of the book's many appendixes (pp. 637–642). One might compare this lack of emphasis on aggregate trends with the 28 pages devoted to an essentially negative review of compensatory education programs.

On page 291 of the main text, Herrnstein and Murray present a table entitled "Reductions in the Black–White Difference on the National Assessment of Educational Progress," which is based on summary data from the early 1970s through 1990 (Mullis, Dossey, Foertsch, Jones, & Gentile, 1991) and which I have reproduced here as Table 1. Herrnstein and Murray reported that across math, science, and reading examinations, and at ages 9, 13, and 17, the Black–White difference declined by an average 0.28 standard deviation between 1969–1973 and 1990. They described these changes as presenting "an encouraging picture" (p. 291). After adding a summary of changes in the SAT "from 1.16 to .88 standard deviation in the verbal portion of the test and from 1.27 to .92 standard deviation in the mathematics portion of the test" (p. 292), Herrnstein and Murray concluded that there has been a "narrowing of approximately .15 to .25 standard deviation units, or the equivalent of two to three IQ points overall" (p. 292). Apparently, Herrnstein and Murray tempered their arithmetic with cautionary data from their fifth appendix when they decided that changes of 0.28, 0.28, and 0.35 standard deviations suggest a range of 0.15–0.25 *SD* units. Then, in an endnote, they discounted this range by a factor of 0.6 or 0.8—to account for the imperfect relationship between SAT or NAEP tests and IQ—to come up with the estimated change of 2–3 IQ points. They acknowledged that if one relied on the SAT alone, the data would suggest a narrowing of 4 IQ points, "but only for the population that actually takes the test" (Note 57, p. 721).

Table 1

Reductions in Black–White Differences on the National Assessment of Educational Progress

Participant and subject area	Black–White difference in standard deviations		
	1969–1973	1990	Change
9 year olds			
Science	1.14	0.84	−0.30
Math	0.70	0.54	−0.16
Reading	0.88	0.70	−0.18
Average	0.91	0.69	−0.21
13 year olds			
Science	0.96	0.76	−0.20
Math	0.92	0.54	−0.38
Reading	0.78	0.40	−0.38
Average	0.89	0.57	−0.32
17 year olds			
Science	1.08	0.96	−0.12
Math	0.80	0.42	−0.38
Reading	1.04	0.60	−0.44
Average	0.97	0.66	−0.31
Overall average	0.92	0.64	−0.28

Note. From Mullis et al. (1991), as reported by Herrnstein and Murray (1994), who stated that "the computations assume a standard deviation of 50." From *The Bell Curve: Intelligence and Class Structure in America*, by Richard J. Herrnstein and Charles Murray, p. 291. Copyright 1994 by Richard J. Herrnstein and Charles Murray. Reprinted with the permission of The Free Press, a division of Simon & Schuster.

Even while acknowledging the trends toward convergence in test scores, Herrnstein and Murray were quick to point out that some of the trend was due to declining scores among Whites, rather than increasing scores among Blacks, and to add that it would be foolhardy to extrapolate the observed trends into the future. Whatever the specific estimate of test-score convergence between Blacks and Whites, one would be hard-pressed to find any acknowledgment of it once Herrnstein and Murray started drawing conclusions and making recommendations.

The more I thought about their findings, the more curious seemed Herrnstein and Murray's treatment of the NAEP data, because those data contain the only set of test scores discussed by these authors that consistently covers an unselected sample of the general population. If one applied their range of discount factors to their estimate of the test-score convergence in the NAEP data alone, the estimated closure would lie between 2.5 and 3.4 points, which is not bad for aggregate change in an immutable quantity over a 20-year period. However, there is more to the story: The footnote in Herrnstein and Murray's table declares that the authors "assume a standard deviation of 50." I recalled some variation in the standard deviations of the NAEP test scores across tests, from year to year and between Blacks and Whites, so I went back to the source.

My research proved to be a cautionary lesson regarding what the book jacket of *The Bell Curve* cited as the "relentless and unassailable thoroughness" of Herrnstein and Murray's analysis. To begin with, several of the numbers in the table are simply wrong. There are no fewer than five copying or multiplication errors in age- and test-specific entries in the body of the table, and these lead to other errors in average differentials and in measures of change. In the end, the effect of these errors is small; the overall average change is .29, rather than .28.

But this is the least of their problems with the NAEP data. In their "relentless and unassailable thoroughness," Herrnstein and Murray evidently confined their reading to a one-page summary of change in the test-score differences (p. 11), plus a footnote on page 1 from Mullis et al. (1991), stating that "each scale was set to span the range of student performance across all three ages in that subject-area assessment and to have a mean of 250.5 [*sic*] and a standard deviation of 50." However, a series of tables in the appendix provides details of the test-score distributions for each population year by year, including their standard deviations, which are typically much less than the value of 50 adopted by Herrnstein and Murray. The difference is mainly due to the

Table 2

Revised Estimates of Reductions in the Black–White Difference on the National Assessment of Educational Progress

Participant and subject area	Black–White difference in standard deviations		
	1969–1973	1990	Change
9 year olds			
Science	1.42	1.04	−0.38
Math	1.06	0.82	−0.24
Reading	0.98	0.79	−0.19
Average	1.15	0.88	−0.27
13 year olds			
Science	1.28	1.01	−0.27
Math	1.48	0.87	−0.61
Reading	1.07	0.58	−0.49
Average	1.28	0.82	−0.46
17 year olds			
Science	1.17	1.04	−0.13
Math	1.29	0.68	−0.61
Reading	1.28	0.71	−0.57
Average	1.25	0.81	−0.44
Overall average	1.23	0.84	−0.39

Note. Author's computations using data from Mullis et al. (1991) and standard deviations for each age group in 1990.

incorporation of variation by age in the larger overall value, whereas the Black–White comparisons should have been conditioned on age, just as Herrnstein and Murray attempted to condition on age in their regression analyses of the effects of the AFQT. The effect of choosing too large a standard deviation was to understate both the initial Black–White differences and the changes in test scores over time in standard deviation units.

Table 2 shows the change in test scores using the estimated standard deviations of the total population of each age in 1990 as the unit of measure. Mullis et al. (1991) did not provide standard deviations for test scores in science and mathematics in the 1970 time period, and for this reason I based the comparisons on the population standard deviations for 1990. In science and mathematics, though not in reading, the variability of the tests declined across time. Both the initial differences between Blacks and Whites in science and mathematics and the changes in those test-score differences would be somewhat smaller if the changes had been normed on the standard deviations for 1970.

Using the revised standard deviations raises the overall average convergence from .29 to .39 *SDs*. On the basis of Herrnstein and Murray's assumptions, this calculation raises the implied convergence in IQ between Blacks and Whites to a range between 3.5 and 4.7 points. I wonder whether Herrnstein and Murray would have waxed so eloquent about immutability and ineducability if they had acknowledged aggregate changes in test-score differentials of this magnitude in the general population over the past 2 decades.

In one important respect, Herrnstein and Murray were surely right: It is most dangerous to project trend lines unthinkingly. Yet another set of NAEP assessments—for 1992—became available after *The Bell Curve* went to press, and these data appear to confirm that the trend toward convergence in Black and White test scores was reversed after 1986–1988 (Mullis et al., 1994). For example, Figure 1 shows trends in the average (mean) NAEP scores of Blacks and Whites at age 13 in reading, science, and mathematics.[4] The years of greatest convergence are not entirely clear because there are no reading scores for 1986 and no science or math scores for 1988. It does appear that sometime in the middle to late 1980s, the convergent trend ended, and Black–White gaps returned to levels of the early 1980s.

Was There No Change at All?

Immediately after the publication of *The Bell Curve* in October 1994, most commentary on Black–White test-score differences focused either

[4] Similar trends appear at age 17 and, to a lesser degree, at age 9.

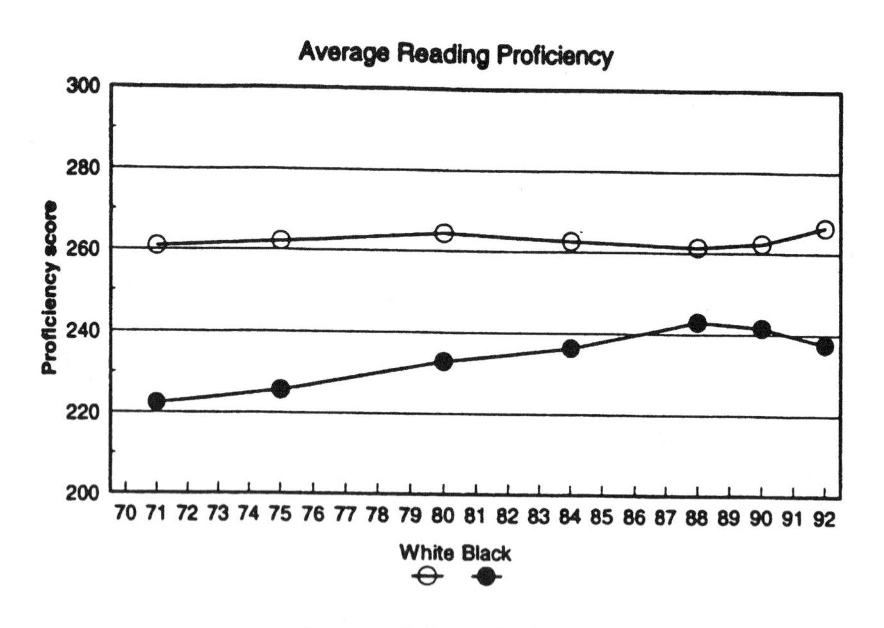

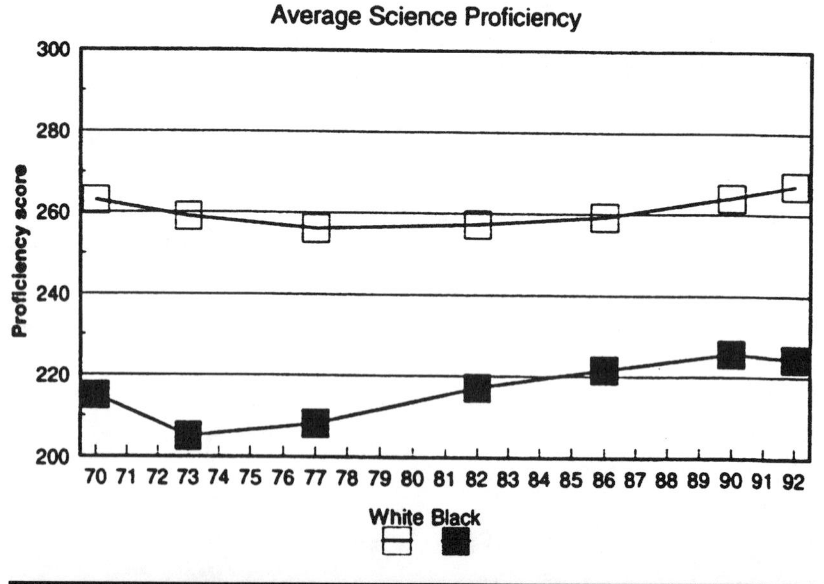

Figure 1

Trends in National Assessment of Educational Progress scores at age 13 among Blacks and Whites.

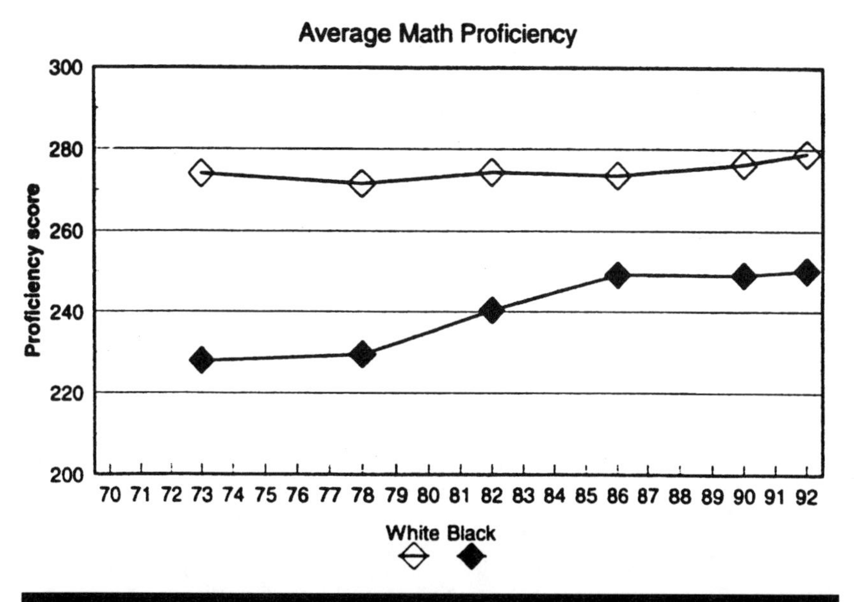

Figure 1 (*Continued*)

on the specious genetic arguments of the book or on its review of compensatory education programs. There was almost no reaction to the treatment of aggregate trend data. One significant exception was a letter signed by 52 academics, which appeared on the editorial page of the *Wall Street Journal* under the heading "Mainstream Science on Intelligence" (Arvey et al., 1994). The letter writers purported to outline conclusions "regarded as mainstream among researchers on intelligence, in particular on the nature, origins, and practical consequences of individual and group differences in intelligence" (p. 18). Among the 25 conclusions, Items 19 and 20 bear on change in Black–White test-score differentials:

> 19. There is no persuasive evidence that the IQ bell curves for different racial-ethnic groups are converging. Surveys in some years show that gaps in academic achievement have narrowed a bit for some races, ages, school subjects and skill levels, but

> this picture seems too mixed to reflect a general shift in IQ levels themselves.
> 20. Racial-ethnic differences in IQ bell curves are essentially the same when youngsters leave high school as when they enter first grade. However, because bright youngsters learn faster than slow learners, these same IQ differences lead to growing disparities in amount learned as youngsters progress from grades 1 to 12. As large national surveys continue to show, black 17-year-olds perform, on the average, more like white 13-year-olds in reading, math and science, with Hispanics in between. (p. 18)

I thus looked further at the NAEP series to learn whether my reading of the trends—and that appearing in published reviews—might have been mistaken.

Figure 2 summarizes trends in White–Black differences in NAEP proficiency scores in the major subject-matter series: reading, science, and mathematics. For each subject, I have used the same scale to show trend lines in mean test-score differences by age, but I have arrayed the data by birth year, rather than by year of assessment. For example, in the upper panel, the reading assessment covers the cohorts of 1954–1975 at age 17, the cohorts of 1958–1979 at age 13, and the cohorts of 1962–1983 at age 9. With this arrangement of the data, it is possible to compare White–Black differences in performance levels across ages by reading the graph vertically at a given birth year. For example, reading performance was assessed for the cohort of 1962 both at age 9 and age 13, and those performance levels might also be compared with that of the adjacent cohort of 1963, measured at age 17.

A first observation about the data in Figure 2 is that there is no distinct pattern to the within-cohort comparisons in any subject. That is, for members of the same (or adjacent cohorts, the Black–White differences are only occasionally larger at age 17 than age 13 and at age 13 than age 9. In other cases, the White–Black differences are largest at age 9 or age 13. Thus, I find no substantial or consistent support in the NAEP data for the claim of Arvey et al. (1994) that there are "growing disparities in amount learned as youngsters pro-

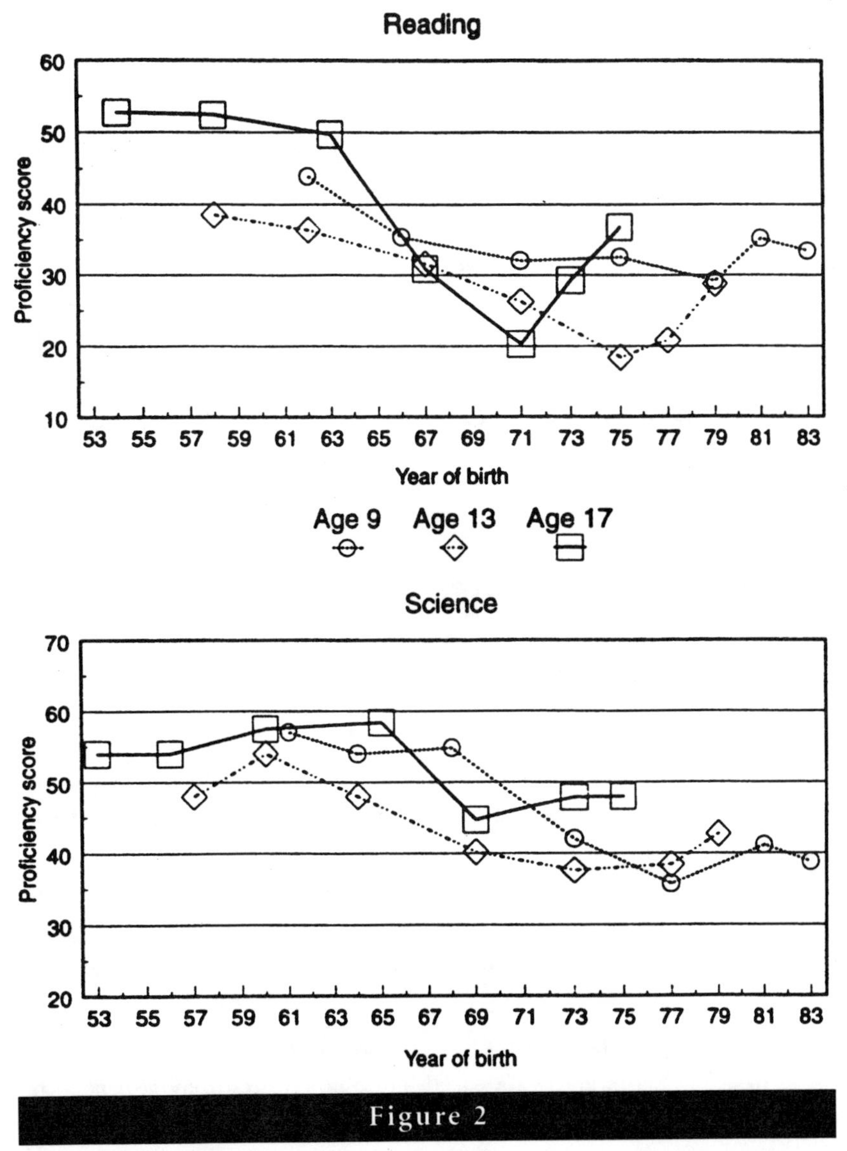

Figure 2

White–Black Differences in National Assessment of Educational Progress proficiency scores by subject, age, and year of birth.

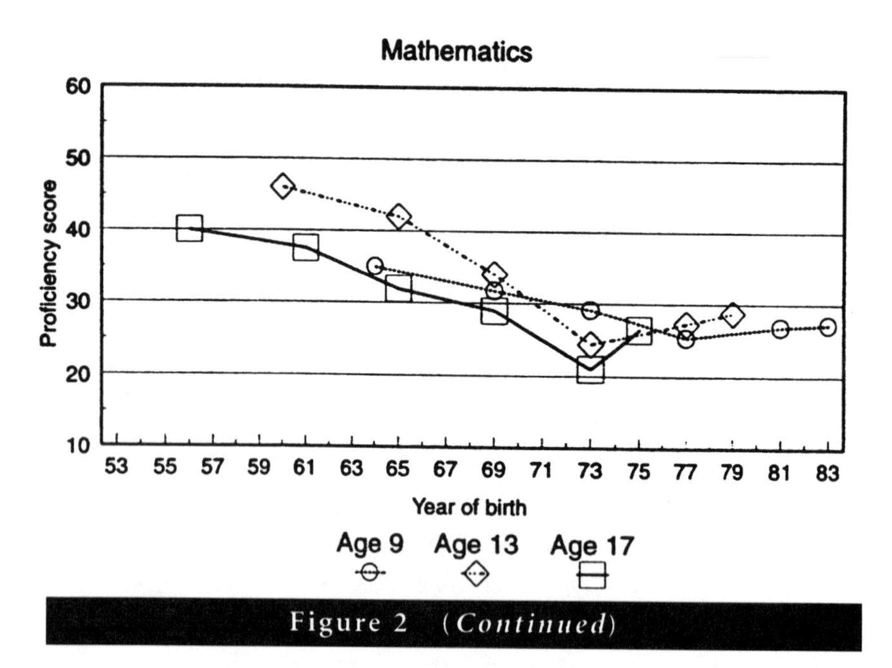

Figure 2 (*Continued*)

gress from grades 1 to 12" (p. 18). To be sure, one might observe smaller differences in the NAEP series at age 17 because low scorers tend to drop out, but available data provide no evidence about the effects of attrition.

More important, Figure 2 clearly shows the major declines in White–Black differences in reading, science, and mathematics achievement. This arrangement of the data suggests that the declines may have occurred in particular cohorts, beginning before age 9. In reading, for example, the major gains for Blacks occurred in the cohorts of 1962–1971. In science, the gains appeared, although not entirely consistently, for cohorts born after 1960. In mathematics, the Black gains appeared for cohorts born from 1956 to 1973. In science, but not in reading or mathematics, Black gains were in part a consequence of declining test scores among Whites. I hesitate to connect these test-score changes too closely to changes in IQ, although Herrnstein and Murray showed no hesitation in doing so. However, in my opinion, the trends have not been "too mixed" to reflect a general shift in test-score differentials among cohorts born from the late 1950s to the early 1970s.

Mean achievement scores provide important, but limited, information about levels of achievement in the population. Such a measure of central tendency may be insensitive to changes in the shape or dispersion of the distribution. For this reason, I have examined changes over time across the entire achievement distributions among Blacks. For example, Figure 3 shows selected percentile points of the NAEP mathematics distributions for Black children at ages 9, 13, and 17 from 1971 through 1992. The test scores are reported in the metric of "proficiency levels" for the 5th, 10th, 25th, 50th (median), 75th, 90th, and 95th percentiles at each age and year. I have used line patterns and symbols for the data points that emphasize the rough symmetry in the distributions; the same lines and marker shapes are used for the corresponding pairs of percentile points below and above the median: 5 and 95, 10 and 90, and 25 and 75. Thus, it is possible to follow changes in both the level and the shape of the distributions across time. One should bear in mind that the writers of the *Wall Street Journal* letter could not have observed the final (1992) data points in each series.

Figure 3 shows that at age 9, mathematics performance improved steadily throughout the distribution from 1978 to 1990, but especially between 1982 and 1990. At age 13, mathematics performance grew between 1978 and 1986 but leveled off thereafter.[5] Growth appears to have been more rapid in the lower half of the distribution than in the upper half. At age 17, there was steady growth from 1978 to 1990 throughout the distribution, but performance fell between 1990 and 1992 at the top and bottom of the distribution. During the 1978 to 1986 period, growth appears to have been faster at the bottom than at the top of the distribution.

I have also examined detailed displays of change in the distributions of reading and science proficiency, and the pattern of findings is similar

[5] Unfortunately, the published series do not include percentile points from the initial mathematics assessment of 1973. The convergence between the average (mean) performance levels of Black and White students in mathematics first appears between 1973 and 1978 at ages 9, 13, and 17 (Mullis et al., 1994, pp. A63–A71).

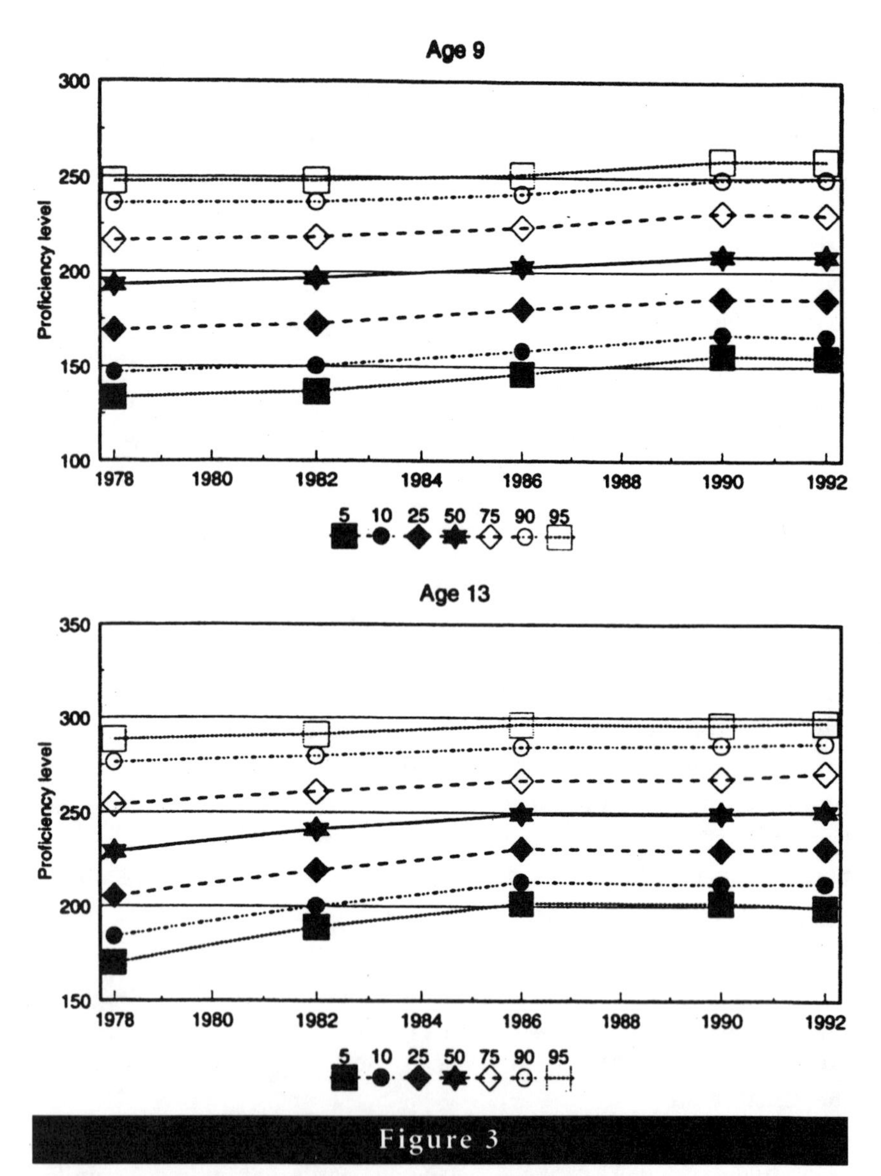

Figure 3

National Assessment of Educational Progress mathematics trends assessment: Percentiles of the distribution among Blacks by age, 1978–1992.

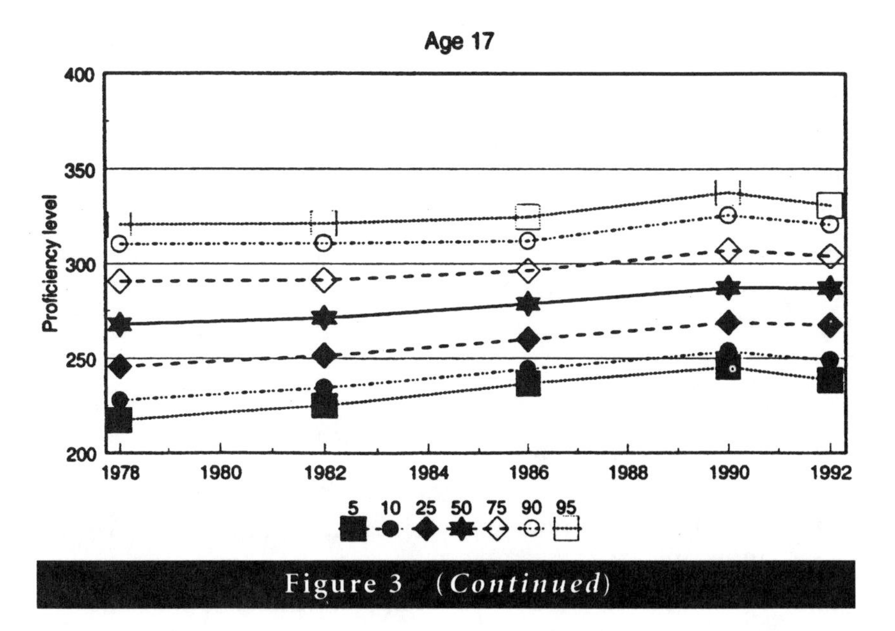

Figure 3 (*Continued*)

to that in mathematics. There is much more to the NAEP trends than the statement by Arvey et al. (1994) that "gaps in academic achievement have narrowed a bit for some races, ages, school subjects and skill levels" (p. 18). Almost all of the growth in Black scores led to convergence in the performance of Blacks and Whites. Although growth in Black performance did not occur between every assessment from the 1970s to 1990, growth did occur throughout most of this period and throughout the entire distributions of performance levels. Thus, I think that the statement in the *Wall Street Journal* substantially understates the extent and pervasiveness of change in Black achievement test scores. Indeed, I doubt that many readers would independently describe the NAEP series in the same terms as those of Arvey et al. (1994).

There is clearly a great deal of opportunity for research on the sources of Black–White test-score convergence during the period from 1970 to the middle 1980s and on the sources of the subsequent slowdown or reversal. One might think, for example, of the reduced enthusiasm for compensatory education after the first Reagan administration took office in 1980 and of the length of time required for its effects to take hold. There also remains the possibility that some part of the convergence or of its reversal may be explained as methodological artifacts of the NAEP design. On the other hand, relative to the larger body of evidence on change in test scores, it seems hard to believe that the NAEP assessments are especially vulnerable to methodological error. I am more inclined to think that both the convergence and its subsequent reversal are real and that both suggest the mutability of Black–White test-score differences, even if the mechanisms of change are now poorly understood.

CONCLUSION

An increasing array of evidence suggests that Black–White differences in scores on cognitive tests have been reduced for cohorts born after the middle 1960s. Whereas several test-score series have shown some signs of convergence, the NAEP series in reading, science, and mathematics, which cover ages 9, 13, and 17, are more nearly representative of the general population than other testing programs. As Smith and O'Day (1991) summarized the findings, between 1970 and the middle to late 1980s, initial test-score differences in reading were reduced by 50%, those in mathematics were reduced by 25–40%, and those in science were reduced by 15–25%. However, there is cause for concern in the last two rounds of NAEP data, because the gains of the 1970s and early 1980s may have begun to erode. A preliminary report of NAEP findings from 1994 suggested that the recent divergent trend may have ended, but it has not been reversed. From 1992 to 1994 there were no significant changes in achievement differences between Blacks and Whites at ages 9, 13, or 17 (Campbell, Reese, O'Sullivan, & Dossey, 1996).

Of what importance is convergence in achievement test scores of Blacks and Whites? Herrnstein and Murray (1994) argued that IQ, or *g*, is the key source of variability in adult social and economic success. In so arguing, they followed a strong tradition in psychology. For example, referring to occupational standing, Jensen (1986) wrote that "although *g* cannot account for all the variance in occupational level, it accounts for more than any other measurable sources of variance, independent of *g*, that we have been able to discover" (p. 318). If that were the case,

I should be most concerned about the strength of the link of IQ with test series like the NAEP assessments, and I should also look for more direct evidence about trends in IQ differentials between Blacks and Whites.

On the other hand, there is increasing evidence that IQ, or *g*, is neither the sole nor necessarily the most important cognitive factor in adult success. Much of this evidence comes from new analyses of the National Longitudinal Study of Youth (NLSY), the same data analyzed by Herrnstein and Murray (1994) in *The Bell Curve.* For example, the Numerical Operations (NO) and Computational Speed (CS) components of the Armed Services Vocational Aptitude Battery (ASVAB) are not closely related to the IQ factor measured by the four components of the ASVAB that make up the AFQT (Herrnstein & Murray, 1994, pp. 580–583); yet Goldberger (1995) and Heckman (1995) have found that NO and CS are at least as important as the AFQT in the determination of earnings. Also, Corcoran (1996) found great variation in the importance of the several components of the ASVAB in determining educational, economic, and social success. That is, the several outcomes analyzed in *The Bell Curve* appear to respond differentially to the several components of the ASVAB, and the differential responses are not explained by the closeness of the components to a general ability factor. These findings, I believe, suggest the importance of the array of cognitive tests across which Black performance has begun to converge toward that of Whites, whether or not those tests may be said to reflect IQ, or *g*. It is unfortunate that there are not more longitudinal data in which the effects of a full range of test performances can be assessed across a broad array of life outcomes.

REFERENCES

Alwin, D. F. (1991, October). Family of origin and cohort differences in verbal ability. *American Sociological Review, 56,* 625–638.

Armor, D. (1992, Summer). Why is Black educational achievement rising? *The Public Interest, 108,* 65–80.

Arvey, R. D., et al. (1994, December 13). Mainstream science on intelligence. *Wall Street Journal,* p. 18.

Berliner, D. C., & Biddle, B. J. (1995). *The manufactured crisis: Myths, fraud, and the attack on America's public schools.* Reading, MA: Addison-Wesley.

Campbell, J. R., Reese, C. M., O'Sullivan, C., & Dossey, J. A. (1996). *NAEP 1994 trends in academic progress* (National Center for Education Statistics, Office of Educational Research and Improvement, U.S. Department of Education). Washington, DC: U.S. Government Printing Office.

Ceci, S. J. (1991). How much does schooling influence general intelligence and its cognitive components? A reassessment of the evidence. *Developmental Psychology, 27,* 703–722.

Coleman, J. S., Campbell, E. Q., Hobson, C. J., McPartland, J., Mood, A. M., Weinfeld, F. D., & York, R. L. (1966). *Equality of educational opportunity* (Office of Education, U.S. Department of Health, Education, and Welfare). Washington, DC: U.S. Government Printing Office.

College Entrance Examination Board, Advisory Panel on the Scholastic Aptitude Test Score Decline. (1977). *On further examination.* New York: Author.

Corcoran, J. (1996). *Beyond* The Bell Curve *and* g: *Rethinking ability and its correlates.* Unpublished senior honors thesis, Department of Sociology, Harvard University, Cambridge, MA.

Fischer, C. S., Hout, M., Sanchez Jankowski, M., Lucas, S. R., Swidler, A., & Voss, K. (1996). *Understanding inequality in America: Beyond* The Bell Curve. Princeton, NJ: Princeton University Press.

Flynn, J. R. (1984). The mean IQ of Americans: Massive gains 1932 to 1978. *Psychological Bulletin, 95,* 29–51.

Flynn, J. R. (1987). Massive IQ gains in 14 nations: What IQ tests really measure. *Psychological Bulletin, 101,* 171–191.

Goldberger, A. S. (1995, December). *Abilities, tests, and earnings.* Paper presented at the MacArthur Foundation Conference on Meritocracy and Inequality, Madison, WI.

Grissmer, D. W., Kirby, S. N., Berends, M., & Williamson, S. (1994). *Student achievement and the changing American family.* Washington, DC: RAND Institute on Education and Training.

Hauser, R. M., Jaynes, G. D., & Williams, R. M., Jr. (1990, Spring). Explaining Black–White differences. *The Public Interest, 99,* 110–119.

Heckman, J. J. (1995, October). Lessons from *The Bell Curve. Journal of Political Economy, 103,* 1091–1120.

Herrnstein, R. (1990a, Spring). On responsible scholarship: A rejoinder. *The Public Interest, 99,* 120–127.

Herrnstein, R. (1990b, Winter). Still an American dilemma. *The Public Interest, 98,* 3–17.

Herrnstein, R. J., & Murray, C. (1994). *The bell curve: Intelligence and class structure in American life.* New York: Free Press.

Humphreys, L. G. (1988). Trends in levels of academic achievement of blacks and other minorities. *Intelligence, 12,* 231–260.

Jaynes, G. D., & Williams, R. M., Jr. (Eds.). (1989). *A common destiny: Blacks and American society* (Committee on the Status of Black Americans, Commission on Behavioral and Social Sciences, National Research Council). Washington, DC: National Academy Press.

Jencks, C. S., & Crouse, J. (1982). Aptitude vs. achievement: Should we replace the SAT? In W. Schrader (Ed.), *New directions for testing and measurement, guidance, and program improvement* (pp. 33–49). San Francisco: Jossey-Bass.

Jensen, A. R. (1986, December). *g*: Artifact or reality? *Journal of Vocational Behavior, 29,* 301–331.

Jones, L. V. (1984, November). White–Black achievement differences: The narrowing gap. *American Psychologist, 39,* 1207–1213.

Korenman, S., & Winship, C. (1995). *A reanalysis of* The Bell Curve. Unpublished manuscript, University of Minnesota.

Koretz, D. (1986). *Trends in educational achievement* (Congress of the United States, Congressional Budget Office). Washington, DC: U.S. Government Printing Office.

Lemann, N. (1995, August). The structure of success in America: The untold story of how educational testing became ambition's gateway—And a national obsession. *Atlantic Monthly,* pp. 41–60.

Loehlin, J. C., Lindzey, G., & Spuhler, J. (1975). *Race differences in intelligence.* San Francisco: Freeman.

Menard, S. (1988, September). Going down, going up: Explaining the turnaround in SAT scores. *Youth & Society, 20*(1), 3–28.

Miller, L. S. (1995). *An American imperative: Accelerating minority educational advancement.* New Haven, CT: Yale University Press.

Morgan, R. (1991). Cohort differences associated with trends in SAT score averages. (College Board Report No. 91–1). Princeton, NJ: Educational Testing Service.

Mullis, I. V., Dossey, J. A., Campbell, J. R., Gentile, C. A., O'Sullivan, C., & Latham, A. (1994). *NAEP 1992 trends in academic progress: Achievement of U.S. students in science, 1969–70 to 1992; mathematics, 1973 to 1992; reading, 1971 to 1992; and writing, 1984 to 1992* (National Center for Education Statistics. Office of Educational Research and Improvement, U.S. Department of Education). Washington, DC: U.S. Government Printing Office.

Mullis, I. V., Dossey, J. A., Foertsch, M. A., Jones, L. R., & Gentile, C. A. (1991). *Trends in academic progress: Achievement of U.S. students in science, 1969–70 to 1990; mathematics, 1973 to 1990; reading, 1971 to 1990; and writing, 1984 to 1990* (National Center for Education Statistics, Office of Educational Research and Improvement, U.S. Department of Education). Washington, DC: U.S. Government Printing Office.

Murray, C., & Herrnstein, R. (1992, Winter). What's really behind the SAT-score decline? *The Public Interest, 106,* 32–56.

Neal, D. A., & Johnson, W. R. (1996). The role of pre-market factors in black–white wage differences. *Journal of Political Economy, 104,* 869–895.

Northwestern University. (1987). *Statistical summary and interpretation* (1987 Midwest talent search). Evanston, IL: Center for Talent Development.

Osborne, T. R., & McGurk, F. C. (1982). *The testing of Negro intelligence (Vol. 2).* Athens, GA: Foundation for Human Understanding.

Shuey, A. M. (1966). *The testing of Negro intelligence.* New York: Social Science Press.

Smith, M. S., & O'Day, J. (1991). Educational equality: 1966 and now. In D. Verstegen & J. Ward (Eds.), *Spheres of justice in education: The 1990 American Education Finance Association yearbook* (pp. 53–100). New York: HarperCollins.

Solomon, R. J. (1983). *Information concerning mean test scores for the Graduate Management Admission Test (GMAT); Graduate Record Examination (GRE); Law School Admission Test (LSAT): Preliminary Scholastic Aptitude*

Test (PSAT); and Scholastic Aptitude Test (SAT) for the National Commission on Excellence in Education. Princeton, NJ: Educational Testing Service.

Wainer, H. (1985). *Some pitfalls encountered while trying to compare states on their SAT scores: Page and Fiefs as an example* (pp. 85–62). Princeton, NJ: Educational Testing Service.

Wainer, H. (1987, January). *Can we accurately assess changes in minority performance on the SAT?* Princeton, NJ: Educational Testing Service.

Wigdor, A. K., & Garner, W. R. (Eds.). (1982). *Ability testing: Uses, consequences, and controversies, Part I. Report of the committee* (Committee on Ability Testing, Assembly of Behavioral and Social Sciences, National Research Council). Washington, DC: National Academy Press.

Zajonc, R. B. (1976). Family configuration and intelligence. *Science, 192,* 227–236.

Zajonc, R. B. (1986). The decline and rise of scholastic aptitude scores: A prediction derived from the confluence model. *American Psychologist, 41,* 862–867.

Zwick, R. (1992, Summer). [Special issue] on the National Assessment of Educational Progress. *Journal of Educational Statistics, 17*(2), 93–232.

10

Exploring the Rapid Rise in Black Achievement Scores in the United States (1970–1990)

David W. Grissmer, Stephanie Williamson, Sheila Nataraj Kirby, and Mark Berends

Available evidence indicates that sustained increases in comparable age-specific test scores have occurred in many Western nations during the current century (Flynn, 1987). This evidence points to gradual increases across generations rather than abrupt changes. It is difficult to pinpoint the causes, empirically; gradual, long-term changes are always difficult to account for because of their simple correlation with time. In addition, the similarity of direction across countries and the paucity of reliable and comparable data also make it difficult to pinpoint any single cause or combination of causes that can explain such a phenomenon.

In this chapter, we describe and hypothesize about a fairly abrupt increase in achievement scores that occurred for students in the United States between 1970 and 1990. The data show, for instance, that the U.S. population of Black students aged 9, 13, and 17 years increased verbal and mathematics scores in the range of 0.4–0.8 *SD*s over a relatively short period. The score gains were registered on tests with iden-

This work was supported by funds from the Lilly Endowment and the Center for Research on Educational Diversity and Excellence (CREDE), which is funded by the U.S. Department of Education.

tical items given every 2–4 years to representative samples of all racial–ethnic groups across the United States. But only minority scores show dramatic increases in this time period. Such a large, sustained increase in achievement scores for millions of children in a relatively short period of time is rare and perhaps unprecedented.

Part of the evidence for the "Flynn effect" comes from a variety of tests, some more specifically designed to tap IQ and some that test more subject-specific or learned knowledge similar to the tests discussed in this chapter. Therefore, the Flynn effect probably reflects a variety of factors that can act to increase learned or subject-specific knowledge as well as "pure" IQ. As such, the evidence cited here can contribute to understanding the secular rise in scores from a variety of tests. However, it cannot be stated that increases in the tests cited here necessarily arise from increases in IQ.

THE NATIONAL ASSESSMENT OF EDUCATIONAL PROGRESS

The U.S. Department of Education has given the National Assessment of Educational Progress (NAEP) tests[1] (Johnson & Carlson, 1994) in reading and mathematics to nationally representative samples of 9-, 13-, and 17-year-old students approximately every 2–4 years between 1971 and 1994. The items on the tests have stayed the same since 1971, removing the need for and uncertainty associated with periodic re-norming.

Scores on achievement tests reflect genetic endowment, family characteristics and environment, the quality of schools and communities, the level of educational and social investment in children and families,[2] and social and educational policies governing access to schools, jobs, and health care. Genetic endowment within a circumscribed population changes very slowly; as such, dramatic changes in test scores over short

[1]The NAEP tests are documented for each application.

[2]We group the maintenance of minimal levels of nutrition, sanitation, and other basic health-related factors (e.g., birth weight) that have been shown to affect achievement levels under "social investment."

periods of time are likely to be the result of corresponding changes in families, communities, schools, the level of social and educational investment, or social and educational policies. Therefore, these scores can provide some evidence about environmental effects on achievement scores.

Cross-sectional studies of student achievement show strong associations between family and racial–ethnic characteristics and test scores. For example, children in households with high parental educational attainment and income tend to score higher on tests. Other characteristics that have a positive effect are smaller family size, older age of mother at birth of the child, and more stimulating home environment (which in itself is a product of some of the previously mentioned factors). School and community characteristics have also been shown to be predictors of achievement scores. The specific mechanisms through which these family, school, and community characteristics work to foster higher achievement are still somewhat elusive. These so-called proximal processes (Bronfenbrenner & Ceci, 1994) range from theories of how different environments can result in more or less permanent differences in brain development patterns or in emotional development to simple theories of different exposure, access, and learning opportunities. Until these mechanisms are more precisely identified, research on student achievement will lack a key element that would allow educators to identify and implement effective and efficient social and educational policies aimed at increasing achievement.

An important issue regarding trends in test scores is their permanence. Evidence has indicated that some interventions, although effective in the short term, tend to be less so in the long run. Indeed, the research seems to indicate that it is easier to achieve long-term changes in other positive outcomes (e.g., high school completion, labor force participation, or avoidance of criminal involvement) than in achievement scores (Gomby, Larner, Stevenson, Lewit, & Behrman, 1995). It is important, therefore, to determine whether gains in scores are permanent or temporary. The NAEP scores offer some important evidence concerning this question because they encompass three groups of children of different ages, so that it can be determined whether the pattern of gains that began at age 9 was followed 4 years later with gains at age 13 and 4 years later with gains at age 17. Unfortunately, the associated data collected along with the scores are quite inadequate for analyzing reasons for changes in scores. Ideally, one would like to have, for each child, measures of family characteristics and home environments as well as school and community characteristics. The NAEP is quite limited in the measures it collects, missing rather basic information (Berends & Koretz, 1996). As a result, although the NAEP scores have been extensively used in research in assessment, they have been less often used to study broad policy issues concerning the well-being of children, the quality of schools, or the effectiveness of educational and social investments and policies.

Although the NAEP scores and their associated data have severe limitations, analyses of these scores is important because they offer perhaps the best broad measures of the social, economic, and educational environments in which children are being raised. Without such efforts to sort out trends in important aspects of children's well-being and their family, community, and school environments, professionals and the public may fail to understand or to measure the effectiveness of social and educational policies and programs over the last 25 years.

The last 25 years of American educational and social policy have been largely directed at changing the environment for disadvantaged children and families. As such, this period represents a major "experiment" in determining whether changing environmental influences through governmental policy can affect the well-being and performance of children perceived to be at risk of educational failure. There exists a widespread perception that this experiment was a failure—that it even produced counterproductive results (Herrnstein & Murray, 1994)—which has led to attempts both to scale back and to restructure many of these programs. Much of the evidence supporting this view has been based on trends in scores on the Scholastic Aptitude Test (SAT), a highly misleading indicator of how the average achievement of all American students is changing (Grissmer, Kirby, Berends, & Williamson, 1994). It is important to examine whether NAEP scores are consistent with these perceptions or tell a different story.

Trying to sort out the relative contributions of families, schools, and social and educational policies and programs to student achievement over the last 25 years is a complex exercise for several reasons. Explaining trends is difficult, first, because several factors perceived to affect student achievement have all changed dramatically: the family environment, demographic mix of students, school quality, public policies directed toward providing equal educational opportunity, and public investment in schools and social programs. Second, assessing the effect of public policies and investment is problematic partly because empirical evidence indicates that family and demographic changes probably have the largest effects on test scores; thus, family and demographic effects on student achievement need to be taken into account before assessments are made of the effect of public policies and investment.

In this chapter, we use the NAEP reading and mathematics scores from 1971 to 1992 together with a new method of compensating for missing family variables to explore the following issues:

- Whether differential changes have occurred over time by racial–ethnic groups
- Whether such trends hold for both younger and older children
- Whether these trends show regional differences
- Whether changing family and racial or demographic composition can account for these changes
- Whether the pattern of differential changes by racial–ethnic group fits the national emphasis on social and educational programs aimed toward low-income and minority groups

The factors that have changed for minority-group populations in the United States in the last 25 years are not dissimilar to factors that have changed more slowly for populations in many Western nations during the current century. Rising income and education; access to better health care, schools, and housing; early childhood education programs; and better nutrition were all objectives of the Civil Rights and Great Society legislation enacted and programs funded in the 1960s.

Much of this legislation and funding began to be implemented in the late 1960s and 1970s.

In addition, evidence indicates that real public resources devoted to children (in 1988 dollars) increased from about $1,300 per child to almost $3,000 per child between 1960 and 1988 (Fuchs & Reklis, 1992), much of it in the form of increased real school spending. Evidence also indicates that this increased spending for regular students was disproportionately directed toward and would be expected to benefit lower income and minority-group students (Rothstein & Miles, 1995).[3] There was an attempt in the United States, therefore, over a relatively short period of time, to make large investments and improve the lives of minority-group families and children. For this reason, it is important to determine whether achievement rose during this period and whether it is possible empirically to link any gain in scores with this investment. If not, the relation between test-score performance and much of what are considered improved living conditions is weakened. If a link can be demonstrated, it becomes plausible that improved living conditions across a variety of dimensions can help explain the Flynn effect.

National NAEP Score Trends

Figure 1A and 1B show simple differences in standard deviation units for Black and non-Black students' reading and math scores from the early 1970s to 1992.[4] Overall, the data show small changes for non-Black students but very large gains for Black students for each age group. The Black gains were smaller for age 9 than for ages 13 and

[3]For instance, experimental data have indicated that although smaller class sizes in early grades benefit all students, the effects are twice as large for minority-group students (Mosteller, 1995).

[4]We use a consistent measure—proportion of a standard deviation—throughout to measure differences in test scores. The standard deviation unit is estimated using the square root of the mean variance of the first and last data point. A measure that is also commonly used in reporting test scores is the percentile, which shows the relative standing of a particular score and indicates the proportion of children scoring lower than that score. A 0.10 *SDs* difference in test scores is approximately 3.4 percentile points for most children. Therefore, two groups of children whose average scores differ by 0.10 *SDs* would indicate that one group scores, on average, 3.4 percentile points higher than the other group.

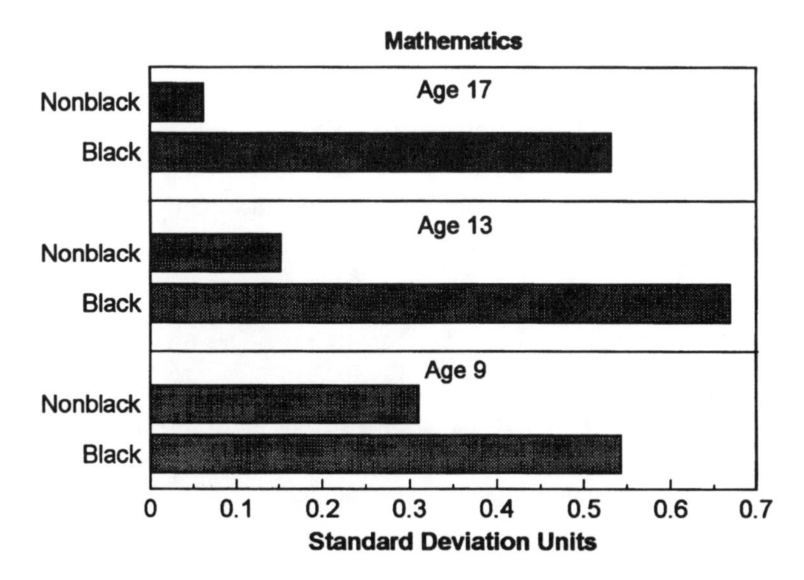

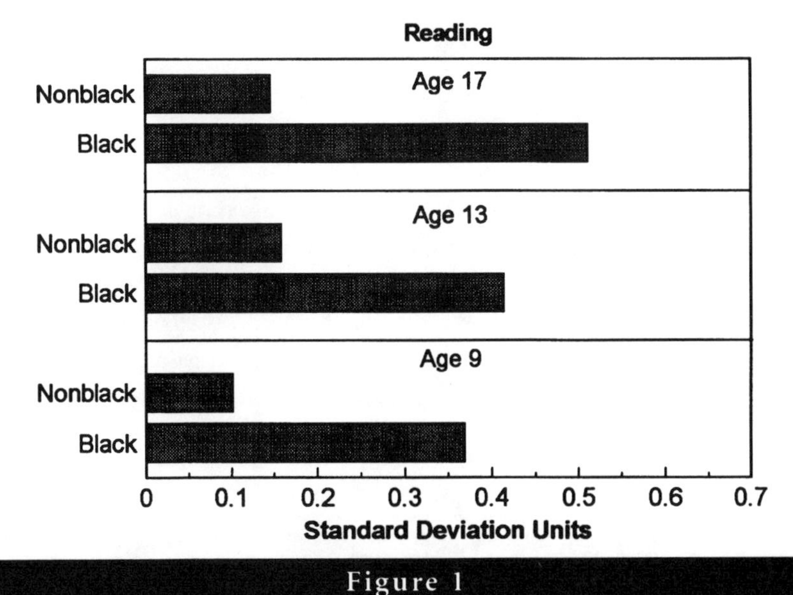

Figure 1

Change in National Assessment of Educational Progress mathematics scores between 1973 and 1992 and reading scores between 1971 and 1992 by race and age.

17[5]—a point that we discuss later. Figure 2 shows similar data for Black, Hispanic, and non-Hispanic White students.[6] Hispanic gains tend to fall in between Black and non-Hispanic White gains. Figure 3 shows the decline in the gap between minority-group scores and non-Hispanic White scores. Gains made by minority-group students have resulted in a closing of the gap between the groups by one fourth to one half.

Trends Among Different Age Groups

Figure 4 shows time series data by age group for all years in which the tests were administered.[7] For Black students, the gains were not uniform over time, tending to occur mostly within a shorter period of time. However, the period of rising scores is different for each age group, and some groups also show later declines in scores.

Regional Trends

Figure 5 presents changes in NAEP reading and mathematics scores for Blacks and non-Blacks disaggregated by region.[8] In almost every region and in both tests, Blacks made larger gains than Whites over the 20-year period. The regional pattern of gains is different across the three age groups for reading. For example, 17-year-old Blacks made the largest gains in the South, whereas those aged 13 made the largest gains in the West. For the youngest age group, the largest gains were in the Midwest region, with the West showing the smallest gains. Non-Blacks did better in the South and West, showing the smallest gains in the Midwest region. In math, Blacks experienced enormous gains, especially

[5]The NAEP scores are for a "student" sample rather than an age-group sample. There is little difference between the two at ages 9 and 13 because all children attend school. However, the data at age 17 are biased to the extent that school dropout rates change over time. In general, school dropout rates have declined, especially for Blacks, so that Black gains at age 17 are likely to be understated if dropouts tend to have lower scores.

[6]These data begin with 1975 because that is the first year Hispanic students were identified.

[7]Standard deviation units are the square root of the mean variance of the first and last data points for all racial–ethnic groups combined.

[8]Standard deviation units used here are the square root of the mean national variance for the first and last data points for each racial–ethnic group.

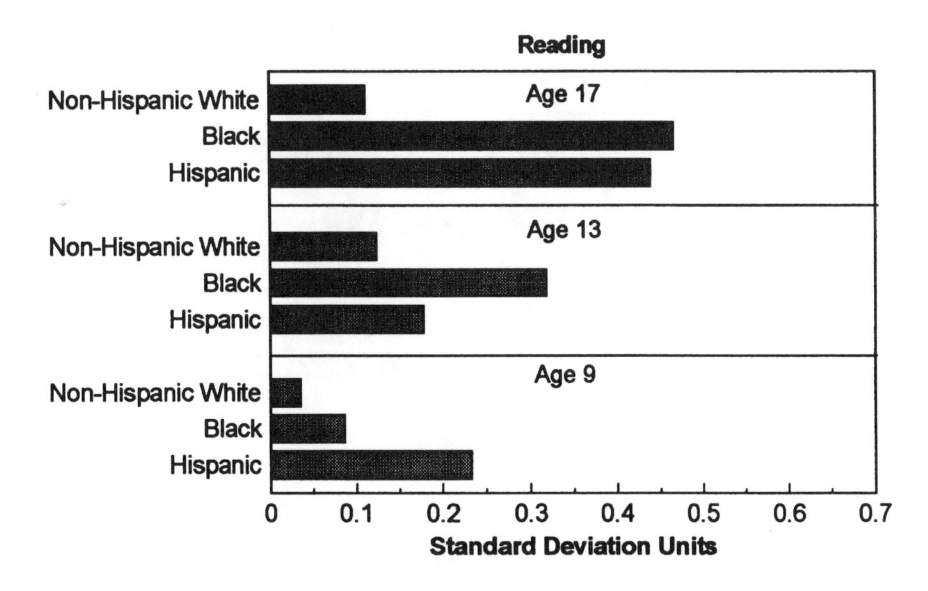

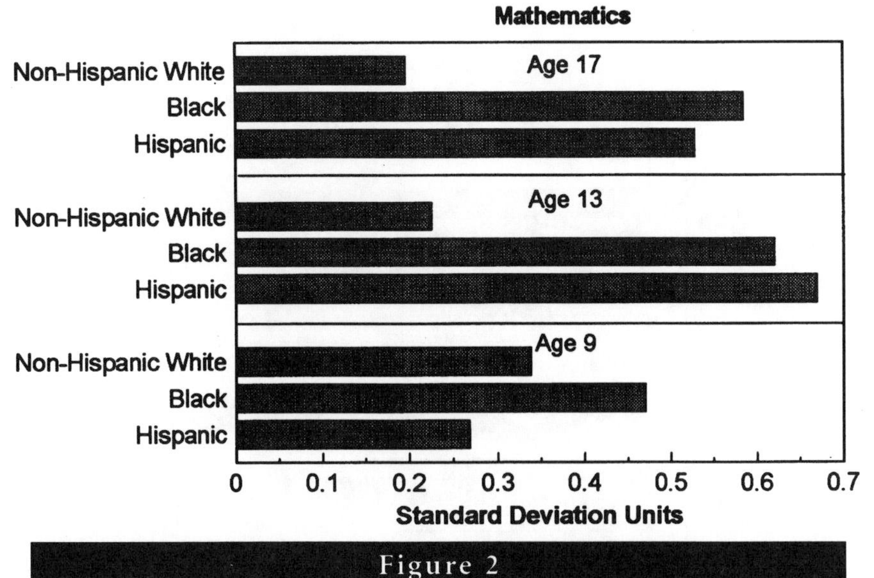

Figure 2

Change in National Assessment of Educational Progress mathematics scores between 1978 and 1992 and reading scores between 1975 and 1992 by race–ethnicity and age.

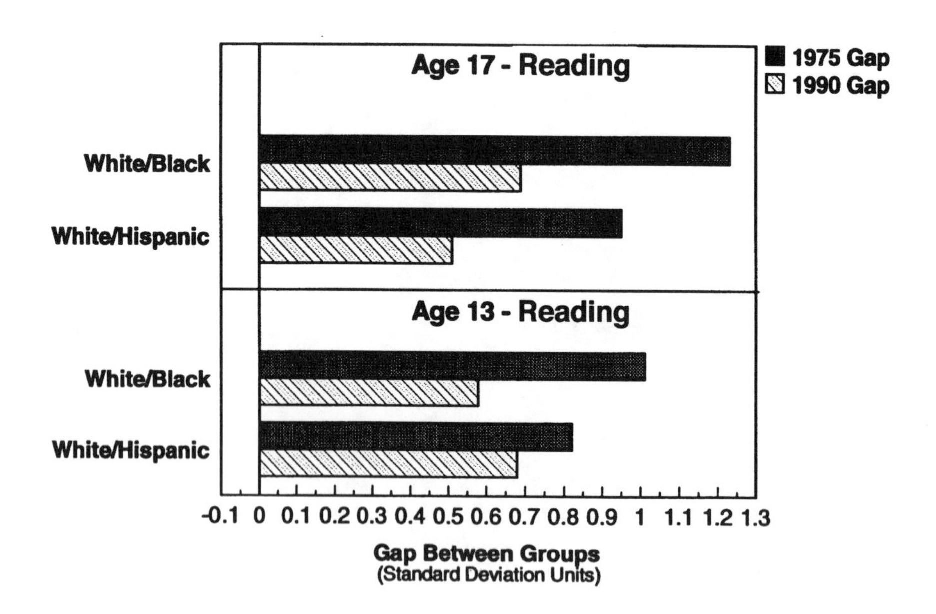

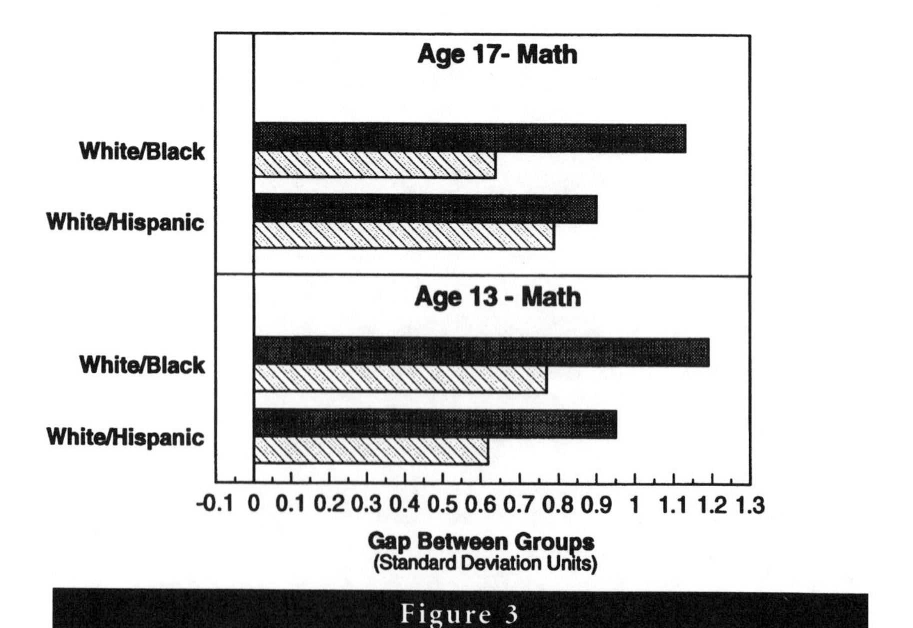

Figure 3

Gap in National Assessment of Educational Progress mathematics scores in 1978 and 1990 and reading scores in 1975 and 1990 by racial–ethnic group.

in the South, although the gains were much smaller in the youngest age group. In contrast, non-Blacks showed the largest gains for the youngest age group, and gains in the South and West were almost equal to those of the Blacks.

An interesting question is whether the Black gains were cohort specific or period specific. A cohort effect would show that gains occurred first for 9-year-old students, then 4 years later for 13-year-old students, and in another 4 years for 17-year-old students. A cohort effect is more developmentally based in that it assumes that a score at any age is a cumulative result of environmental conditions from birth.[9] For instance, if better prenatal care resulted in higher birth weights, one would expect to see this cohort effect for all age groups 9, 13, and 17 years after the implementation of such a prenatal program. A period-specific effect would occur for all age groups at the same point in time. For instance, a misnormed test would affect all groups in the same year.[10]

Analyses of NAEP data using a cohort perspective offer several useful insights. Figure 6 presents the NAEP reading and math scores for

[9] A developmental approach to explaining test scores at a given age would assume that family environments, the quality of schools and communities, and social and educational policies and investments made over the child's entire lifetime up to the time of test taking were relevant. For example, if one assumed that desegregating schools is likely to affect test scores, and beneficially so for some subset of the student body, then a policy completely desegregating schools in a single year would be expected to have a different pattern of effect on 9-, 13-, and 17-year-old students' time series test scores. One would hypothesize that the full effect of this policy action would be greatest for children who experienced the effects of desegregated schools throughout their schooling. For example, a fourth grader who was in desegregated schools only in the fourth grade would be unlikely to benefit as much from this policy change as a child who was in desegregated schools for the third and fourth grades or from the first to the fourth grade. Thus, a plausible hypothesis might suggest a pattern of increases over the four consecutive cohorts and then a flattening out of the effect. Whether this increase is linear or nonlinear depends on assumptions made about the relative importance assigned to first grade versus second versus third versus fourth grade in determining fourth-grade scores. For instance, a more developmentally based hypothesis might suggest that school quality in earlier grades is more important than in later grades and that very small gains would be seen until children were in desegregated schools over the whole period prior to the test. However, some might hypothesize that the learning occurring closest to the test application should be weighted more heavily and that large gains would be seen in the first year of testing with much smaller gains in the other 3 years. A final hypothesis might assign equal importance to all grades, resulting in a linear increase. Thus, in modeling the effects of family, schools, communities, and social and educational policies, one needs to account for the program differences existing over a cohort's lifetime and make assumptions about the relative importance of environmental changes at different ages and grades.

[10] Koretz first made this distinction with respect to test-score data (Koretz, 1987).

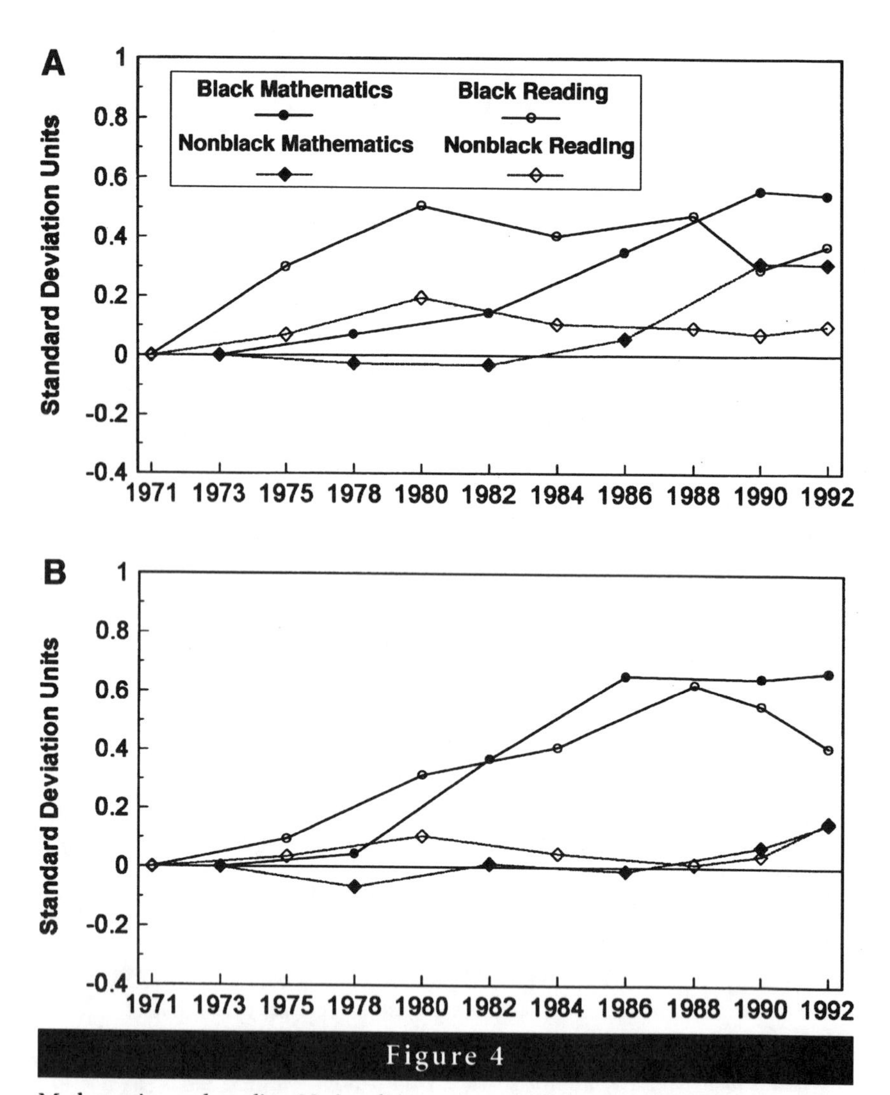

Figure 4

Mathematics and reading National Assessment of Educational Progress scores for 9-year-olds (A), 13-year-olds (B), and 17-year-olds (C) by race.

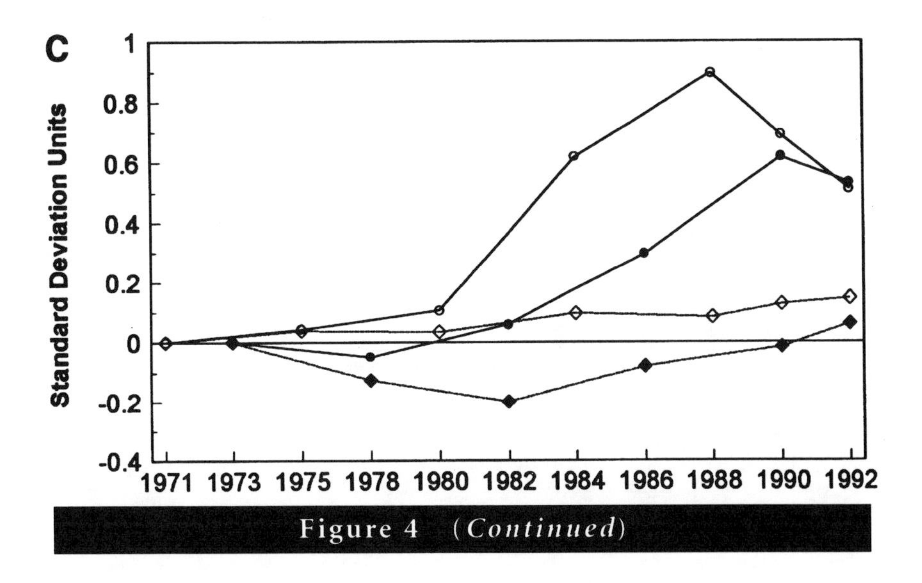

Figure 4 (*Continued*)

Blacks by entering school cohort. These data allow one to compare scores of a single cohort at ages 9, 13, and 17. The data for Blacks display an interesting pattern of fairly stable scores for cohorts entering school before 1968, rapid gains for cohorts entering approximately between 1968 and 1978, and little or no gain or some decline for cohorts entering school in 1980 and after. For instance, the data for reading scores show that the 1976 cohort scores were about 0.4 *SD*s higher than the 1968 cohort scores at ages 9 and 13 and almost 0.8 *SD*s higher for 17-year-olds. Therefore, the 1976 cohort showed sizable reading gains at each age that these students took the test, compared with the 1968 cohort. The cohorts after 1980 show no additional gains in scores and some decline. However, the sustained gains are in the 0.4 *SD*s range. The reading scores display a pattern that would be expected from a cohort effect.

The math data also display the pattern of rapidly rising scores for the 1975 and 1979 cohorts compared with the 1968 cohort. Increases occurred for all age groups in these cohorts, although they were much

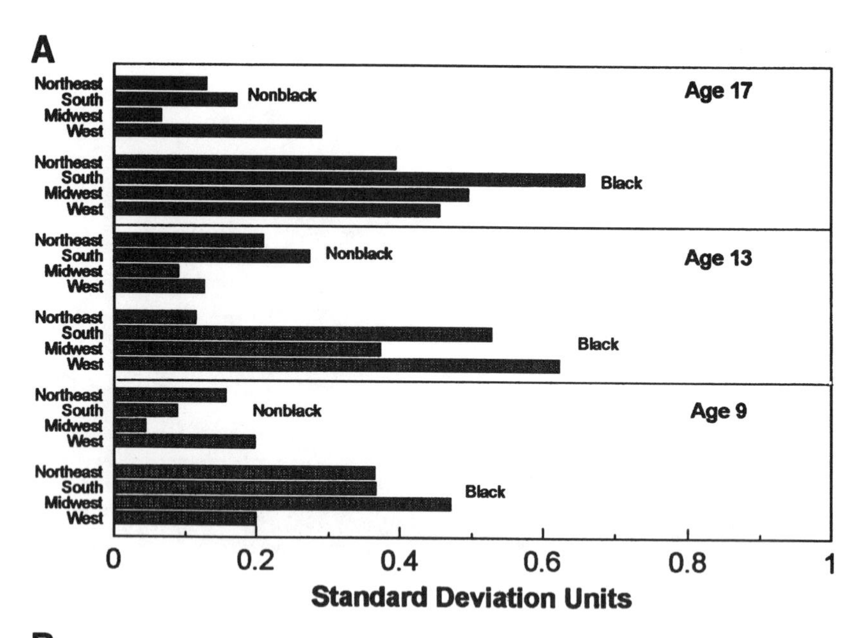

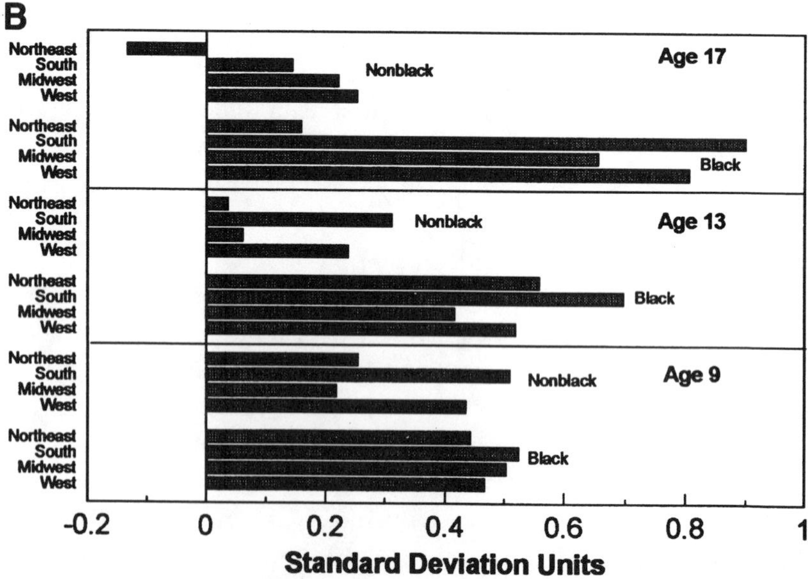

Figure 5

Change in National Assessment of Educational Progress reading scores between 1971 and 1992 (A) and math scores between 1978 and 1990 (B) by region, race, and age.

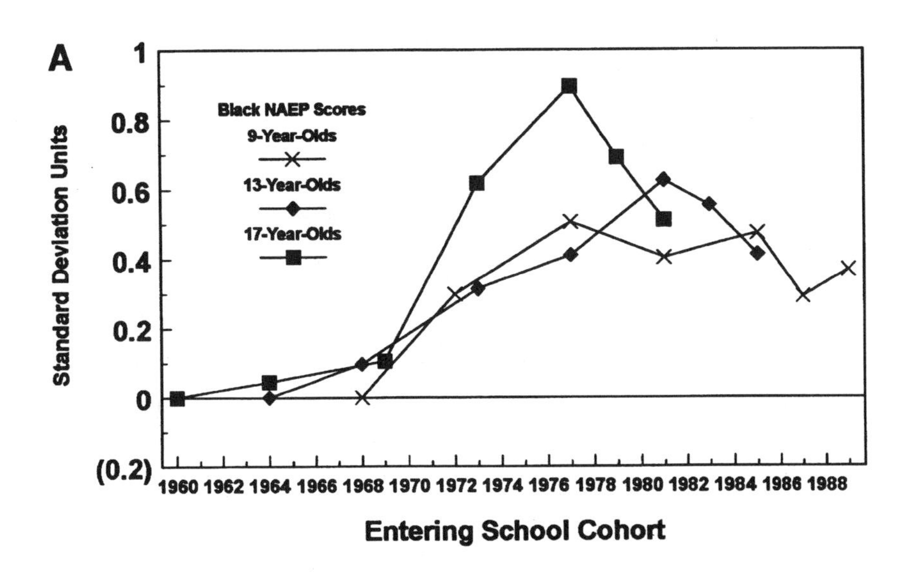

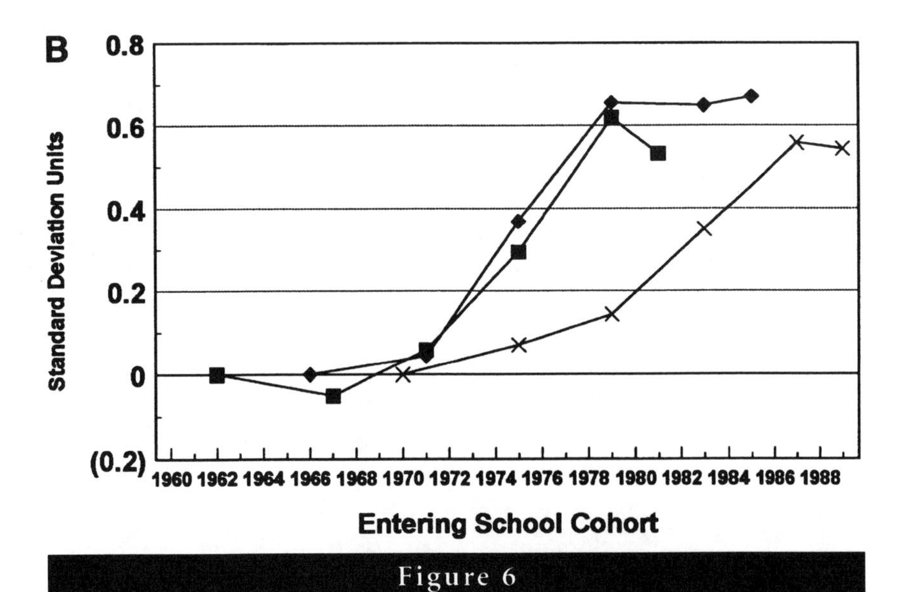

Figure 6

Change in National Assessment of Educational Progress reading scores (A) and mathematics scores (B) by entering school cohort and age: Blacks.

more pronounced for 13- and 17-year-old students, whose scores also stabilized after 1980. However, the 9-year-old students' scores alone continued to increase after 1980, an exception to the cohort effect. For the 1983 and 1987 cohorts, the 9-year-olds' scores increased without corresponding increases in the scores for older students.

The data tend to focus attention on the question of what differences were experienced by 1970 and previous cohorts compared with 1970–1980 cohorts compared with post-1980 cohorts. This appears to be the critical question in explaining the large gains for Black students. Certainly a hypothesis that needs examination is whether these gains corresponded to the implementation of the new programs passed in the late 1960s, often characterized as Civil Rights, "Great Society," or "War on Poverty" legislation. This was also a period of declining class size and of compensatory education funding for the disadvantaged. However, before testing hypotheses about the effects of public investment and policies, it is necessary to understand how the changes that have taken place in American families would be expected to affect scores. Because family characteristics are the strongest correlates with test scores, it is necessary to see what role these changes may have played in changing test scores over this period.

EFFECTS OF THE CHANGING FAMILY ON ACHIEVEMENT SCORES

The method used here to estimate effects is described by Grissmer et al. (1994). In that study, we developed estimates of the net effect of the changing family and demographic environment on student verbal–reading and mathematics test scores over time and a residual that contained the effect of factors not associated with family and demographic changes as well as errors. The method consisted of three steps: (a) developing equations relating student achievement to family and demographic characteristics using two large, nationally representative data sets: the 1980 National Longitudinal Survey of Youth (NLSY) and the 1988 National Education Longitudinal Study (NELS); (b) using these

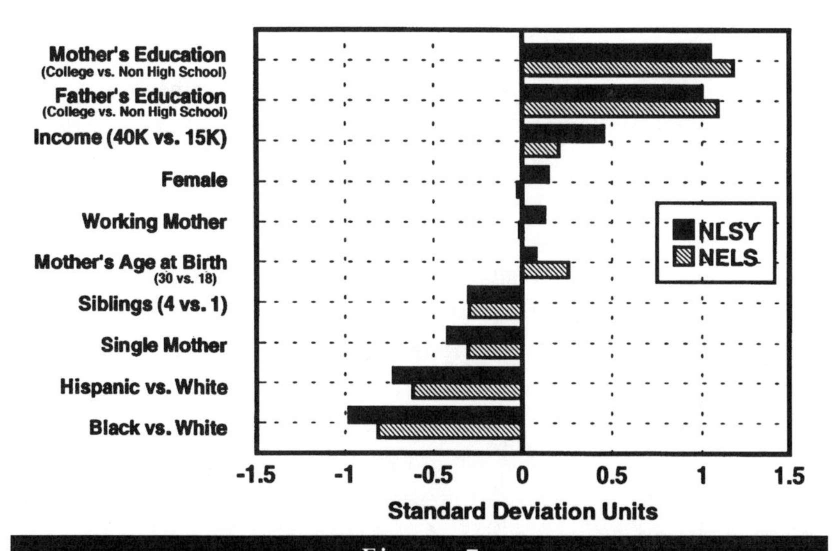

Figure 7

Simple differences in mean mathematics test scores for selected groups, National Longitudinal Survey for Youth (NLSY), and National Education Longitudinal Study (NELS).

equations to predict test scores for each student in a national sample of children (from the Current Population Surveys) in 1970, 1975, and 1990 using their family and demographic characteristics; and (c) comparing the mean differences in these predicted test scores (estimates of the effect of changing family and demographic characteristics) with actual scores from the NAEP. This procedure provides an estimate of how much family and demographic changes contributed to actual changes in test scores, and the residual change in test scores (actual changes minus family and demographic effect) provides an estimate of the effect that factors unrelated to family and demographic characteristics had on changing test scores.

How Much Do Family and Demographic Characteristics Affect Test Scores?

The results from both the NELS and the NLSY showed large differences in test scores associated with family and demographic characteristics and great consistency in the direction and relative significance of these differences. Figure 7 shows simple comparisons of mathematics test scores among youths from families with different characteristics according to the NLSY and NELS.[11]

The figure shows large differences among the average test scores of children living in families with different levels of parental education or of different racial–ethnic background. For instance, children whose mother or father graduated from college scored approximately 1.0 *SD* higher than children whose mother or father did not graduate from high school, and Black and Hispanic youths scored between 0.50 and 1.0 *SD* lower than non-Hispanic White youth.

Somewhere smaller test-score differences are evident among young people living in families with different levels of annual income ($40,000 vs. $15,000), living in families of different sizes (four siblings vs. one sibling), having younger versus older mothers (age 30 vs. age 18 at birth), and living in two-parent versus single-parent families. For instance, children living in two-parent families scored about 0.30–0.40 *SD*s higher than youths living in single-parent families, whereas children in large families scored approximately 0.30 *SD*s lower than children from smaller families. There was little difference in test scores between those with working versus nonworking mothers.

Public debate and the press often focus on these simple comparisons of achievement scores for different family and demographic characteristics and mistakenly attribute the difference in scores between two groups to the particular characteristic in which the groups differ. However, these comparisons and inferences are misleading because the students being compared usually differ in several characteristics, not just the one being cited. For instance, young people in higher income families are also more likely to have parents with higher levels of education and to be nonminority-group members. Therefore, the difference in average test scores between children from high- versus low-income families is probably due to a combination of factors, not income alone. A

[11] The mathematics and verbal–reading test-score differences reported by Grissmer et al. (1994) showed fairly similar patterns and sizes.

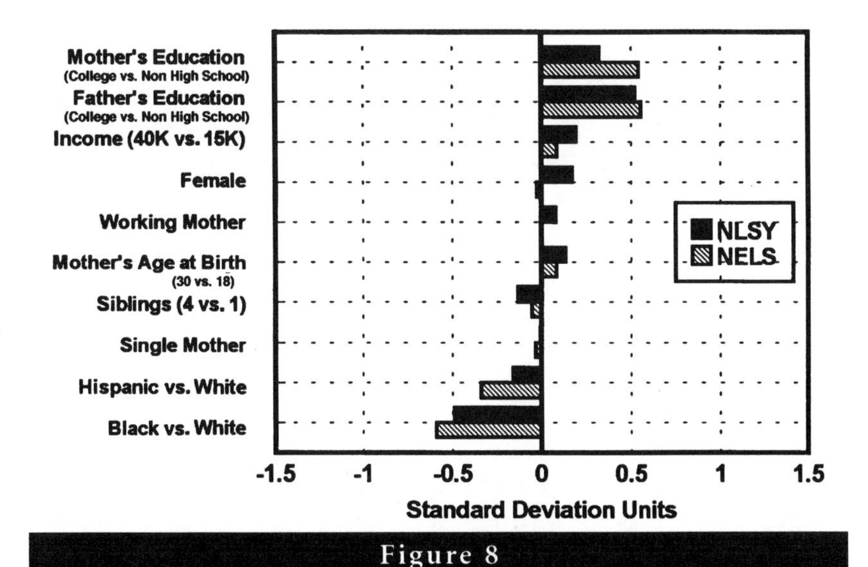

Figure 8

Controlled differences in mean mathematics test scores for selected groups, National Longitudinal Survey for Youth (NLSY), and National Education Longitudinal Study (NELS).

better measure of the effect of income on test scores would be a controlled comparison of two groups of young people who have similar family characteristics except for income. This type of comparison would be helpful for other characteristics as well. Figure 8 summarizes these controlled comparison differences for mathematics scores.[12]

Figure 8 shows that the net effect of each factor is considerably smaller than in the simple comparisons in Figure 7. However, the controlled differences remain statistically significant for almost all characteristics. For example, youths whose parents were college graduates scored about 0.50 *SDs* higher than youths who were otherwise similar but whose parents did not graduate from high school. In addition, controlling for other family characteristics, the difference between

[12]These effects are derived by using the estimates from our multivariate model of student achievement. Multivariate models allow one to examine the effect of a particular characteristic, holding constant other important variables.

Blacks and non-Hispanic Whites was 0.50 *SDs*, and the difference between Hispanics and non-Hispanic Whites was somewhat smaller. Youths with different levels of family income or different family sizes showed much smaller differences in test scores. Controlled test-score differences due to family structure and labor force participation of mother appeared to be negligible. These results suggest that the simple differences between scores of youths in single- versus two-parent families arose from other differences in family characteristics, such as family income, parental education, or previous family environment, rather than from the structure of the one- versus two-parent family itself.

How Much Would Changing Families and Demographics Change Test Scores?

We used the estimates from the multivariate models (which formed the basis for Figure 8) to predict the changes in test scores that would be expected due to the changes in family and demographic characteristics that occurred between 1970–1975 and 1990. We found that 14- to 18-year-olds living in U.S. families in 1990 would be expected to score *higher*, not lower, on tests compared with youths in families in 1970. The size of the shift in mean scores is approximately 0.20 *SDs*. This means that youths in 1990 would be expected to have higher scores by about 7 percentile points than their counterparts in 1970 on the basis of combined changes in demographic and family characteristics. It should be emphasized that these findings estimate average effects, taking into account all American families with 14- to 18-year-olds.

Our analysis suggests that the most important family influences on student test scores are the level of parental education, family size, family income, and the age of the mother when the child was born. Of these variables, the two that have changed most dramatically in a favorable direction are parental education levels and family size. Children in 1990 are living with better educated parents and in smaller families. These factors are the primary reasons that changes in family characteristics lead to predictions of higher test scores. For example, 7% of mothers of 15- to 18-year-old children in 1970 were college graduates compared with 16% in 1990, whereas 38% did not have high school degrees in 1970 compared with only 17% in 1990. Similar but somewhat smaller changes occurred in the educational attainment of fathers. Changes in family size were also dramatic. Only about 48% of 15- to 18-year-old children lived in families with one or no siblings in 1970 compared with 73% in 1990.

Our analysis indicates that average family income changed little over the period from 1970 to 1990 (in real terms), so it would not be expected to affect average test scores. However, the decline in family size coupled with unchanged average family income means that family income per child actually increased from 1970 to 1990.

One change that has had a slight negative effect on test scores is the small decline in the average age of mother at birth of child. This is partly due to increased births to younger mothers but also to the decline in family size. The effect of the large increase in working mothers and single-parent families is more complex (discussed in more detail below). Our equations imply that the large increase in working mothers would—if all other things were equal—have a negligible or small positive effect on youths' test scores. However, mother's labor force participation was measured when the youth was approximately 14 years old, so our results may not apply to younger children. In the case of the increase in single mothers, our models imply no negative effects from the changed family structure alone. However, such families tend to have much lower income levels, so the predictions for youths in these families incorporate a negative impact due to increasing numbers of poor, single-parent families.[13]

We turn now to the results by racial–ethnic group between 1975 and 1990. Figure 9 (mathematics and reading) show the estimated family effects separately for non-Hispanic Whites, Blacks, and Hispanics as well as for the total youth population between 1975 and 1990. Higher mathematics scores in 1990 would be expected for 13- and 17-year-olds for each racial–ethnic group on the basis of changing family characteristics.

[13] A more technical discussion of these complex effects is given in our main report (Grissmer et al., 1994, chapter 5).

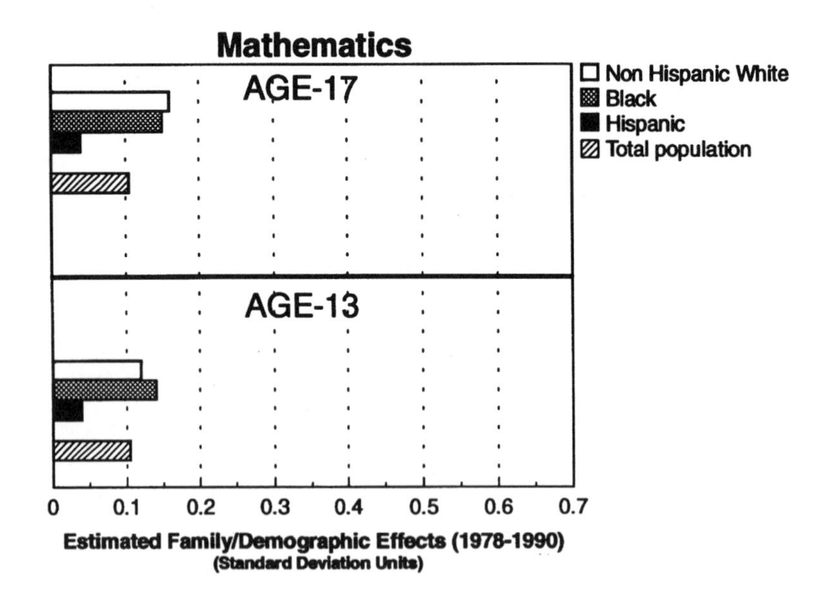

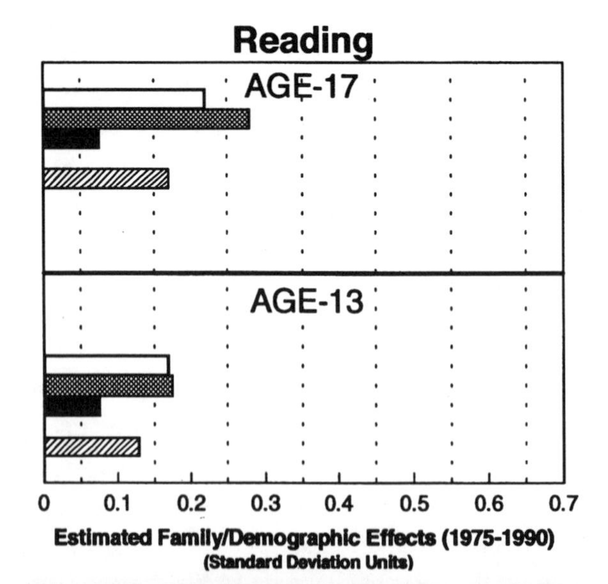

Figure 9

Estimated family and demographic effects on mathematics test scores between 1978 and 1990 and reading test scores between 1975 and 1990 by racial–ethnic groups.

The data show that non-Hispanic White and Black youths have similar predicted family gains of approximately 0.15 *SD*s, but Hispanic youths show smaller gains of approximately 0.05 *SD*s. Verbal–reading score comparisons show slightly higher gains than do those for mathematics, although the pattern is similar by racial–ethnic groups. The positive changes in the Black family in terms of increased parental education and reduced family size are greater than those for non-Hispanic White families, but there were offsetting increases in births to younger and single mothers. The smaller gains for Hispanic youths are explained by smaller increases in parental education, falling family income, and smaller reductions in family size compared with Black families. This is probably due to the continuing immigration of large numbers of Hispanic families into the population, many of whom may have lower levels of educational achievement and fewer labor market skills than previous waves of immigrants (Borjas, 1990).

How Much of the Test-Score Changes Can Be Accounted for by Changes in Family and Demographics?

We compared our projected family–demographic effects on test scores with actual trends in NAEP test scores over similar time periods and for similar age groups to see how much of the changes might plausibly be attributed to changes in family and demographic characteristics. We subtracted the predicted change in test scores (due to family–demographic effects) from the actual change in NAEP scores to compute a residual effect. Figure 10 (mathematics and reading) shows these residuals. The data for mathematics show no residual gain for non-Hispanic White students, indicating that these students' gains in test scores could be accounted for entirely by family effects. However, there are large positive residuals for Hispanic and Black students, suggesting that changing family characteristics alone cannot explain the large gains made by these students. In fact, changing family characteristics account for only approximately one third of the total gain. The residual term does, of course, include errors arising from faulty specifications of the family model, as well as correlations between family and nonfamily

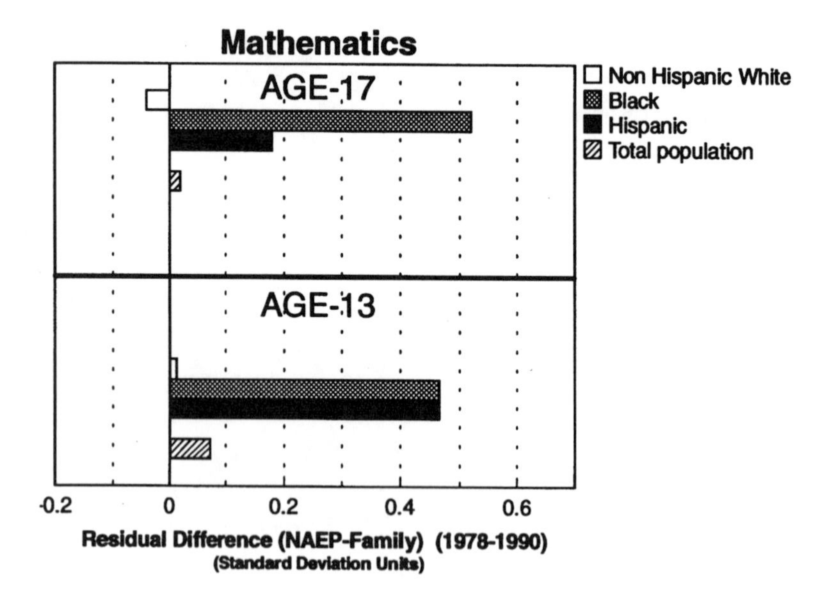

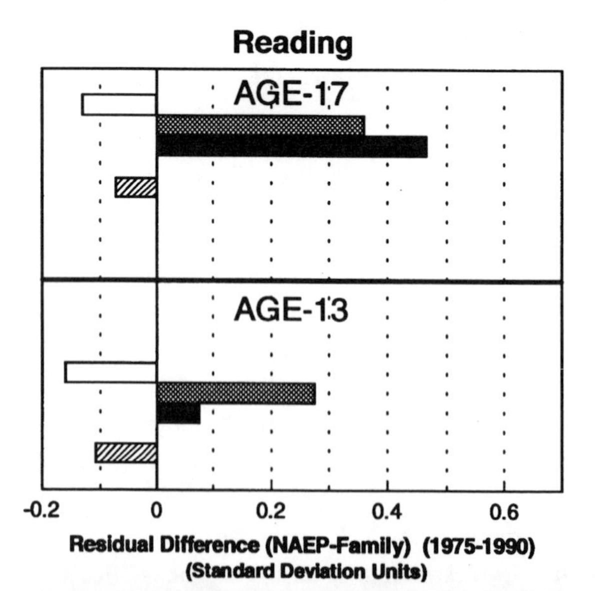

Figure 10

Residual differences between National Assessment of Educational Progress and family effects on mathematics test scores (1978–1990) and reading test scores (1975–1990) for different racial–ethnic groups.

factors. Therefore, these data do not constitute empirical data for "nonfamily" effects.[14]

The data generally indicate smaller residual gains for verbal–reading scores than for mathematics, but they still show substantial Black and Hispanic residual gains not accounted for by family effects. The verbal–reading data also show that non-Hispanic White students have a small negative residual in both age groups, indicating that their NAEP gains were not as large as would be expected from family changes.

U.S. TEST-SCORE TRENDS: TESTING FURTHER HYPOTHESES

We explore two directions for explaining the pattern of residuals in Figure 10. This pattern of residuals shows large score gains for Black and Hispanic students not explained by family effects but no gains for White students that could be attributed to factors outside the family. This pattern would be the expected one if either (a) public resources and policies were disproportionately directed toward Black children or (b) programs and policies directed toward all students disproportionately benefited minorities.

There is certainly little doubt that significant increases in national resources disproportionately directed toward minority-group and low-income families were made between the late 1960s and 1990. There were also dramatic changes in national policies on civil rights, affirmative action, and equal opportunity education, which might be expected to benefit minority-group families and children. However, the coincidence of rising test scores and the implementation of these programs and policies does not constitute empirical evidence that one

[14] Our language of "family–demographic" and "nonfamily" effects is somewhat imprecise. What we mean by *family effects* are the effects of the variables relating to family–demographics in our equations. However, this does not imply that all changes due to these variables originate in the family or are free of government policy influence. For instance, parental education levels are the product of educational policies from past generations; birth control methods certainly have affected family size; and welfare policy has an impact on family income levels. However, a more accurate distinction may be between variables that act primarily through affecting the family characteristics as opposed to variables that might affect intrafamily factors that are not captured in our models (locus of control) or have their impact through schools and communities.

likely caused the other. It suggests that a more refined analysis that attempts to link the timing and regional pattern of score changes with the timing and regional pattern of program and policy implementation would be worthwhile. We discuss some initial results of this analysis below.

A second approach to explaining the residual patterns shown in Figure 10 is to make different assumptions concerning the effects of changing families. The mathematical models estimated to explain children's cognitive and social outcomes or to evaluate intervention programs typically incorporate linear family variables. The exception to this is the frequent inclusion of nonlinear income terms. However, linear models may not fully capture the more complex family phenomena suggested by the child development and clinical literature. For example, the developmental literature indicates that risks may not cumulate linearly but may compound in a nonlinear or exponential manner for children facing multiple risks (Rutter, 1988; Sameroff, Seifer, Zax, & Greenspan, 1987). Linear models do not capture this compounding effect, which may be particularly crucial for explaining changing test scores of lower scoring students. For example, children facing multiple risks may do much worse than would be predicted by simple linear models, whereas removing some of these risks may result in much larger gains than would be predicted by linear models.

A possible supporting explanation of our residual pattern, therefore, is that large test-score gains occurred among Black students because these students were much more likely to be at "multiple risk" in their families and that removal of some of this risk (through policies and programs, perhaps) resulted in test-score increases much larger than would be predicted from linear models of family effects. White students would not have experienced such gains because the incidence of multiple risks for White students was much lower than for Black students.

Exploring the Effects of Nonlinear Variables and Multiple Risks

There are several factors besides multiple risk suggested by the developmental and clinical literature whose effects can be incorporated only by a nonlinear specification. For example, in the case of mother's age at birth of child, one could argue that there would be stronger effects for teenage mothers. For family size, the decline in resources available within the family, assuming an equal distribution among members, would be nonlinearly related to the number of siblings.

The effects of several variables tend to be dependent on the family context (i.e., the coefficient of a variable depends on the value of other variables). The effect of being in a single-parent family may be very different for children in high- versus low-income households or in large versus small families. Similarly, the effect of a mother working could be quite different in small, higher income families than in large, lower income families. Finally, one can postulate an interaction between working- and single-mother variables. Working may be more of a choice in two-parent families; alternately, the lack of working in a single-parent family may be an indicator of dysfunction that extends to parenting.

The literature describing the role of protective factors and resiliency in children (Masten, 1994; Rolf, Masten, Cicchetti, Nuechterlein, & Weintraub, 1990) has not been fully incorporated into models of student achievement, although the role of resiliency factors has been explored in explaining the wide differences in children's outcomes from similar "poor" environments. For instance, Werner and Smith (1982) pointed out that children born with lower birth weights or with birth complications are more "resilient" in higher income than lower income families. This phenomenon can be modeled with simple interaction terms between income and health at birth. The literature also indicates that girls may be more resilient to stress than boys, suggesting the need for interaction terms or separate estimation of gender models. Clearly, the concept of resiliency is essential for modeling risks facing children and needs careful conceptualization. Another interesting hypothesis derived from developmental literature and related to the concept of resiliency is that individual children in families can be targeted to "carry" the family stress and anxiety in dysfunctional families. This concept also may help explain the quite different outcomes of children in the same family (Dunn & Plomin, 1990).

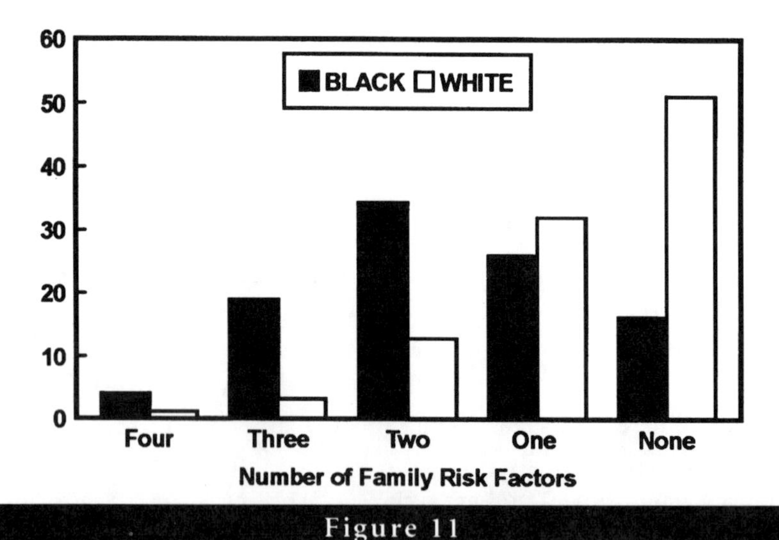

Figure 11

Proportion of children under different amounts of risk by race.

There are many different ways in which these contextual effects can be operationalized and tested. One is through nonlinear and interaction terms. Another method involves assigning each individual a multiple risk index on the basis of the number of risks he or she faces. We first establish a threshold for risk, which is admittedly somewhat arbitrary, for each significant family factor in our regressions. The thresholds are currently set so that approximately 15–30% of children at the bottom of each distribution are defined as being at risk. For example, the risk threshold for education is having a mother without a high school education; for age of mother at birth, it is being born to a teenage mother; for family size, the risk factor is living in a household with four or more siblings; and for income, it is being in the lowest third of the distribution: approximately $23,000 (in 1987 dollars). We create a dichotomous variable for each variable, assigning 1 if the risk exists and 0 if the risk is not present.

We then define a multiple risk index for each individual by adding the separate risk indexes. An individual at risk because of income, mother's education, mother's age at birth, and family size would receive a score of 4 on this multiple risk index. Figure 11 shows the distribution

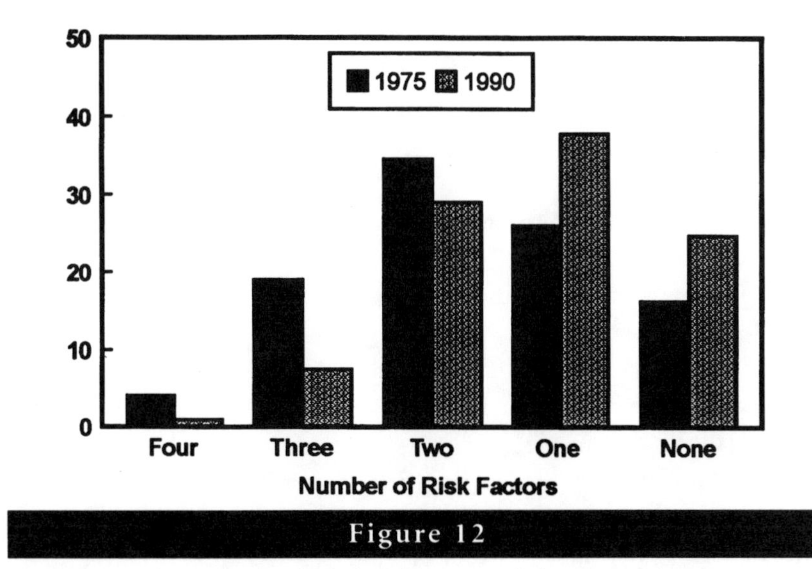

Figure 12

Comparison of proportion of Black children at risk for 1975 and 1990.

of individuals by multiple risk factor by race for 1970 data. The data show that Black students lived in families with a higher number of risk factors than non-Black children. Over 20% of Black children lived in families with three or four risk factors compared with 5% of non-Black children. However, Figure 12 compares the proportion of Black children in 1990 living with the same risks. The data show that the proportion of Black youths living in families with two or more risks has declined markedly over this 20-year period. This dramatic reduction in multiple risks might help explain the large minority-group gains if multiple risks were taken into account in family equations.

When we used the multiple risks equation rather than linear equations to predict family effects, we found that Black family effects increased about 0.03–0.05 *SDs*. This means that somewhat more of the increase in scores may be family related, and somewhat less may be nonfamily related. Therefore, multiple risk and contextual effects can explain a part, but only a small part, of the residuals in Figure 10. A large part of the residuals is left to be explained by factors other than those included in the family equations.

Exploring the Effects of Reduced Class Size and Desegregation

We have tested for the relationship between achievement scores and policies of desegregation and increased educational investment using the NAEP scores, disaggregating the age-specific scores regionally and by race from 1971 to 1992, which gives approximately 84 data points. Regional disaggregation is the lowest level of disaggregation allowed by the NAEP sample design.

This method depends on regional and time series differences occurring in the independent variables. Figures 13 and 14 show the regional pattern of change over time for the two variables of current interest. Our measure of desegregation for each year is the percentage of Black students in schools with 90% or more minority students in that region.[15] The data show that massive desegregation occurred in the South very rapidly over a 7-year period between 1968 and 1975. The trends in the other regions were more gradual and far less dramatic.

We believe a plausible hypothesis is that the potential effect of the dramatic desegregation in the South will be fully seen only in cohorts that attended desegregated schools during their entire school attendance up to the time of the test. This means that any differences in test scores caused by desegregation should be seen first in comparisons between 9-year-olds who took the test in 1971 (who entered school before desegregation occurred) and those who took it in 1975 (who were more likely to have experienced desegregated schools for all 4 years); for 13-year-olds taking the test in 1975, compared with those taking the test in 1980 and later; and for 17-year-olds taking the test in 1980, compared with those who took it in 1985 and later. This pattern of earlier gains for 9-year-old Black students and later gains for 13- and 17-year-old Black students is approximated by the data in Figures 7, 8, and 9. Our analysis may provide an initial hypothesis for this pattern of gains.

We use one measure of pupil:teacher ratio aggregated regionally (see Figure 14). The data show that class size decreased more rapidly

[15] Desegregation data are based on data collected by the Office of Civil Rights, U.S. Department of Education.

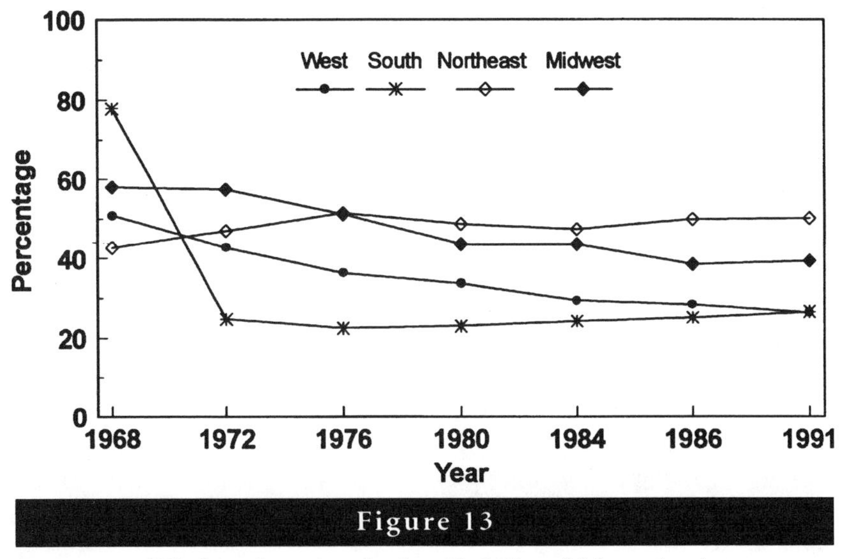

Figure 13

Percentage of Black students in schools with 90% or higher minority-group enrollment by region.

in the 1970s than in the 1980s and that the regional patterns were somewhat different. In particular, the data show increases in class sizes in the West, mostly due to large class size increases in California beginning in the 1980s. However, the largest declines in class size occurred in regions that have a greater proportion of Black students. These variations regionally and over time can allow testing of hypotheses with respect to NAEP scores; again, the data must be reestimated to obtain average pupil:teacher ratios over the school career by region for each NAEP age group.

We use pupil:teacher ratio rather than per-pupil expenditure for several reasons. First, experimental research has shown that class sizes have significant effects on student achievement. A carefully designed experiment revealed significant positive effects of lower class sizes in the early grades (Mosteller, 1995). Reducing class sizes from about 22 to about 14 resulted in test-score increases overall of approximately

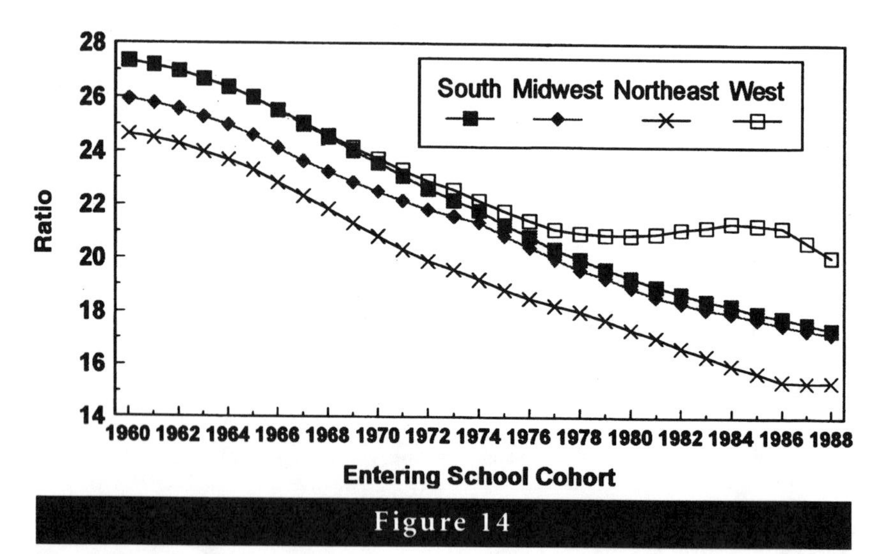

Figure 14

Pupil:teacher ratio by region.

0.20–0.25 *SDs* by Grade 4. Effects for minority-group students in the first two grades were twice that: 0.40–0.50 *SDs*.

Second, the effect of per-pupil expenditures depends largely on how the money is spent. Without more detailed data and given the aggregated level of our analysis, we cannot hope to capture the true effect of such expenditures. For instance, hiring more teachers to reduce class size may be more effective than raising teacher salaries in increasing student achievement.

Third, in a recent paper, Rothstein & Miles (1995) showed that the commonly used per-pupil expenditure data adjusted with the consumer price index overstates the increase in educational expenditures for regular students between 1967 and 1991 by between 60 and 75%. Therefore, pupil:teacher ratio may be a better measure of actual increased resources than is expenditure data.

Our simple regression analysis of the race-by-region data shows that desegregation is statistically significant and that pupil:teacher ratio is highly statistically significant for Blacks' scores—and in the expected direction. For White students, the pupil:teacher ratio is highly significant, although the coefficient is less than one half that for Blacks. The coefficients for pupil:teacher ratio are approximately equal to those of the experimental results cited earlier. For Whites, the desegregation coefficient, although negative, is statistically nonsignificant, suggesting that desegregation had little or no effect on the achievement of White students. This preliminary analysis provides some empirical evidence that the timing and regional pattern of desegregation and reduced class size are consistent with a hypothesis that the investment in schools and equal opportunity policies may be partly responsible for increasing Black students' test scores.

CONCLUSION

There is little doubt that achievement test scores for Black children increased greatly between the early 1970s and 1992, whereas scores for non-Hispanic White students registered small gains. The Black students' gains were made in cohorts starting school beginning in the late 1960s and continuing until the late 1970s. The gains were for both reading and math, have persisted across age groups, and have been sustained for the most part. The score increases were somewhat larger in the South, but they occurred across all regions. Cohorts entering school after 1980 have not made further gains, and some evidence points to slight deterioration in scores for those entering after 1980. The Blacks' gains cannot be accounted for by estimated gains from changes in family characteristics.

There is also little doubt that the family, school, and community environments improved for Black students in this period. National policies to desegregate schools, address poverty, and increase investment in education were implemented in this period. Our initial regression analyses of two factors suggest that the timing and regional pattern of the score gains were consistent with the timing and regional pattern of desegregation in schools and of the decline in class size. Clearly, much work needs to be done to refine and extend these analyses. However, it is difficult to suggest hypotheses that could account for such dramatic and relatively rapid gains in scores other than those concerning the variables connected with the additional government resources and changing policies of this era.

Our analysis also suggests that directing resources toward children from multiple risk families may be much more effective than directing them toward children from families with only a single or no risk. Part of the dramatic gains in Blacks' scores may have occurred because large numbers of Black children in the 1960s were in multiple risk families, so that policies that removed some of this risk had unexpectedly large gains. These data certainly would be consistent with a partial explanation of the Flynn effect arising from changes in policies and economic conditions that occurred in Western countries throughout the century. What occurred in the United States for minority-group populations in the 1970s and 1980s was a relatively rapid shift in resources and policies in an attempt to compensate for previous economic and social discrimination. These changes are probably not unlike what occurs more slowly for entire populations of nations and thus might be expected to lift scores gradually for whole countries.

REFERENCES

Berends, M., & Koretz, D. M. (1996). Reporting minority students' test scores: How well can the National Assessment of Educational Progress account for differences in social context? *Educational Assessment, 3,* 249–285.

Borjas, G. J. (1990). *Friends or strangers: The impact of immigrants on the U.S. economy.* New York: Basic Books.

Bronfenbrenner, U., & Ceci, S. J. (1994). Heredity, environment, and the question "How?"—A first approximation. In R. Plomin & G. E. McClearn (Eds.), *Nature, nurture, and psychology* (pp. 313–324). Washington, DC: American Psychological Association.

Dunn, J., & Plomin, R. (1990). *Separate lives: Why siblings are so different.* New York: Basic Books.

Flynn, J. R. (1987). Massive IQ gains in 14 nations: What IQ tests really measure. *Psychological Bulletin, 101,* 171–191.

Fuchs, V. R., & Reklis, D. M. (1992). America's children: Economic perspectives and policy options. *Science, 255,* 41–45.

Gomby, D. S., Larner, M. B., Stevenson, C. S., Lewit, E. M., & Behrman, R. E.

(1995, Winter). The Tennessee study of class size in the early school grades. *Future of Children, 5*(2). (Available from the Center for the Future of Children, The David and Lucille Packard Foundation, Los Altos, CA.)

Grissmer, D. W., Kirby, S. N., Berends, M., & Williamson, S. (1994). *Student achievement and the changing American family* (Rep. No. MR-488-LF). Santa Monica, CA: RAND Corporation.

Herrnstein, R. J., & Murray, C. (1994). *The bell curve: Intelligence and class structure in American life.* New York: Free Press.

Johnson, E. G., & Carlson, J. E. (1994, July). *The NAEP 1992 technical report.* Washington, DC: U.S. Department of Education, National Center for Education Statistics.

Koretz, D. (1987). *Educational achievement: Explanations and implications of recent trends.* Washington, DC: U.S. Congressional Budget Office.

Masten, A. S. (1994). Resilience in individual development: Successful adaptation despite risk and adversity. In M. C. Wang & E. W. Gordon (Eds.), *Education resilience in inner-city America: Challenges and prospects* (pp. 3–25). Hillsdale, NJ: Erlbaum.

Mosteller, F. (1995, Summer/Fall). The future of children: Long-term outcomes of early childhood programs. *Future of Children, 5*(2). (Available from the Center for the Future of Children, The David and Lucille Packard Foundation, Los Altos, CA.)

Rolf, J., Masten, A. S., Cicchetti, D., Nuechterlein, K. H., & Weintraub, S. (1990). *Risk and protective factors in the development of psychopathology.* Cambridge, England: Cambridge University Press.

Rothstein, R., & Miles, K. H. (1995). *Where's the money gone? Changes in the level and composition of education spending.* Washington, DC: Economic Policy Institute.

Rutter, M. (Ed.). (1988). *Studies of psychosocial risk: The power of longitudinal data.* Cambridge, England: Cambridge University Press.

Sameroff, A. J., Seifer, R., Zax, M., & Greenspan, S. (1987). IQ scores of 4-year-old children: Social environment risk factors. *Pediatrics, 79,* 343–350.

Werner, E. E., & Smith, R. S. (1982). *Vulnerable but invincible: A study of resilient children.* New York: McGraw-Hill.

11

The Shrinking Gap Between High- and Low-Scoring Groups: Current Trends and Possible Causes

Stephen J. Ceci, Tina B. Rosenblum, and Matthew Kumpf

In *The Bell Curve*, Herrnstein and Murray (1994) argued that a cognitive divergence is taking place in the United States. As a result of a genetically induced dysgenesis, they argued, the nation is fast becoming a bipolar society in which the offspring of high-IQ, mostly White, professional parents attend elite colleges and universities and end up marrying the offspring of other high-IQ professionals.[1] According to these authors, the offspring of these pairings are statistically more likely to possess higher IQs than are the offspring of low-IQ pairings. In addition, because low-IQ parents tend to start childbearing at an earlier

[1] Elsewhere in *The Bell Curve*, Herrnstein and Murray (1994) argued that *any* increase in the gap between high- and low-scoring groups is dysgenic, whether the source is genetic or environmental, because the result is equally negative in terms of widening the distribution of scores in the next generation: "We refer to this downward pressure (on IQ scores) as *dysgenesis* . . . it is important not to be sidetracked by the role of genes versus the role of the environment. . . . If women with low scores are reproducing more rapidly than women with high scores, the distribution of scores will, other things equal, decline, no matter whether the women with low scores came by them through nature or nurture" (p. 342). However, on principled grounds it is useful to distinguish between changes owing to genetic trends and those that come about as a consequence of environmental trends, for any lowering of subsequent generations' IQ distributions will have quite different remediation strategies under the two causes (e.g., if a widening of the IQ gap is due to environmental causes, a solution might focus on increasing access to education and financial resources, as opposed to deterring pregnancy among low-IQ groups).

age than do their high-IQ counterparts, they end up having greater numbers of children *across* generations (this is true even if the actual number of offspring remains the same in both groups *within* a given generation). As a consequence of the well-established 15-point racial asymmetry in IQ scores, the negative demographic effect noted by Herrnstein and Murray is claimed to affect Blacks, Latinos, and the socioeconomically disadvantaged disproportionately because of their earlier onset of childbearing. Following are the words of Herrnstein and Murray (1994):

> Mounting evidence indicates that demographic trends are exerting downward pressure on cognitive ability in the U.S. and that these pressures are strong enough to have social consequences.... Blacks and Latinos are experiencing even more severe dysgenic pressures than whites, which could lead to further divergence in future generations.... Putting the pieces together, something worth worrying about is happening to the cognitive capital of the country. (p. 341) The effect is dysgenic when a low-IQ group has babies at a younger age than a high-IQ group.... In the United States women of lower intelligence have babies younger than women of higher intelligence. (p. 351) The higher fertility rates of women with low IQs have a larger impact on the black population than on the white. The discrepancies are so dramatically large that the probability of further divergence seems substantial. (p. 353–354)

In this chapter, we discuss the issue of whether a negative demographic effect on intelligence has been taking place and, if it has, whether it promises to continue into the next century. To address the first part of this dual question requires that one start by asking what counts as evidence of a negative demographic effect on intelligence. Following this, we examine the claim that various groups' IQ scores are becoming more disparate, particularly racial and socioeconomic groups.

THE ESTABLISHED FACTS

There is no dispute among psychometric researchers that Whites outscore Blacks on IQ tests as well as on standardized achievement tests. The gap most commonly reported is approximately 1 *SD*. (On the most widely used individual IQ tests, this translates into a 15- to 16-point gap between Blacks and Whites; Hispanics fall midway between these groups, and Asian Americans score about 3 points, on average, higher than Whites.) Racial and ethnic gaps in IQ and achievement tests scores have existed throughout this century; for example, IQ differences between Blacks and Whites were evident on the first Stanford–Binet IQ test normed in 1932. Even earlier signs of a racial gap of approximately 1 *SD* were apparent on the Army Alpha tests administered to recruits during World War 1 (Loehlin, Lindsay, & Spuhler, 1975). These facts are not in dispute among researchers, although their interpretation is open to argument.

We attempt to show in this chapter that there are ample grounds for disagreeing with the claim that the racial gap in IQ is genetically driven, immutable, or widening, especially the last of these claims, which lies at the core of Herrnstein and Murray's (1994) expressed worry—that the gap is widening. The data we recruit to make our case against Herrnstein and Murray's claims generally are achievement test scores reported by Grissmer, Williamson, Kirby, and Berends (chapter 10, this volume), Hauser (chapter 9), and Huang and Hauser (chapter 12), rather than IQ test scores per se. Later, we justify the decision to focus on achievement test scores instead of traditional IQ test scores.

In the remainder of this chapter, we address four questions: (a) What do achievement test scores tell about IQ changes among members of various racial and socioeconomic groups? (b) What are the trends in achievement test scores for racial and socioeconomic groups in the United States? (c) How can one explain the changes? (d) Why is it necessary to think in terms of a galaxy of factors to explain them? We start with the first two of these questions, which are interrelated.

ACHIEVEMENT VERSUS IQ

Our first contention is that trends in achievement test scores are close reflections of trends in IQ scores. Indeed, one is hard pressed to find one trend without the other. Regardless of how one conceptualizes "intelligence" and "achievement," the empirical reality is that trends in one mimic trends in the other (see a review by Neisser et al., 1996). Another way of stating this assertion is to point out that the theoretical distinctions that some make between intelligence and achievement are immaterial to the empirical reality that a good measure of one almost always is highly correlated with a good measure of the other. Psychometric researchers frequently use one score as a stand-in for the other, even though the content of IQ tests may at times appear quite dissimilar from that of achievement tests (e.g., matrices, mazes, or puzzles vs. paragraphs about political ideas or problems requiring scientific analysis).

If one doubts the assertion that IQ scores are highly correlated with a reliable measure of achievement, one can do the following experiment. Administer two batteries of tests to a random, stratified sample: one, an achievement battery (e.g., tests of mathematics, reading comprehension, or scientific analysis), and the other, a widely used IQ battery (e.g., the 10 subtests from the Wechsler Adult Intelligence Scale–III–Revised). Next, distill from the achievement test battery a single summative score that captures the covariances among the various math, reading, and science scores. This is traditionally accomplished by taking the first principal component from the correlation matrix.[2] This summary score of the sample's achievement test scores will be closely related to the IQ scores for the same sample (Ceci, 1996). It is rare to find a measure of achievement that does not closely correlate with an IQ score. Generally, the correlation between scores on a battery of diverse achievement tests and IQ scores for the same sample approaches the correlation of IQ with itself, when corrected for reliability (i.e., the .8 –.9 range). This level of correlation justifies our decision to focus on

[2]This is operationalized as the maximum (linear) variance that can be accounted for, independent of any type of factor rotation, in the matrix of correlations among the various test scores.

achievement test scores as harbingers of IQ trends. If there are trends for achievement batteries, one can confidently assume that IQ scores on the sample would follow similar trends.

To some, the close relationship between tests that are avowedly achievement based, including subject matter that is explicitly taught in schools (e.g., science, reading, and math tests) and so-called intelligence tests, is attributed to the role that intelligence is alleged to play in achievement: Students with high IQ achieve more in school precisely because they possess sufficient intelligence to do so. To others, however, this close relationship between achievement and intelligence reflects the manufacturing of hard distinctions where none may exist; IQ is claimed to be a form of achievement, responsive to the same constellation of variables (schooling, parenting, genetics) that influence other forms of achievement. Whatever one's view on this topic (we are "agnostic"), the empirical reality allows the use of one as a statistical surrogate for the other in the confidence that the two trends mimic each other closely. Therefore, in what follows we challenge Herrnstein and Murray's (1994) claim of a widening racial and socioeconomic gap in intelligence test scores by focusing on trends in achievement test scores. Elsewhere in this volume, these trends in achievement test scores have been ably documented by Hauser (chapter 9), Huang and Hauser (chapter 12), and Grissmer et al. (chapter 10). These researchers have independently arrived at highly similar analyses and conclusions to ours, as we demonstrate.

RACIAL TRENDS IN ACHIEVEMENT TEST SCORES

As already mentioned, 1 *SD* (15–16 IQ points) separates American Blacks and Whites, 0.2 *SD*s (3 IQ points) separates East Asians and Whites, the former scoring higher. On batteries of achievement tests, a similar racial–ethnic gap has existed, that is, until recent times. As seen in the chapters by Hauser and Grissmer et al. (chapters 9 and 10), a gap of approximately 1 *SD* existed for achievement test scores, which showed up across a broad variety of tests as recently as 1970; however, these authors take issue with Herrnstein and Murray's (1994) claim of a negative downward trend. Although the racial gap has stubbornly held at 1 *SD* as long as records have been kept, there began to appear signs in the early 1970s of a turnaround of sorts, or convergence between the races in achievement test scores. Black students closed between one third and one half the gap in achievement test scores with White students in the course of little more than 15 years! By 1986, the gap between Whites and Blacks on one of the most respected achievement test batteries (the National Assessment of Educational Progress, NAEP) had dwindled to roughly half its magnitude of 1970, depending on the particular test used and the age group studied. By any measure, this is an enormous improvement, and it is due to gains by Black students rather than losses of White students.

By the late 1980s, all indications of test score convergence between Black and White students had ceased and a slight divergence in their scores began again. Because there are multiple achievement tests on the NAEP (math, science, reading, and writing) as well as multiple age groups taking these tests (9-, 13-, and 17-year-olds) and multiple grades that cross-cut each age group (e.g., on some trend data eighth graders include not only 13-year-olds but somewhat older and younger students as well), a detailed picture of trends is complicated because the magnitude of racial convergence varies somewhat by the specific achievement test, grade, and age group discussed. Summing across all of these categories, it appears that the racial convergence in test scores ceased by the late 1980s. The trend to converge does not appear to have continued in the 1994 NAEP trend data, which came out in November, 1996. These latest NAEP trend data (National Center for Education Statistics, 1996) show no systematic further changes in racial means, if we restrict the discussion to statistically reliable differences. (After 1988, however, there is hint of a potential divergence in the scores of Blacks and Whites; however this suggestion of divergence has not yet reached statistically significant levels.) Recently, Williams and Ceci (1997) summarized these newest analyses as follows.

> Comparing the gap between Black and White students' scores in these most recent NAEP data with those in the 1990 report revealed that 8 of the 12 trends showed a slight divergence be-

> tween White and Black students' test scores but that these differences are not statistically reliable. (In addition, the remaining 4 of the 12 contrasts showed signs of further convergence, but these also were not significant.) . . . Clearly, the next release of NAEP data in 1998 will be extremely interesting: Will the hint of a potential reversal (not statistically significant as of yet) apparent in the 1994 data continue and reach conventional levels of statistical significance? Will there be a reversal of the gains made during the 1970s and the 1980s? These questions are ones to which educators and policymakers should be attuned. . . . If the trends in IQ test scores were to mimic the achievement test score trends just described, one would expect a similar reduction in the historically stubborn racial intelligence-score gap, indexed by the two most widely used IQ tests (i.e., the Stanford–Binet and the Wechsler series). Specifically, Black students (who showed the greatest gains on the NAEP achievement test scores in the 1970s and 1980s) might also exhibit a comparable gain in IQ, thus closing the one-standard deviation gap by approximately one half. . . . In contrast, because the gains made by Blacks ended by 1988, and actually have shown signs of reversing direction since then, it might be expected that the later cohort's IQ scores will show a commensurate gap-widening. Again, this assumes that trends in achievement test scores mimic trends in IQ scores, regardless of the conceptual distinctions one wishes to draw between achievement and intelligence. (pp. 1229–1230)

Even Herrnstein and Murray documented signs of a convergence. They mentioned three studies that showed that the Black–White IQ gap seemed to be converging, not diverging,[3] and they noted further that in their own analysis of the NAEP data, they, too, found a convergence of math and verbal fluency measures of Blacks and Whites,

[3]However, they attempted to moderate this conclusion by noting statistical concerns about one of the three studies.

especially for the 17-year-olds. Although the magnitude of the convergence in their analysis was less than what Grissmer et al. and Hauser reported (this volume),[4] there is no denying that for all nine achievement tests included in the table on page 291 of *The Bell Curve* (i.e., science, math, and verbal fluency, for each of three age groups, 9-, 13-, and 17-year-olds), there was convergence, not divergence. Herrnstein and Murray acknowledged this result but then proceeded to gainsay it in the following way:

> As the table indicates, black progress in narrowing the test score discrepancy with whites has been substantial on all three tests and across all of the age groups. The overall average gap of .92 standard deviation in the 1969–1973 tests had shrunk to .64 standard deviation by 1990. The gap narrowed because black scores rose, not because white scores fell. Altogether, the NAEP provides an encouraging picture. (p. 291) The question that remains is whether black and white test scores will continue to converge. If all that separates blacks from whites are environmental differences and if fertility patterns for different socioeconomic groups are comparable, there is no reason why they shouldn't. The process would be very slow, however . . . reaching equality sometime in the middle of the twenty-first century. . . . If black fertility is loaded more heavily than white fertility toward low-IQ segments of the population, then at some point convergence may be expected to stop, and the gap could begin to widen again. (p. 293)

Herrnstein and Murray, troubled by the earlier onset of childbearing by Black teenagers, clung to a "dysgenesis" hypothesis even though there appears to be no scientific evidence in the direction of a continued divergence of test scores between Black and White youngsters. On the

[4]Herrnstein and Murray arrived at a somewhat different conclusion regarding the size of the convergence in the NAEP data, arguing that approximately a third of the racial gap had been closed by the late 1980s. As Hauser (this volume) correctly notes, Herrnstein and Murray arrived at their lower estimate of convergence by using the wrong measure of variance in their calculations. The appropriate (smaller within-group) standard deviations lead to the higher rates of convergence reported by both Grissmer et al. and Hauser.

basis of findings of Hauser and Grissmer et al. (chapters 9 and 10, this volume), if Black fertility is loaded downward, it is difficult to see why all indicators in the 1970s and 1980s pointed in the direction of a narrowing of the racial gap that was, as recently as 25 years ago, approximately twice as large as it is presently. It is interesting that Blacks' gains on achievement tests occurred predominantly during the time period that the teenage pregnancy rate of unmarried Black women rose dramatically (Bronfenbrenner, McClelland, Wethington, Moen, & Ceci, 1996).

WHAT DO TRENDS IN ACHIEVEMENT TEST SCORES REVEAL ABOUT IQ CHANGES?

On the basis of trends in achievement test scores reported by both Grissmer et al. and Hauser, we are led to predict a similar trend in IQ scores. Specifically, we anticipate that if the Black cohorts that showed the greatest gains on the NAEP achievement test scores in the 1970s and 1980s were administered IQ tests today, they would exhibit a comparable gain in IQ, closing the gap of 1 *SD* between Blacks and Whites by about half. (Again, note that the closing of the racial gap has not been the result of declining test scores of Whites, but of gains in the achievement test scores of Blacks.) These cohorts of high-scoring Blacks are easily identified (e.g., the 17-year-olds of 1986 would be about 29 years old today), so that it would not be hard to test our prediction of a racial convergence of IQ scores among the groups that showed the largest gains in achievement test scores in the 1970s and 1980s.

Furthermore, because the convergence in test scores of Blacks and Whites had run its course by the late 1980s and the scores had begun to diverge again, it should be possible to show that the IQ scores of the latter cohort (those who were in school in the late 1980s and early 1990s) were diverging again, in parallel with the resumption in the decline of Blacks' test scores around 1988. These predictions are based on our argument that trends in IQ scores and achievement test scores mimic each other, despite whatever conceptual distinctions one wishes to draw between them. If we are correct in this prediction, it will represent the first time this century that the stubborn 1-*SD* gap between Blacks and Whites has been bridged, leading to optimism that similar improvements might be possible if the conditions that fostered the gains of Black students in the 1970s and 1980s were reinstated.

CHANGES IN IQ SCORES OF DIFFERENT SOCIOECONOMIC GROUPS

We can add to Grissmer et al.'s and Hauser's important statistical evidence of convergence the documentation provided by Flynn of another form of convergence. Flynn (in press; see also 1984) showed that the claim of a widening gap in the IQ scores of the rich and poor is based more on illusion than reality. Following is an account of how he did this.

Herrnstein and Murray's (1994) dysgenesis–divergence hypothesis leads to the expectation of a growing tendency for "good" genes for IQ to rise to the top of the occupational scale and for "bad" genes to fall to the bottom. Specifically, it leads to the expectation that, like the racial gap, the IQ gap between the children of the upper and lower income groups has been diverging over time. Flynn's (in press) data come from the Stanford-Binet normative sample tested in 1932 and the Wechsler Intelligence Scale for Children (WISC) samples tested in 1948, 1972, and 1989; he analyzed the standardization samples used to norm these two tests between 1932 and 1989. He measured the difference between the mean IQs of children whose parents were in the top third versus the bottom third in terms of occupational status. For White students in the United States, this gap was over 12 points in 1932, fell to 10 points in 1948, and has not altered through the WISC-III sample of 1989. When the most recent Stanford–Binet and WISC-III samples are pooled, the gap stands at little more than 9 points.[5] Therefore, it appears

[5] Flynn (in press) showed that the gap between socioeconomic status (SES) groups for all races combined increased between 1972 and 1989 from 11.64 to 12.85 points. However, this is due to increased immigration plus an increase in non-Whites, which together doubled the number of non-Whites in the recent samples. Because non-Whites have a mean IQ that is lower than that of Whites and tend to be concentrated in low-status occupations, such a demographic trend automatically increases the all-races class IQ gap. This has nothing to do with the divergence thesis of *The Bell Curve,* which predicts a widening class IQ gap on the basis of dynamics within groups who have been in the United States throughout the century but were said to differ in genetic talent.

that the divergence hypothesis cannot be supported by the trends concerning SES and IQ.

CHANGES IN THE TOP AND BOTTOM GROUPS' SCORES

Finally, we are in the process of analyzing changes in Preliminary Scholastic Aptitude Test (PSAT) scores over the same period that the NAEP trends were collected, 1970–1990 (Williams & Ceci, 1997). Our interest is in whether another type of divergence has occurred, namely, whether the PSAT scores of the top and bottom quintiles of the distribution have widened. (Unlike the Scholastic Aptitude Test [SAT], the PSAT has been given to virtually all high school juniors since 1965, irrespective of their intentions to attend college, thus making it a superior source for this analysis.) Preliminary evidence indicates that the scores of the highest and lowest quintiles of students are not widening but, in fact, are converging. As one piece of supporting evidence, in 1961 the verbal gap between the top 25% and bottom 25% of students' PSAT scores was 15.2 points; this same gap in 1995 was only 14.0 points. Similarly, for the math gap the trend was slightly downward, going from 15.34 to 15.00 points over the past 35 years. We have yet to determine whether this convergence is similar for all racial and economic groups; however, the finding that the best students have not outpaced the lowest scoring students in PSAT scores leads to the expectation that a similar convergence may be occurring between the IQ scores of the top and bottom scorers—again, because scores on achievement tests such as the PSAT and IQ tests are so highly correlated. We are fairly confident in dismissing the dysgenesis–divergence hypothesis, a central theme in Herrnstein and Murray's.

HOW CAN ONE EXPLAIN THE CONVERGENCES?

It is easier to say how one cannot explain the racial convergence, followed by divergence, than to say how one can explain it. Certainly one cannot explain the closing of the racial gap during the 1970s and early

1980s in terms of genetics, for the changes occurred too quickly for genetic shifts to be causal; often, the changes in scores occurred within a single generation, whereas genetic sources can operate only across generations. In addition, there is no plausible genetic force that could be invoked to have caused, first, a convergence (from the early 1970s through the mid-1980s), followed by a divergence. An increase in teenage parents among Blacks and Whites would, if anything, argue in the opposite direction, insofar as the White teenage birth rate crept steadily closer to the Black rate throughout the entire 1980s (Bronfenbrenner et al., 1996). But as we demonstrated, the White scores were fairly static, and the Black scores rose considerably over this period. Of course, it is possible that competing and potentially offsetting forces operated simultaneously over this period: one, a genetically based force with downward consequences, and the other, an environmental force with upward consequences (see chapter 13, this volume). In such a situation, it is not possible empirically to disentangle the elements. All of the observable data we have considered here reflect the net effect of such competing forces, which, as seen, is in the upward direction. To whatever degree there are genetic downward forces at work, therefore, there seems to be no reason to think the population of the United States is getting dumber or more divergent.

A number of factors could be involved in producing the convergence in test scores, each contributing to part of it, with no single factor explaining it all. Grissmer, Kirby, Berends, and Williamson (1994) focus on the increased educational spending that occurred over the period of rapid score increases by Black students, especially in programs that targeted minority children (e.g., Title 9, school lunch programs, busing). Over a period of roughly 25 years (1967–1991), there has been more than a 60% increase in real dollars for education (Rothstein & Miles, 1995), but only 25% of this spending found its way into regular or mainstream classrooms (i.e., much was mandated for special education programs). The spending that found its way into regular classrooms was probably on the order of a 1.0–1.5% increase per year in real dollars over a 25-year period. This is a nontrivial increase, however, and it leads to the conclusion that money did seem to matter, especially for children enrolled in programs that experienced the largest increases.

The desegregation of all-Black schools could also be involved in the rise in Black students' scores. For this factor to be important, of course, it would be necessary to show that there were not comparable test-scores gains for Black students who remained in segregated schools, a conclusion that has been rejected by some (see Armor, 1992).

Finally, there has been an enormous increase in parental educational attainment by Black and Hispanic parents over the same period in which there were rapid score gains by Black youngsters (Grissmer et al., 1994). It is well known that parental educational level is tied to children's educational attainment (Bronfenbrenner et al., 1996); if parents are getting better educated, and Black parents are disproportionately becoming better educated, one ought to expect an elevation of Black children's scores relative to those of other students. Cook and Evans (1997) and others have estimated that approximately one fourth of the racial gap was closed as a result of the particularly large gains made by Black parents in their educational levels over the period in question (i.e., over three fourths of the gap closing was not as a result of changes in families or schools). The remaining 75% of the gap closing could be due to a variety of factors that are as yet undisclosed.

CONSIDERING A GALAXY OF FACTORS

Both Hauser and Grissmer (chapters 9 and 10, this volume) have alerted scholars to the steady increase in Black children's test scores during the 1970s and 1980s. Moreover, they have sounded the alarm: The convergence seems to be waning with downward trends beginning to appear by the late 1980s in Blacks' test scores. No single factor can explain this rise and subsequent fall in test scores; there is a need to think in terms of a galaxy of factors. Some of these factors are valenced upward, whereas others are valenced downward. Unless one thinks in these terms, one is apt to become mired in seeming contradictions in comparisons of various racial groups and birth cohorts.

One factor that has been valenced upward during the 1970–1988 period is the mean number of years of schooling completed. Since the time that the first Stanford–Binet IQ test was normed in 1932, the mean educational attainment rose by over 4 years, essentially from an 8th-grade education to just beyond a 12th-grade education (Ceci, 1991). As the parents of the 1950s and 1960s became better educated, their children reaped the cognitive benefits, such as by being exposed to more complex words and syntax and to greater use of hypotheticals. Some of this increase was a continuation of the century-long trend toward increased education, but some reflected the greater availability of education following the G. I. Bill of Rights after World War 2.

At the same time that this upward trend in years of completed education occurred, there was a downward trend in the quality of that very education. The best evidence for this downward trend can be seen in Hayes, Wolfer, and Wolfe's (1996) recent analysis of the "dumbing down" of American textbooks following World War 2. These researchers showed that American (but not British) schoolbooks were systematically reduced in their difficulty immediately following World War 2. They gave reasons for this change, but for the present purposes they are not important. What is important is these authors' parallel finding that the same cohorts who were exposed to the "dumbed down" readers and science texts during their elementary and secondary school years later experienced a decline in verbal SAT scores (but not math SATs). The decline was fairly specific; it affected the cohorts in an almost linear fashion such that for each year a cohort was exposed to dumbed down books, that cohort lost additional SAT points. This dumbing down of books affected the SAT patterns of even the highest scoring group, those with a verbal score of over 700; their numbers were diminished despite the greatly expanded pool of test takers.

It is only through a juxtaposition of the downward and upward valenced factors that one can understand why, on balance, Black and Hispanic students closed nearly half of the test-score gap that separated them from White students. Sometimes the two valenced sets of factors coexisted even though the positive ones more than compensated for the negative ones. It is only by simultaneously thinking about negative and positive factors that one can explain the turnabout in Black scores in the late 1980s (e.g., there appeared to be large social changes for this cohort in poverty rates).

Other factors, both beneficial as well as baleful to the development of IQ, can be identified during this epoch of asymmetrical test-score rise among racial groups. Poverty rates for children under the age of 6 skyrocketed during the 1980s, although rates held steady for other birth cohorts and even fell for some (Bronfenbrenner et al., 1996). Divorce rates, illegitimacy rates, and high school dropout rates affected some racial groups and birth cohorts more than others (e.g., despite the upward trend in completing high school among White and Black students, Hispanic students lagged significantly). If social scientists hope to explain past shifts in IQ among some groups and predict future trends, it will be important to think in terms of multiple factors that yield a net score that may differ for different groups over time.

REFERENCES

Armor, D. (1992, Summer). Why is Black educational achievement rising? *The Public Interest.*

Bronfenbrenner, U., McClelland, P., Wethington, E., Moen, P., & Ceci, S. J. (1996). *The state of Americans: This generation and the next.* New York: Free Press.

Ceci, S. J. (1991). How much does schooling influence general intelligence and its cognitive components?: A reassessment of the evidence. *Developmental Psychology, 27,* 703–722.

Ceci, S. J. (1996). *On intelligence: A bioecological treatise on intellectual development* (Expanded ed.). Cambridge, MA: Harvard University Press.

Cook, M. D., & Evans, W. N. (1997). *Families or schools? Explaining the convergence in White and Black academic performance.* Unpublished manuscript, U.S. Naval Postgraduate College.

Flynn, J. (1984). Banishing the spectre of meritocracy. *Bulletin of the British Psychological Society, 37,* 256–259.

Flynn, J. (in press). IQ trends over time: Intelligence, race, and meritocracy. In S. Durlauf, K. Arrow, & S. Bowles (Eds.), *Meritocracy and equality.* Princeton, NJ: Princeton University Press.

Grissmer, D. W., Kirby, S. N., Berends, M., & Williamson, S. (1994). *Student achievement and the changing American family* (Rep. No. MR-488-LF). Santa Monica, CA: RAND Institute on Education and Training.

Hayes, D. P., Wolfer, L. T., & Wolfe, M. F. (1996). Schoolbook simplification and its relation to the decline in SAT-verbal scores. *American Educational Research Journal, 33,* 1–18.

Herrnstein, R. J., & Murray, C. (1994). *The bell curve: Intelligence and class structure in American life.* New York: Free Press.

Loehlin, J. C., Lindsay, G., & Spuhler, J. (1975). *Race differences in intelligence.* New York: Freeman.

National Center for Education Statistics. (1996). *National assessment of educational progress (NAEP), 1994 long-term trend assessment.* Washington, DC: U.S. Department of Education.

Neisser, U., Boodoo, G., Bouchard, T. J., Boykin, A. W., Brody, N., Ceci, S. J., Halpern, D. F., Loehlin, J. C., Perloff, R., Sternberg, R. J., & Urbina, S. (1996). Intelligence: Knowns and unknowns. *American Psychologist, 51,* 77–101.

Rothstein, R., & Miles, K. H. (1995). *Where's the money gone? Changes in the level and composition of educational spending.* Washington, DC: Economic Policy Institute.

Williams, W. M., & Ceci, S. J. (1997). Are Americans becoming less alike? Trends in race, class, and ability differences in intelligence. *American Psychologist, 52,* 1226–1235.

12

Trends in Black–White Test-Score Differentials: II. The WORDSUM Vocabulary Test

Min-Hsiung Huang and Robert M. Hauser

There are a great many studies of the cognitive test performance of Blacks and Whites, but few can be used with confidence to measure differential performance trends (chapter 9, this volume). One key resource, the series on Black and White schoolchildren from the National Assessment of Educational Progress, began only about 1970. Another series of data on verbal ability, which also began in the 1970s, provides clues about much longer running trends in Black–White differences. These data are from the General Social Survey (GSS) of the National Opinion Research Center (NORC), which has regularly administered the same 10-item verbal ability test (WORDSUM) to adult household members of all ages since the early 1970s (Davis & Smith, 1994). Because there are repeated, usually annual administrations of WORDSUM to cross-sectional samples of about 1,000, it is possible to identify age and cohort effects on performance by assuming that there are no period effects. Using the GSS data, we can estimate trends in verbal ability among Blacks and Whites who were born from 1909 to 1974. These data show a consistent pattern of convergence between the test scores of Blacks and Whites throughout the period covered by the GSS cohorts. The convergence is not fully explained either by changes in social background or by changes in educational attainment across cohorts.

The 10 GSS vocabulary items were chosen from "Form A," one of two parallel, 20-item vocabulary tests selected by Thorndike. Each form contained two vocabulary test items from each of the levels of the vocabulary section of the Institute for Educational Research Intelligence Scale: Completion, Arithmetic Problems, Vocabulary, and Directions (R. L. Thorndike, 1942). Form A was developed by Thorndike in response to the need for a brief test of intelligence that could be administered in social surveys (R. L. Thorndike & Gallup, 1944), and it was also used in an attempt to study the feasibility of an aptitude census (R. L. Thorndike & Hagen, 1952). Form A was later used by Miner (1957) in his monograph *Intelligence in the United States*, in which he attempted to assess the intellectual ability of the U.S. population using a national household sample. Alwin (1991) used the GSS WORDSUM data from 1974 to 1990 in an analysis that demonstrated that changes in family configuration could not account for the decline of verbal ability in the Scholastic Aptitude Test (SAT-V).

For each of 10 WORDSUM items, GSS respondents are asked to choose the one word out of five possible matches that comes closest in meaning to the word in capital letters. Figure 1 gives a set of sample items that are similar to those in WORDSUM.[1] The GSS conducts personal interviews, during which each item is handed to the respondent on a preprinted card. Before 1988, WORDSUM was administered to the full GSS sample, but only every other year. Since 1988, it has been administered to two thirds of the sample in each survey year in an alternate-forms design. From 1974 to 1994, WORDSUM was completed by 11,160 Whites and 1,418 Blacks who were between the ages of 20 and 65 at the survey date and who also provided valid data concerning years of schooling, number of adults, number of siblings, and structure of the family of orientation.

Miner (1957) argued that vocabulary tests are highly correlated with

[1]We have selected these items at random from the list provided by Miner (1957, p. 53). The National Opinion Research Center has requested that we not reveal the specific items used in WORDSUM.

Item	1	2	3	4	5
a. LIFT	1. sort out	2. raise	3. value	4. enjoy	5. fancy
b. CONCERN	1. see clearly	2. engage	3. furnish	4. disturb	5. have to do with
c. BROADEN	1. efface	2. make level	3. elapse	4. embroider	5. widen
d. BLUNT	1. dull	2. drowsy	3. deaf	4. doubtful	5. ugly
e. ACCUSTOM	1. disappoint	2. customary	3. encounter	4. get used	5. business
f. CHIRRUP	1. aspen	2. joyful	3. capsize	4. chirp	5. incite
g. EDIBLE	1. auspicious	2. eligible	3. fit to eat	4. sagacious	5. able to speak
h. CLOISTERED	1. miniature	2. bunched	3. arched	4. malady	5. secluded
i. TACTILITY	1. tangibility	2. grace	3. subtlety	4. extensibility	5. manageableness
j. SEDULOUS	1. muddled	2. sluggish	3. stupid	4. assiduous	5. corrupting

Figure 1

Illustrative vocabulary test items. Adapted from data of Miner (1957, p. 53).

tests of general intelligence. He reviewed about 36 studies in which a vocabulary measure had been correlated with a measure of general intelligence and found a median correlation of .83. Miner also noted that whereas Wechsler had originally excluded a vocabulary test from the Wechsler–Bellevue scales because "he felt it might be unfair to illiterates and those with a foreign language background" (pp. 28–30), he later decided that it was an excellent measure of general intelligence. Finally, Miner noted that the median correlation of verbal tests with full-scale IQ tests is larger than the median correlation Wechsler reported between the Wechsler–Bellevue full-scale IQ and 15 other measures of general intelligence (Wechsler, 1944). More recently, Wolfle (1980, p. 110) reported that the correlation between the GSS vocabulary test and the Army General Classification Test (AGCT) was .71.

Despite these indications of validity, we urge caution in the use of WORDSUM. First, the test is very short, so its reliability is low. The internal consistency reliabilities are .712 among Whites and .628 among Blacks. In the 1994 GSS, in addition to WORDSUM half the sample was administered 8 of the 14 similarity (abstract reasoning) items from the Wechsler Adult Intelligence Scale–Revised (WAIS-R; Wechsler, 1981).[2] The correlations between WORDSUM and the WAIS-R similarity scores were .394 for Whites and .409 for Blacks, without correction for unreliability. After correction for internal consistency reliability, the correlations were .589 among Whites and .608 among Blacks. Second, as we discuss later in more detail, despite the common use of vocabulary tests in IQ instruments and the high correlations between tests of vocabulary and of general intelligence, there is evidence of divergent trends between IQ and verbal ability in the past several decades. Therefore, even if verbal ability is a valid indicator of general intelligence, neither the overall trend in WORDSUM nor the specific trends in WORDSUM among Blacks and Whites need follow the trends found in more general tests of ability.

Figure 2 shows the distributions of vocabulary test scores for Blacks

[2]Hauser and Mare (1993) added the similarity items in order to calibrate them against WORDSUM for use in supplementary telephone interviews with brothers and sisters of the GSS respondents.

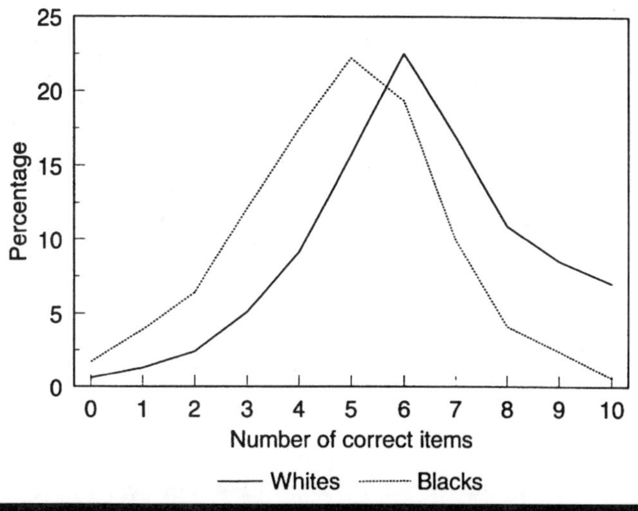

Figure 2

WORDSUM distributions of Blacks and Whites: General Social Survey, 1974–1994.

and Whites in the GSS from 1974 to 1994. The mean numbers of correct items are as follows: M_w = 6.23 (SD_w = 2.09) for Whites, and M_b = 4.78 (SD_b = 1.93) for Blacks. From the display of relative frequencies, it is clear that there is a ceiling effect on the scores for Whites in the general population. Similarly, there are ceiling or floor effects on the number of correct answers in other subpopulations (e.g., persons with many or few years of schooling). For this reason, our analyses of the WORDSUM data are based on a two-sided tobit specification, which compensates for censoring at both ends of the distribution on the assumption that the true distribution of test scores is Gaussian (Maddala, 1983).

METHODOLOGICAL ISSUES

Before attempting a trend analysis of the GSS data, we carried out several methodological analyses. Only 460 GSS respondents (3.7%) refused to answer any of the WORDSUM items, and we ignored these participants' data throughout the analysis. Those refusing had completed slightly fewer years of schooling (M = 10.5 vs. 12.9 years), but they were also more variable in schooling (SD = 3.6 vs. 2.8 years). We considered whether item nonresponses—other than complete refusal—should be treated as incorrect answers or whether total scores should be adjusted for the number of items answered. We compared the internal consistency reliabilities among complete responses and among incomplete responses, treating nonresponses in the latter group as errors. These were virtually the same, .62 and .64. Furthermore, the correlations between educational attainment and WORDSUM score among the complete versus incomplete responses were identical. Therefore, we assumed that item-specific nonresponses were erroneous responses.

We also experimented with a range of corrections for guessing by assuming that respondents who missed easy items may have answered harder items correctly by guessing. Under each guessing scenario, we recoded the harder "correct" answers as erroneous and examined the correlations of WORDSUM scores with educational attainment, occupational status (Stevens & Cho, 1985; Stevens & Featherman, 1981), and, for 1994 GSS respondents only, scores on the WAIS-R similarity items. We found that the correlations with external criteria were lowered by corrections for guessing, so we did not introduce such a correction.

Finally, we carried out logistic regression analyses of each item, by race and by birth cohort, to look for differential item functioning (DIF). We carried out each analysis with and without adjustment of total test scores for unreliability, and in each case we controlled age at test administration. In the absence of DIF, the odds of a correct response to an item would be the same for two groups at every ability level, as indicated by the total test score. Uniform DIF occurs when the odds of answering an item are equally greater or equally lower at every ability level. Nonuniform DIF occurs when the difference between groups in the odds of a correct answer are not the same across ability levels (Swaminathan & Rogers, 1990).

In looking for DIF, we modified the usual model by including dummy variables for age at testing as covariates. For Whites, we based the intercohort comparisons on 10-year birth cohorts from 1910–1919 to 1960–1969. Among Blacks, we collapsed the 1910–1929 birth cohort to increase sample size. The usual logistic regression model for predicting the probability of a correct response to an item is

$$P(u = 1) = e^z/(1 + e^z), \quad (1)$$

where

$$z = \tau_0 + \tau_1\theta + \tau_{2j}w_j + \tau_{3j}(\theta \times w_j). \quad (2)$$

The variable w_j indicates membership in the jth group, and θ is the observed ability of an individual. In this case, θ is the respondent's total WORDSUM score, w_j is the respondent's birth cohort, and $\theta \times w_j$ is the product of the two independent variables, w_j and θ. An item shows uniform DIF if $\tau_{2j} \neq 0$ for some j and $\tau_{3j} = 0$ for all j, and it shows nonuniform DIF if $\tau_{3j} \neq 0$ for any j, whether or not $\tau_{2j} = 0$. Because WORDSUM varies with age at testing, we also included dummy variables for respondent's age in the regression. Thus, the model becomes

$$z = \tau_0 + \tau_1\theta + \tau_{2j}w_j + \tau_{3j}(\theta \times w_j) + \tau_{4i}x_i, \quad (3)$$

where x_i is a dummy variable for membership in the ith 10-year age group. We estimated this model separately for Whites and Blacks, and we looked for evidence that items had become easier or harder for successive cohorts. We used conventional statistical tests with $p = .01$, supplemented by the Bayesian information criterion (BIC; Raftery, 1995), and we also looked for systematic variation across cohorts in the parameters for DIF. By all standards, we found little evidence of nonuniform DIF, and in only one case, a relatively difficult item among Whites, did the Bayesian criterion indicate there was reliable evidence of uniform DIF. However, there were four items among Whites and two items among Blacks in which there was nominally significant uniform DIF. Among Blacks, both items had become successively more difficult, and among Whites, two items had become more difficult and two had become less difficult.

We also estimated a model similar to Equation 3, in which we pooled the data for Blacks and Whites and specified effects of racial–ethnic group corresponding to those of cohorts in Equation 3. In comparing Blacks and Whites, when total WORDSUM scores were controlled, a few items were slightly easier for Whites than for Blacks with the same WORDSUM score. After adjustment for internal consistency reliability, those effects were reversed: Five items were significantly easier for Blacks than for Whites with the same estimated true scores. One other item is then significantly easier for low-scoring Blacks and significantly harder for high-scoring Blacks. There does not appear to be any relationship between item difficulty and DIF. Given these equivocal findings, it would probably be a good idea to test the sensitivity of our regression analyses to the elimination of selected items. Because of the small number of items in WORDSUM, we have not done so. However, these findings provide yet more reason for caution in the interpretation of WORDSUM trends.

Finally, we looked for external evidence that the difficulty of the WORDSUM items may have changed across time. Between 1921 and 1967, four studies were published that contained frequency counts and ranks for English words (Kučera & Francis, 1967; E. L. Thorndike, 1921, 1931; E. L. Thorndike & Lorge, 1944). Unfortunately, the several sets of rankings are of uncertain comparability. E. L. Thorndike (1921) compiled a list of the 10,000 most frequent words "in a count of about 625,000 words from literature for children; about 3,000,000 words from the Bible and English classics; about 300,000 words from elementary school text books; about 50,000 words from books about cooking, sewing, farming, the trades, and the like; about 90,000 words from the daily newspapers; and about 500,000 words from correspondence" (p. iii). E. L. Thorndike (1931) reported "extensive additional counts from over 200 other sources including about 5,000,000 words" (p. iii), and he extended the list to 20,000 words. E. L. Thorndike and Lorge (1944) added information from counts of an additional 4.5 million words. Kučera and Francis (1967) did not build on the work of Thorndike and Lorge but analyzed "a collection of statistical information obtained from analysis of *The Standard Corpus of Present-Day Edited American English*, a computer processible corpus of language texts assembled at Brown

University during 1963–64" (p. xvii). The corpus of more than 1,000,000 words was selected from the press (reportage, editorials, and reviews); books about religion, skills and hobbies, and popular lore; belles lettres and biography, learned and scientific writings, humorous works, and several categories of fiction writings. It appears that the Thorndike–Lorge series was cumulative, beginning with a focus on children's reading material and later extending to more general collections of text. For that reason, that series is not entirely appropriate for an assessment of trend. On the other hand, the Kučera–Nelson corpus is almost entirely made up of adult-oriented text, and it is therefore not strictly comparable to the Thorndike–Lorge series.

For each WORDSUM item, we looked up or estimated the approximate rank of each stimulus word, of its synonym, and in most cases of a plausible distracter. We had no basis for interpreting changes in the relative frequency of key words within a given item. Kučera and Francis (1967) identified the 50,000 most common words, and in many cases words that had appeared in the earlier lists were much less common in the 1967 list. We do not know whether differences between rankings in that list and the earlier lists are due to true temporal change or to other differences in the selection of text. In any event, we found that the ranks of WORDSUM items have been either stable or decreasing. That is, if frequency of usage in written text is an indicator of difficulty, WORDSUM has become somewhat more difficult across time, independent of any other change in verbal ability in the general population. That tendency occurs to some degree across the lists of 1921, 1931, and 1944, and it appears strongly when we include the 1967 list in the comparison. However, we do not believe that changes in item difficulty, if they are real, could account for decreasing differences between the test scores of Blacks and Whites.

TREND ANALYSES OF WORDSUM

We estimated three models of intercohort trends in verbal ability in the 1974–1994 GSS data. In each case, we analyzed the total number of correct answers to the 10 WORDSUM items using a two-sided tobit

Table 1

Baseline, Social Background, and Education Models of Vocabulary Test Scores: General Social Survey

	Baseline		Social background		Education	
Variable	Coefficient	*SE*	Coefficient	*SE*	Coefficient	*SE*
Intercept	3.86	0.32	3.86	0.30	5.33	0.28
Sex (Men = 1)	0.02	0.12	−0.02	0.11	0.03	0.10
Race (White = 1)	2.46	0.31	1.69	0.28	1.36	0.26
Age						
20–29	—	—	—	—	—	—
30–39	0.47	0.06	0.47	0.06	0.28	0.05
40–49	0.51	0.08	0.53	0.07	0.29	0.07
50–59	0.29	0.10	0.30	0.09	0.01	0.08
60–65	0.18	0.13	0.22	0.11	−0.07	0.11
Birth year						
1909–1919	—	—	—	—	—	—
1920–1929	0.21	0.35	−0.12	0.32	−0.41	0.29
1930–1939	0.61	0.33	0.27	0.30	−0.34	0.28
1940–1949	0.85	0.33	0.14	0.30	−0.77	0.28
1950–1959	0.63	0.33	−0.19	0.30	−1.13	0.28
1960–1969	0.65	0.34	−0.56	0.31	−1.41	0.29
1970–1974	1.37	0.49	−0.13	0.45	−0.87	0.41

Interaction of sex and race	−0.23	0.13	−0.18	0.11	−0.36	0.10
Interaction of birth year and race						
1909–1919	—	—	—	—	—	—
1920–1929	−0.38	0.37	−0.22	0.33	−0.05	0.30
1930–1939	−0.91	0.35	−0.92	0.31	−0.54	0.29
1940–1949	−0.89	0.34	−0.84	0.30	−0.31	0.28
1950–1959	−0.87	0.33	−0.93	0.30	−0.35	0.28
1960–1969	−1.08	0.34	−0.89	0.31	−0.35	0.28
1970–1974	−2.05	0.51	−1.78	0.46	−1.21	0.42
Father's education			0.05	0.01	0.01	0.01
Mother's education			0.10	0.01	0.04	0.01
Father's occupational status			0.02	0.00	0.01	0.00
Number of siblings			−0.09	0.01	−0.05	0.01
Not living with both parents at age 16			−0.00	0.06	0.09	0.05
Lived in foreign country at age 16			−0.98	0.10	−0.96	0.09
Farm background			−0.44	0.04	−0.30	0.04
Lived in the South at age 16			−0.45	0.04	−0.34	0.04
Missing father's education			−0.43	0.06	−0.09	0.06

Table continues

Table 1 (*Continued*)

	Baseline		Social background		Education	
Variable	Coefficient	*SE*	Coefficient	*SE*	Coefficient	*SE*
Missing mother's education			−0.64	0.06	−0.25	0.06
Missing father's occupational status			0.19	0.08	0.04	0.07
No schooling					0.39	0.66
Years of school						
1					−3.19	1.01
2					−2.16	0.73
3					−2.39	0.39
4					−2.50	0.30
5					−2.39	0.25
6					−1.50	0.20
7					−1.57	0.15
8					−1.24	0.09
9					−1.00	0.10
10					−0.75	0.08
11					−0.68	0.08
12					—	—
13					0.40	0.06

14	0.75	0.06
15	1.06	0.08
16	1.67	0.06
17	1.89	0.10
18	2.07	0.10
19	2.57	0.15
20+	2.32	0.14

Note. Entries for reference groups in sets of more than two dummy variables are marked by a dash. Thus, main effects of race pertain to the Black–White difference in the 1909–1919 cohort; main effects of cohorts pertain to contrasts among Blacks, relative to the 1909–1919 cohort; and Race × Cohort interactions pertain to cohort-specific Black–White differences. Age effects are expressed relative to ages 20–29, and education effects are expressed relative to the completion of 12 years of school.

specification. The estimated coefficients and their standard errors are shown in Table 1. The baseline model includes sex, race, age, birth cohort, and interactions between race and sex and between race and birth cohort[3]:

$$E[y] = \alpha + \sum_{1}^{I} \beta_i x_i + \sum_{1}^{J} \gamma_j w_j + \delta_1 z_1 + \delta_2 z_2 + \delta_3(z_1 z_2) + \sum_{I}^{J} \lambda_j z_1 w_j, \qquad (4)$$

where y is the number of correct WORDSUM items, α is the intercept, the x_i are dummy variables for age groups, the β_i are age effects, the w_j are dummy variables for birth cohorts, the γ_j are cohort effects, z_1 is a dummy variable for race, z_2 is a dummy variable for sex, the δs are effects of sex and race, and the λ_j are effects of race by cohort interactions.

The model is intended to describe trends in verbal ability among Blacks and Whites, free of the confounding influences of temporal changes in the age composition of the population and of the association of age with test scores. If observations from a single cross-sectional sample were available, one could not conceivably separate the effects of chronological ages from those of birth cohorts. Because the GSS provides repeated cross-sectional measures in samples of the same cohorts at different ages, one can estimate distinct effects of chronological age and year of birth. However, the model requires a strong identifying assumption, namely, that there are no period effects on test scores. That is, we assumed that there were no effects on test scores, specific to the year of the survey, above and beyond the combination of age and birth cohort effects that pertain to persons in each survey year.

[3]The number of cases for Whites born in 1909–1919, 1920–1929, 1930–1939, 1940–1949, 1950–1959, 1960–1969, and 1970–1974 were 543, 1,304, 1,660, 2,542, 2,860, 1,555, and 187, respectively. The counts for Blacks were 60, 186, 298, 426, 587, 342, and 28, respectively. The small number of cases in the oldest and youngest cohorts of Blacks implies that the findings about them should be interpreted with caution.

In the case of one item, we were particularly concerned about the assumption of no period effects because a large corporation began using one of the stimulus words in the name of a product line in the late 1970s. Because we expected this change of usage to increase knowledge of the meaning of the stimulus word, we carried out a separate analysis of trends in knowledge of this word, in which we estimated period and age effects but not cohort effects. We found a possible trend toward increasing item difficulty, but it began before and ended shortly after initiation of the commercial use of the word. Therefore, we do not think the commercial use of the word affected its validity for our purposes.

As shown in Table 1, effects have been defined relative to Black women aged 20–29 who were born from 1900 to 1919. In the oldest cohort, Whites answered almost 2.5 more items correctly than Blacks. There was almost no overall gender difference in the WORDSUM score, and White men scored insignificantly lower than the combined effects of race and gender would predict. In the baseline model, persons aged 30–39 and 40–49 each obtained scores about 0.50 words higher than persons aged 20–29. Persons aged 50–59 had only about half the advantage of 30- to 49-year-olds, relative to those at ages 20–29, and persons aged 60–65 had a small but statistically insignificant advantage over the youngest age group. There was an irregular upward progression in the coefficients for birth years (γ_j) and an irregular downward progression in the effects of interactions between birth years and race (λ_j). These patterns imply that verbal ability scores have been increasing among Blacks and decreasing among Whites.

The social background model adds eight social background characteristics to the baseline model of Equation 4: father's educational attainment, mother's educational attainment, father's occupational status, number of siblings, nonintact family (at age 16), foreign residence (at age 16), farm background, and Southern residence (at age 16). In addition, the social background model includes three dummy variables that flag missing values on father's education, mother's education, and father's occupation. The education model adds respondent's years of schooling to the social background model. Years of education are coded into a series of 21 dummy variables with 12 years of education as the reference group. Comparisons among estimates from these models may help to explain the effects of social background and respondent's years of education on Black–White differences and their trends over time. That is, changes across models in the effects of birth cohorts among Blacks and Whites reveal the degree to which the trends can be explained by the makeup of Black and White cohorts in terms of social background and educational attainment.

The effects of age in the three models are also of interest in their own right. They have been estimated with 20- to 29-year-olds as the reference group. In the baseline model and, again, when social background was controlled, test scores increased by about half a point at ages 30–39 and 40–49 and by about half that amount at ages 50–59 and 60–65. However, this age pattern is partly a consequence of the variation in schooling by cohort in the GSS data. When years of schooling were also controlled, there remained a smaller increase in test scores at ages 30–39 and 40–49, about 0.3 year relative to persons aged 20–29. However, the two older age groups no longer had any advantage relative to the youngest age group. This age pattern appears to contradict suggestions in the research literature that relatively crystallized abilities, such as vocabulary knowledge, increase indefinitely with age.

Figure 3 shows the intercohort trends in WORDSUM scores of Blacks and Whites born from 1909 to 1974, as estimated in the baseline model. Black–White differences in test scores have been decreasing.[4] In the most recent birth cohort, 1970–1974, the Black–White difference was 16.4% as large as in the earliest birth cohort, 1909–1919.[5] The differences declined irregularly across birth cohorts, but they never increased substantially from one cohort to the next. Relative to the oldest cohort, they are 84%, 63%, 64%, 64%, 56%, and 16% as large in suc-

[4]In the baseline model, the likelihood-ratio test statistics for intercohort differences in test scores and for the Race × Cohort interactions are both highly significant.

[5]Again, because of the small number of cases in the youngest cohort, this observation should not be taken too seriously.

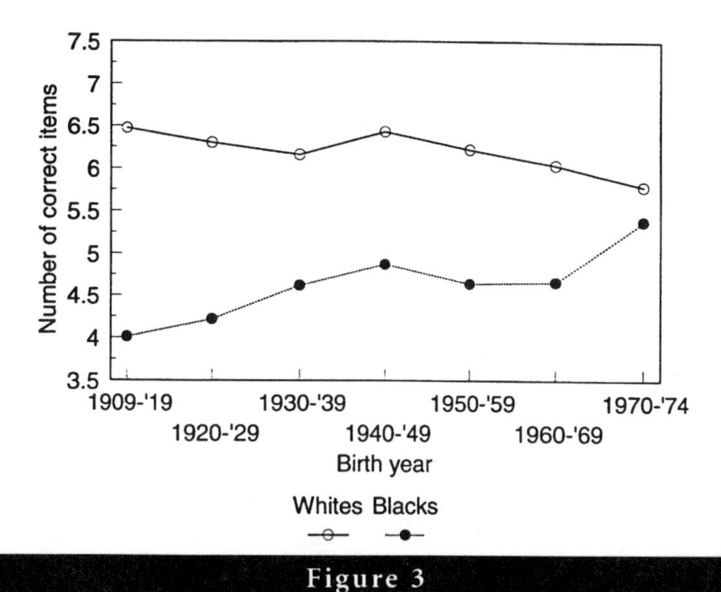

Figure 3

Baseline model of WORDSUM trends: General Social Survey, 1974–1994.

cessive cohorts.[6] About two thirds of the convergence can be attributed to an upward trend in test scores among Blacks, from 4.0 in the oldest cohort to 5.4 in the youngest. About one third of the convergence can be attributed to a downward trend among Whites, from 6.5 in the oldest cohort to 5.8 in the youngest. Most of the White decline took place in cohorts born after 1950, whereas the growth in Black test scores took place between the cohorts of 1909 and 1949 and those born after the 1960s.

Figure 4 shows the trends in test scores for Blacks and Whites when social background variables were added to the baseline model. Because there have been persistent but declining differences in social background between Black and White cohorts, one might expect the initial differences between Black and White test scores to be smaller and the convergence to be less rapid when social background is controlled. This

[6]These estimates and others cited later refer specifically to differences between White and Black women. Because the Race × Gender interactions are negligible, the estimates are similar for differences between White and Black men.

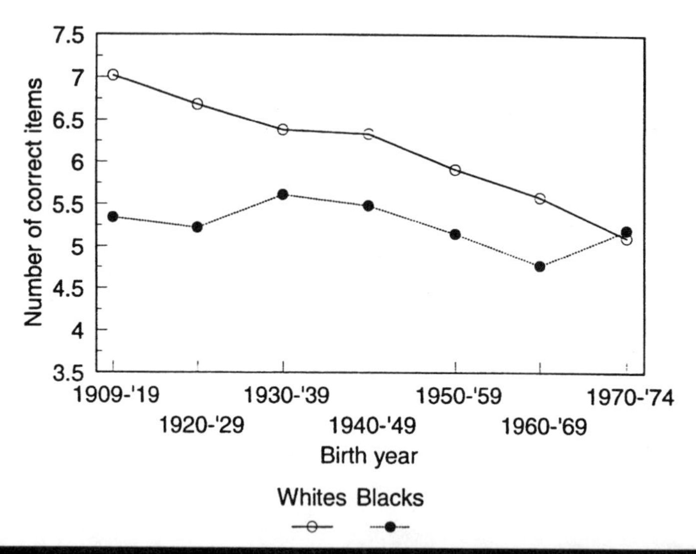

Figure 4

Social background model of WORDSUM trends: General Social Survey, 1974–1994.

appears to be the case.[7] The initial difference between Blacks and Whites in the cohort of 1909–1919 is 1.7 points in the social background model, whereas it was 2.5 points in the baseline model. Moreover, convergence is greater than in the baseline model. Black–White differences in WORDSUM disappear in the 1970–1974 birth cohort when social background variables are controlled. The convergence appears to have been driven largely by a continuous decline in the verbal ability of Whites, whereas the test scores of Blacks varied irregularly from one cohort to the next. That is, the estimates suggest that verbal ability was relatively stable among Blacks with the same social background throughout most of this century and that it declined steadily among Whites of constant social background throughout the century. Black–White differences in social background account for part of the initial difference between average test scores, and improvements in the social

[7]In the social background model, the likelihood-ratio test statistics for intercohort differences in test scores and for the Race × Cohort interactions are both highly significant.

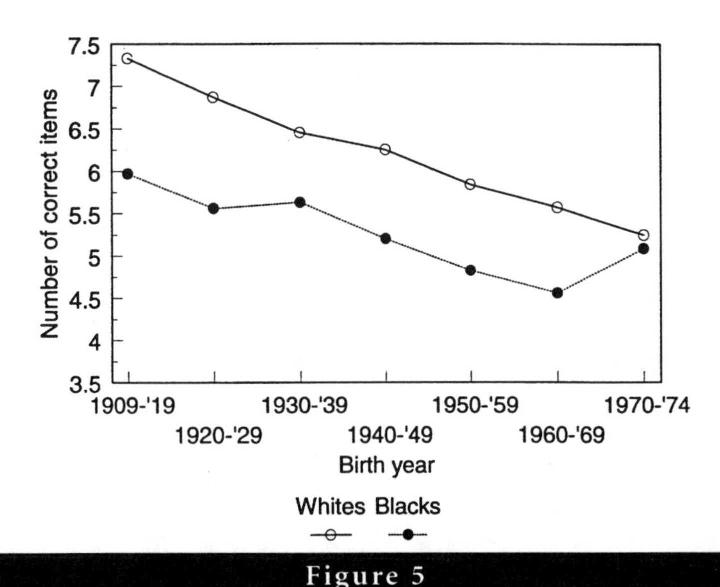

Figure 5

Education model of WORDSUM trends: General Social Survey, 1974–1994.

background of Blacks relative to Whites account for part of the convergence.

Figure 5 shows the intercohort trends in test scores when years of schooling as well as social background are added to the baseline model.[8] The effect of schooling, as one would expect, is to bring average Black and White test scores closer together. In the oldest cohorts, the differential is more than one correct answer on the WORDSUM test (half a standard deviation), and in all but the youngest cohort, the differential is slightly less than one correct answer. Thus, much of the Black–White difference in vocabulary scores is associated with differences in social background and schooling. They account for more than a 1-point test-score difference in the oldest cohort and for slightly less than a half-point difference in the next-to-youngest cohort.

The data also suggest that changes in social background and school-

[8]In the schooling model, the likelihood-ratio test statistic for intercohort differences in test scores is highly significant. However, the Race × Cohort interactions are only marginally significant ($p < .03$); thus, the cohort trend lines for Blacks and Whites are almost parallel.

ing contributed to the convergence in test scores. For example, in the baseline model, the change in the Black–White difference between the oldest and youngest cohorts was 2.1 points; in the social background model, it was 1.8 points; and in the schooling model, it was 1.2 points. Moreover, there are also signs of convergence in test scores that are above and beyond the salutary effects of social background and schooling. That is, in the schooling model, the Black and White test-score series are almost parallel, but there are additional signs of convergence in the youngest cohort. Finally, among Blacks and Whites, the net trends shown in Figure 5 are close to linear. Both groups show an overall decline in vocabulary knowledge, excepting the relatively high performance of the youngest cohort of Blacks. Thus, as successive cohorts completed higher nominal levels of schooling, those levels of schooling may have become less selective with respect to a basic level of verbal proficiency.[9] Some might argue that this finding indicates a decline in school quality, but we prefer to reserve such a conclusion for situations in which there is direct evidence that the effects of schools have declined.

SOCIAL BACKGROUND, SIZE OF SIBSHIP, AND TREND IN VERBAL ABILITY

When social background is controlled, the trend in verbal ability is downward among both Blacks and Whites. For this reason, it would be easy to ignore the positive effects of intercohort changes in social background on verbal ability. Table 2 shows mean levels of the eight social background variables used in the preceding analysis. There have been very large, and for the most part unidirectional, shifts in the composition of successive cohorts on six of the eight background variables.[10] For this reason, we have also estimated the effect of changes in social background on verbal ability.

[9]Also, recall that there may have been some increase in the difficulty of WORDSUM items, as indexed by word frequency.

[10]See Hauser and Featherman (1976) for additional evidence of intercohort change in social background.

Average levels of maternal and paternal educational attainment have grown rapidly, by 5 years or more among Whites and by 7 years or more among Blacks, even as the variability in parental schooling has declined. For example, the growth in schooling of the fathers of Blacks—7.7 years—is 1.7 times its standard deviation in the cohort of 1909–1919 and 2.5 times its standard deviation in the cohort of 1970–1974. Among Whites, father's occupational status has grown by almost as much as its standard deviation in the oldest cohort; among Blacks, father's occupational status has grown by close to half its standard deviation in the older cohorts.

The number of siblings has declined rapidly, from 4.7 to 2.4 among Whites and from 6.0 to 4.2 among Blacks, and there has also been a decline in the variability of sibship size. However, relative to the standard deviations, the change has not been as large as in the case of parental schooling. Among Whites, the decline in size of sibship was about 1.3 times as large as the standard deviation in the youngest cohort, and among Blacks the decline was less than half as large as the standard deviation in the youngest cohort. Among Whites, the share of children who were not living with both their parents was about 20% for cohorts born before 1959 but began to grow rapidly thereafter. Among Blacks, the share of children raised in nonintact families was much higher than among Whites early in the century. It rose after the end of World War I to 47%, and after a modest decline in the Depression cohorts, it has grown steadily to cover almost two thirds of Black adults.

In cohorts born from 1909 to 1974, very few adult Americans were living in a foreign country at age 16, although more may have been born abroad. There has been little change in this share over time. On the other hand, the share of the population with rural origins has declined dramatically. The percentage with farm background has declined from 38 to 19% among Whites and from 58 to 10% among Blacks. Thus, Blacks in the youngest cohort were much less likely than Whites to be of rural origin. There has been little change among Whites in the share of adults, 26–30%, who lived in the South at age 16, but Southern origin declined rapidly among Blacks, from 79% in the cohort of 1909–1919 to 53% in the cohort of 1970–1974.

Table 2

Means (and Standard Deviations) of Selected Social Background Variables: General Social Survey

Social background variable	Birth cohort						
	1909–1919	1920–1929	1930–1939	1940–1949	1950–1959	1960–1969	1970–1974
Whites							
Father's education	7.6 (4.1)	8.4 (4.0)	9.0 (3.9)	10.4 (3.8)	11.7 (3.8)	12.6 (3.6)	13.4 (3.2)
Mother's education	8.1 (3.8)	8.9 (3.6)	9.7 (3.4)	10.8 (3.1)	11.7 (3.0)	12.3 (2.8)	13.1 (2.6)
Father's occupational status	28.3 (15.4)	28.9 (15.7)	30.4 (16.7)	33.5 (18.5)	37.1 (19.8)	39.8 (20.4)	42.9 (21.6)
Number of siblings	4.7 (3.1)	4.1 (3.2)	3.9 (3.2)	3.5 (2.9)	3.5 (2.6)	3.2 (2.4)	2.4 (1.8)
Not living with both parents at age 16	0.23 (0.42)	0.22 (0.41)	0.22 (0.41)	0.19 (0.39)	0.20 (0.40)	0.27 (0.44)	0.34 (0.47)
Lived in foreign country at age 16	0.02 (0.14)	0.04 (0.18)	0.05 (0.21)	0.03 (0.17)	0.03 (0.17)	0.03 (0.16)	0.01 (0.11)
Farm background	0.38 (0.49)	0.35 (0.48)	0.33 (0.47)	0.27 (0.44)	0.24 (0.42)	0.23 (0.42)	0.19 (0.39)
Lived in the South at age 16	0.27 (0.44)	0.26 (0.44)	0.28 (0.45)	0.29 (0.45)	0.27 (0.44)	0.30 (0.46)	0.30 (0.46)

Blacks														
Father's education	4.4	(4.4)	6.7	(4.1)	7.3	(3.8)	8.7	(4.3)	9.4	(3.8)	11.4	(3.5)	12.1	(2.9)
Mother's education	6.7	(4.4)	8.1	(4.3)	8.2	(3.6)	9.4	(3.5)	10.3	(3.4)	11.9	(3.0)	13.8	(2.6)
Father's occupational status	21.4	(9.0)	21.0	(8.4)	21.8	(8.2)	25.2	(15.2)	24.7	(13.2)	30.5	(18.5)	25.2	(10.2)
Number of siblings	6.0	(4.6)	5.3	(3.8)	6.0	(4.4)	5.4	(3.7)	5.7	(3.7)	5.0	(4.0)	4.2	(4.1)
Not living with both parents at age 16	0.37	(0.48)	0.47	(0.50)	0.41	(0.49)	0.43	(0.50)	0.46	(0.50)	0.51	(0.50)	0.63	(0.49)
Lived in foreign country at age 16	0.00	(0.00)	0.02	(0.13)	0.05	(0.22)	0.06	(0.24)	0.07	(0.25)	0.04	(0.21)	0.01	(0.12)
Farm background	0.58	(0.50)	0.44	(0.50)	0.37	(0.48)	0.27	(0.44)	0.20	(0.40)	0.12	(0.33)	0.10	(0.31)
Lived in the South at age 16	0.79	(0.41)	0.73	(0.45)	0.69	(0.46)	0.57	(0.50)	0.52	(0.50)	0.49	(0.50)	0.53	(0.50)

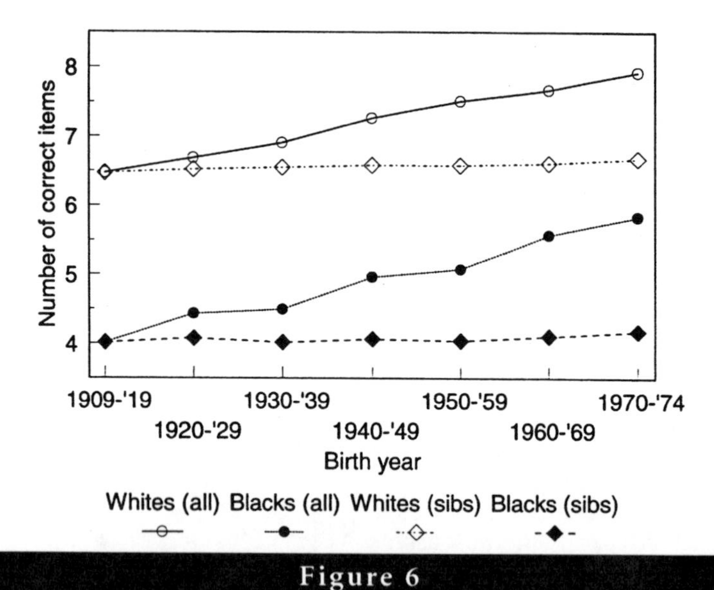

Figure 6

Effects of intercohort changes in social background on WORDSUM trends: General Social Survey, 1974–1994. (sibs = siblings.)

Figure 6 shows the effects of intercohort changes in social background on trend in WORDSUM among Blacks and Whites. That is, it displays the changes in WORDSUM that are implied by changes in social background alone. The estimates were constructed by applying the regression coefficients of the social background model of Table 1 to the means of the social origin variables in Table 2 (including effects of the dummy variables for missing data):

$$\hat{y}_{jg} = \hat{\alpha}_g + \sum_{1}^{K} \hat{\pi}_k \bar{s}_{jkg}, \tag{5}$$

where $\hat{y}_{jg}$ is the predicted WORDSUM score of the *j*th cohort in racial–ethnic group *g*, $\hat{\alpha}_g$ is the intercept of group *g*, $\hat{\pi}_k$ is the estimated effect of the *k*th social background variable, and $\bar{s}_{jkg}$ is the mean of the *k*th social background variable for the *j*th cohort in group *g*. The steeper pair of lines in Figure 6 shows the combined effects of intercohort changes in all eight social background variables among Blacks and Whites. The other two lines show the effects of intercohort changes in the observed mean number of siblings in each cohort. That is, the former series shows the combined effect of all changes in social background on WORDSUM, and the latter series isolates the effect of change in numbers of siblings. We looked separately at the effects of change in size of sibship because of persistent scholarly interest in the effects of that variable on ability (Blake, 1989) and occasional efforts to connect changes in average ability with fertility decline (Alwin, 1991; Zajonc, 1976, 1986; Zajonc & Mullally, 1997).

Intercohort changes in social background have had profound implications for observed levels of verbal ability. Among Whites, the implied growth in WORDSUM is 1.46 items from the oldest to the youngest cohorts, and among Blacks, the implied growth is 1.82 items. That is, improvements in social background across U.S. cohorts born between 1909 and 1974 imply growth of about 0.7 *SD*s in verbal ability among Whites and more than 0.9 *SD*s in verbal ability among Blacks. However, changes in numbers of siblings contributed modestly to these trends. When changes in size of sibship are analyzed jointly with those of other social background characteristics, declining sibship size accounted for an increase of only 0.21 correct WORDSUM items among Whites and 0.16 correct WORDSUM items among Blacks. These effects accounted for only 14% of the effects of social background on trend in WORDSUM among Whites and 9% among Blacks. By contrast, the effects of change in parental schooling were much larger: 0.79 items among Whites and 1.09 items among Blacks. That is, growth in parental schooling accounted for more than half of the salutary effects of intercohort changes in social background on verbal ability.

Why does number of siblings contribute so little to the effects of social background on intercohort change in verbal ability? First, number of siblings is correlated with each of the other social background variables, and these other variables also affect WORDSUM. Thus, the effect of number of siblings was substantially less in the regression analysis than it might have appeared to be if other social background characteristics were ignored. Second, although the declining numbers of siblings seen in the GSS data reflected the sharp fertility decline of the first half of this century, there have also been very large (perhaps even larger) changes in other significant social background variables. Third, those other social background variables, especially parental schooling, are strongly associated with verbal ability.

CONCLUSION

To the evidence that Black–White differences in academic achievement were reduced in cohorts born after the middle 1960s, we can add new evidence of a longer term convergence between Blacks and Whites in verbal ability. Under reasonable assumptions with respect to the effects of age and sex, data collected by the NORC GSS since the middle 1970s suggest a 65-year trend toward convergence between Black and White adults in verbal ability, as indicated by a short test of vocabulary knowledge. Differences between Blacks and Whites in social background and schooling accounted for a large share of the observed differences in test scores. The convergence in verbal ability appears to have been close to complete for persons of similar background and schooling who were born after 1970. Improvements in the social background and schooling of Blacks accounted in part for the convergence in test scores, but other factors also appear to have reduced the test-score difference, especially in recent cohorts.

In combination, the effects of social background on verbal ability and the positive intercohort changes in social background that have occurred throughout the 20th century imply substantial intercohort growth in verbal ability, almost a standard deviation among Blacks and almost three fourths of a standard deviation among Whites. Therefore, if it were possible to generalize from effects on verbal ability to effects on other abilities, one would count improvements in social background among the sources of long-term cognitive growth in the United States and, perhaps, in other nations. The effects of changes in social background are only to a small degree consequences of fertility decline. Rather, they are mainly consequences of improvements in parental schooling, occupational standing, and geographic origin (farm background and Southern origin).

At the same time, a singular aspect of the GSS series is that they show a long-term decline in verbal ability in the White population. A similar trend appears for Blacks and Whites when social background and schooling have been controlled, although the decline among Blacks is less than that among Whites. Are the test-score series from the GSS plausible in light of Flynn's (1984, 1987) findings of rising IQ scores in the United States and in 13 other countries? We believe that the GSS series are not necessarily inconsistent with Flynn's findings, although we certainly hope that it will be possible to validate those findings directly or indirectly with data from other sources. Although the evidence is weak, our examination of word frequency ranks suggested that the difficulty of the WORDSUM items may have increased over the decades. Also, Loehlin, Lindzey, and Spuhler's (1975) review of global trends in intelligence took note of two studies in which verbal scores had declined across time, whereas other ability measures had increased (pp. 135–139). It should also be recalled that in Flynn's (1984) initial report of temporal gains in test scores, much of the discussion concerned the possible inconsistency between IQ gains through 1978 and the post-1963 declines in the SAT-V (pp. 36–39). In his recent review, "IQ Gains Over Time," Flynn (1994) noted that tests may be ordered from those with reduced cultural content, "many of which are indicators of fluid intelligence," to those that measure crystallized intelligence, with "less emphasis on on-the-spot problem solving and more on whether someone has acquired the skills, or general knowledge, or vocabulary we would expect an intelligent person to gain in a normal life" (p. 617). Thus, he placed "pure vocabulary tests" at the extreme of tests of crystallized intelligence. Flynn's reading of the available evidence, worldwide, is as follows:

> IQ gains over time diminish as tests get farther and farther from measuring fluid intelligence. . . . Verbal IQ gains vary from almost nil to 20 points per generation, with 9 as a rough median, and some of this is adult data from military testing. Among the eleven countries that allow a comparison, there is not one in

> which verbal gains match the gains on culture-reduced, or performance, or nonverbal tests and often the ratios run against verbal gains by two or three to one. Where vocabulary gains can be distinguished from verbal gains in general, they rarely match them. (p. 617)

If one accepts this reading, one should not be entirely surprised to find a valid time series of vocabulary measures with a pronounced downward trend. At the same time, the clear distinction between performance on vocabulary tests and on other tests of ability across time is a valuable reminder of the limits of the present findings.

In this context, aside from our specific findings in the GSS, one might want to consider other applications of similar research designs. There is no hope of going as far back in time in the measurement of fluid intelligence as the GSS series permits in the assessment of verbal ability. However, one could cover a considerable span of cohorts by administering more comprehensive tests to a single, large cross-sectional sample, provided only that one truly understood the intrinsic relationship between age and performance on each test. Alternatively, a new series of ability measures in repeated cross-sectional samples, following the GSS model, would permit estimation of both age and cohort effects and could provide both retrospective and prospective assessments of trends and differentials in abilities.

REFERENCES

Alwin, D. F. (1991, October). Family of origin and cohort differences in verbal ability. *American Sociological Review, 56,* 625–638.

Blake, J. (1989). *Family size and achievement.* Berkeley: University of California Press.

Davis, J. A., & Smith, T. W. (1994). *General social surveys, 1972–1994: Cumulative codebook.* Chicago: National Opinion Research Center.

Flynn, J. R. (1984). The mean IQ of Americans: Massive gains 1932 to 1978. *Psychological Bulletin, 95,* 29–51.

Flynn, J. R. (1987). Massive IQ gains in 14 nations: What IQ tests really measure. *Psychological Bulletin, 101,* 171–191.

Flynn, J. R. (1994). IQ gains over time. In R. J. Sternberg, S. J. Ceci, J. Horn, J. Matarazzo, & S. Scarr (Eds.), *Encyclopedia of human intelligence* (pp. 617–623). New York: Macmillan.

Hauser, R. M., & Featherman, D. L. (1976, April). Equality of schooling: Trends and prospects. *Sociology of Education, 49,* 99–120.

Hauser, R. M., & Mare, R. D. (1993). *A General Social Survey mini-module on trends and differentials in cognition* (a proposal to the Governing Board of the GSS). Center for Demography and Ecology, University of Wisconsin, Madison.

Kučera, H., & Francis, W. N. (1967). *Computational analysis of present-day American English.* Providence, RI: Brown University Press.

Loehlin, J. C., Lindzey, G., & Spuhler, J. (1975). *Race differences in intelligence.* San Francisco: Freeman.

Maddala, G. (1983). *Limited-dependent and qualitative variables in econometrics.* Cambridge, England: Cambridge University Press.

Miner, J. B. (1957). *Intelligence in the United States: A survey.* New York: Springer.

Raftery, A. E. (1995). Bayesian model selection in social research. In P. V. Marsden (Ed.), *Sociological methodology 1995* (pp. 111–163). Cambridge, MA: Basil Blackwell.

Stevens, G., & Cho, J. H. (1985). Socioeconomic indexes and the new 1980 census occupational classification scheme. *Social Science Research, 14,* 142–168.

Stevens, G., & Featherman, D. L. (1981). A revised socioeconomic index of occupational status. *Social Science Research, 16,* 364–395.

Swaminathan, H., & Rogers, H. (1990). Detecting differential item functioning using logistic regression procedures. *Journal of Educational Measurement, 27,* 361–370.

Thorndike, E. L. (1921). *The teacher's word book.* New York: Teachers College, Columbia University.

Thorndike, E. L. (1931). *A teacher's word book of the twenty thousand words found most frequently and widely in general reading for children and young people.* New York: Teachers College, Columbia University.

Thorndike, E. L., & Lorge, I. (1944). *The teacher's word book of 30,000 words.* New York: Teachers College, Columbia University.

Thorndike, R. L. (1942). Two screening tests of verbal intelligence. *Journal of Applied Psychology, 26,* 128–135.

Thorndike, R. L., & Gallup, G. H. (1944). Verbal intelligence of the American adult. *Journal of General Psychology, 30,* 75–85.

Thorndike, R. L., & Hagen, E. P. (1952). Analysis of results of field trials to determine the feasibility of an aptitude census. *Human Resources Research Center Research Bulletin,* No. 52-22.

Wechsler, D. (1944). *The measurement of adult intelligence (3rd ed.).* Baltimore: Williams & Wilkins.

Wechsler, D. (1981). *WAIS-R manual: Wechsler Adult Intelligence Scale–Revised.* San Antonio, TX: The Psychological Corporation.

Wolfle, L. M. (1980, April). The enduring effects of education on verbal skills. *Sociology of Education, 53,* 104–114.

Zajonc, R. B. (1976). Family configuration and intelligence. *Science, 192,* 227–236.

Zajonc, R. B. (1986). The decline and rise of scholastic aptitude scores: A prediction derived from the confluence model. *American Psychologist, 41,* 862–867.

Zajonc, R. B., & Mullally, P. (1997). Birth order: Reconciling conflicting effects. *American Psychologist, 52,* 685–699.

PART THREE

The Hypothesis of Dysgenic Trends

13

The Decline of Genotypic Intelligence

Richard Lynn

Concern that the intelligence of the populations in Western nations is deteriorating was first expressed by Francis Galton in 1865 and elaborated in a number of subsequent publications (Galton, 1865, 1869, 1908, 1909). Galton believed that two factors were operating to cause genetic deterioration. The first was the relaxation of natural selection in the form of reduced mortality of the "less fit," which included those with low intelligence. The second was the development in the 19th century of a tendency for intelligent people to have fewer children than unintelligent people. Galton believed that in accordance with the general principles of natural selection, the effect of this tendency must be that the intelligence of the population would deteriorate. It was to counteract this deterioration that Galton (1883) proposed the concept of *eugenics.* His idea was that conscious steps needed to be taken to maintain, and even to improve, the genetic quality of the population with respect to intelligence. This would be done by measures to encourage intelligent people to increase the numbers of their children, which he called *positive eugenics,* and by measures to encourage the unintelligent to have fewer children, which he called *negative eugenics.*

Galton understood that these two factors would bring about a genetic deterioration of the population only if intelligence is to some degree under genetic control. He went to considerable pains to argue that this was the case, for instance, by collecting pedigree data showing the transmission of intelligence in families of eminent lawyers and scientists, and by showing that separated twins had similar levels of intelligence (Galton, 1869, 1874).

In the first half of the 20th century, many psychologists and biologists came to agree with Galton that an inverse association between intelligence and fertility existed in modern populations and that as a result the populations must be deteriorating genetically. They also thought that this was a serious problem for the maintenance of civilization. Those who took this view included Karl Pearson (1912), Sir Ronald Fisher (1929), Sir Julian Huxley (1936), Hermann Muller (1938–1939), Sir Cyril Burt (1952), and Raymond Cattell (1937). The word *dysgenic* was coined to describe the genetic deterioration that these people believed was taking place.

In the second half of the 20th century, support for this position declined. Nevertheless, it is my thesis in this chapter that Galton and the eugenicists were right in their contention that there is an inverse relationship between intelligence and fertility in modern populations and that as a result genotypic intelligence is deteriorating. There are four lines of evidence in support of this contention, consisting of the inverse relations between (a) socioeconomic status (SES) and fertility, (b) intelligence and number of siblings, (c) intelligence and fertility, and (d) educational level and fertility. There is extensive worldwide evidence for these four propositions, which I have reviewed at length in my book *Dysgenics* (Lynn, 1996); a summary of the evidence is given in this chapter.

SOCIOECONOMIC STATUS AND FERTILITY

Early Evidence of Dysgenesis

Evidence for a deterioration of intelligence appeared first in the inverse association between SES and fertility. Studies demonstrating this relationship began to be published around the turn of the century. For instance, Heron (1906) found that among the London boroughs an index of average earnings was negatively associated with average family size. Subsequently, the population censuses of 1910 in the United States and 1911 in Britain provided extensive evidence for this relationship. This association is illustrated in Table 1 for White married women born between 1861 and 1865 in the United States and between 1865 and 1870 in England (Haines, 1989; Kiser, 1970). For both countries, there is a more or less linear trend line of increasing numbers of children with declining SES.

The inference that the inverse relationship between SES and fertility entails the genetic deterioration of intelligence depends crucially on the assumptions that the social classes differ in intelligence and that these differences have a genetic basis. Galton believed that this was so, writing that the "brains of the nation lie in the higher of our classes" (1909, p. 11), and most of the eugenicists in the first half of this century agreed with him.

With the appearance of intelligence tests in the United States at the time of World War 1, constructed by Terman and others, it became clear that the social classes do differ in intelligence. When Terman (1921, p. 188) published a preliminary report of highly intelligent children with IQs of 140 and above, he found that 53% came from SES 1

Table 1

Completed Fertility of White Married Women in the United States and England by Socioeconomic Status

		Socioeconomic status					
Country	Birth cohort	1	2	3	4	5	6
United States	1861–1865	3.0	3.5	3.2	4.3	4.5	4.9
England	1865–1870	2.6	3.4	3.4	4.2	4.5	4.8

Note. Socioeconomic status: 1 = professional, 2 = managerial, 3 = clerical, 4 = skilled worker, 5 = semiskilled worker, 6 = unskilled worker.

(professionals), and 37% came from SES 2 (minor white-collar workers), and 10% were from SES 3 (skilled workers). None came from SES 4 or 5 (semiskilled and unskilled workers). Numerous other studies confirmed the positive association between SES and intelligence. For instance, Johnson (1948) showed a linear increase in mean IQ among World War 1 military personnel, from 96 in SES 5 to 123 in SES 1, and a similar trend among conscripts in World War 2.

The eugenicists believed further that these social class differences have a genetic basis, a contention that has frequently been disputed. For instance, a historian of eugenics, Daniel Kevles, wrote that "there is no evidence that the higher birth rate of lower income groups was polluting the gene pool" (1985, p. 285); Theodosius Dobzhansky wrote that "the occupational and class differentiation is not established on a genetic basis" (1962, p. 248); Milo Keynes said that "there is no class of individuals who are an elite" (1993, p. 23); and Richard Lewontin held that "there is not an iota of evidence that social classes differ in any way in their genes" (1993, p. 37).

The assertion that the social classes do differ genetically with respect to intelligence rests on two arguments. The first is a logical argument advanced by Richard Herrnstein (1971) in the form of a syllogism. The terms of this syllogism were as follows: (a) If intelligence is to some degree inherited and (b) it contributes to SES, then (c) the social classes will inevitably be to some degree genetically differentiated with regard to intelligence. I believe that the evidence for the two premises is massive and that it is not disputed by serious scholars (e.g., Bouchard, 1993; Brody, 1992; Jencks, 1972). I believe also that Herrnstein was right in asserting that these two premises lead inescapably to the conclusion that the social classes must differ genetically. Herrnstein reasserted this conclusion in a later book, *The Bell Curve*, coauthored with Charles Murray (Herrnstein & Murray, 1994).

However, the conclusion that the social classes differ genetically for intelligence does not rest solely on logic. It is supported also by evidence from a series of studies that have shown that the intelligence of adopted children and children in orphanages is positively associated with the SES of their fathers whom they have never seen. The first of these studies was carried out in England by Jones and Carr-Saunders (1927). They found that among a sample of children reared in an orphanage, those who had professional-class fathers had a mean IQ of 107; the mean IQ of the children fell steadily in parallel with the SES of the fathers, reaching 93 among the children of the unskilled.

Six subsequent studies of adopted children in England, the United States, and France confirmed this early result (Capron & Duyme, 1989; Lawrence, 1931; Leahy, 1935; Munsinger, 1975; Skodak & Skeels, 1949; Weinberg, Scarr, & Waldman, 1992). For instance, in the study by Weinberg et al. (1992) of Black and interracial adopted children, the correlations between the children's IQ and the educational level of their biological fathers and mothers (a reasonable proxy for SES) were .28 and .23, respectively. The only explanation for these results is that the parents of different social classes differ genetically for intelligence and transmit their genes to their children. Hence, the inverse relationship between SES and fertility found in the middle years of the 19th century implies that the genes for high intelligence were replicating less successfully than those for low intelligence and that the genotypic intelligence of the populations was deteriorating.

Recent Development of Dysgenic Fertility

The inverse relationship between SES and fertility is a recent phenomenon that appeared in Europe and North America in the early decades of the 19th century. In previous times, the relationship had generally been positive. This was especially the case in polygamous societies in which high-status men had several wives, or even hundreds of them, and large numbers of children. For instance, the Moroccan Emperor Mouley Ismail the Bloodthirsty is credited with having fathered 888 children by his substantial harem (Daly & Wilson, 1983). In polygamous societies of this kind, low-status men frequently had neither wives nor children. In historical Europe, the middle classes typically had higher fertility and lower mortality than the working classes (e.g., Skipp, 1978; Weiss, 1990). The principal reason for this was that the middle classes were better nourished, which improved their fertility and lowered their mortality.

The historically positive relationship between social status and fertility was replaced by a negative relationship in the early 19th century as an effect of the "demographic transition." This was the shift from large families, generally accompanied by high infant and child mortality, to small families with low mortality. The demographic transition has been dysgenic because it took place first among the professional and middle classes. Three stages of the transition can be distinguished. In the first stage, there was a slight reduction in the fertility of the professional and middle classes. This is illustrated for White American women born between 1830 and 1840 in the first row of Table 2. Fertility increased slightly with declining SES and was therefore slightly dysgenic. The magnitude of the dysgenic trend is quantified by the ratio of fertility in SES 5 to that in SES 1 and is given in the column headed "Dysgenic Fertility Ratio." At the beginning of the demographic transition, the dysgenic fertility ratio was quite small at 1.23.

In the second stage of the demographic transition, the dysgenic fertility ratio became greater. This is demonstrated for the United States in the 1861–1865 cohort of married women, for whom the dysgenic fertility ratio reached its maximum of 1.63. In Stage 3, the relationship weakened again but remained negative. This is shown in the 1916–1925 and 1946–1954 cohorts of White American married women, for whom the dysgenic fertility ratio declined to 1.25 and 1.10, respectively. It is not possible to examine this relationship for more recent cohorts because it can be validly examined only for completed fertility, which means for women in their 40s and older. These stages in the evolution of dysgenic fertility have been present in a number of other economically developed nations, including England (Haines, 1989), Australia (Wong, 1980), and France and Norway (Glass, 1967; Haines, 1992). The last three rows of Table 2 give comparable data for Black and Hispanic married American women. The data indicate that dysgenic fertility is greater among Black than among White women, and that of Hispanic women is intermediate.

The major reason that dysgenic fertility appeared in the first half of the 19th century, grew more pronounced in the second half, and has declined but not disappeared in the 20th century probably lies in the

Table 2

Fertility in Relation to Socioeconomic Status in the United States

Sample	Date of birth	Age	Socioeconomic status 1	2	3	4	5	6	7	Dysgenic fertility ratio	Data	Study
White wives	1830–1840	60–70	5.6	5.1	7.8	5.6	7.1	6.9	5.2	1.23	1900 census	Haines (1992)
White wives	1861–1865	45–49	3.0	3.5	3.2	4.3	4.5	4.9	5.2	1.63	1910 census	Kiser (1970)
White wives	1916–1925	35–44	2.4	2.4	2.3	2.6	2.7	3.0	4.0	1.25	1960 census	Preston (1974)
White wives	1946–1954	35–44	2.0	2.0	2.0	2.2	—	2.2	2.5	1.10	1990 survey	Bachu (1991)
Black wives	1916–1925	35–44	2.2	2.6	2.5	3.0	3.2	3.5	5.1	1.59	1960 census	Preston (1974)
Black wives	1946–1954	35–44	1.9	2.4	2.7	2.5	—	2.6	3.9	1.37	1990 survey	Bachu (1991)
Hispanic wives	1946–1954	35–44	2.4	2.6	2.4	2.8	—	2.9	3.6	1.21	1990 survey	Bachu (1991)

Note. Socioeconomic status: 1 = professional, 2 = managerial, 3 = clerical, 4 = skilled worker, 5 = semiskilled worker, 6 = unskilled worker, 7 = farm worker.

increasing use of contraception. In the early 19th century, books on birth control began to appear, such as Richard Carlile's (1826) *Every Woman's Book*, published in London, and Charles Knowlton's (1877) *The Fruits of Philosophy*, published in New York; the latter, despite the opaque title, provided sound advice on the safe period, coitus interruptus, spermicidal chemicals, sponges, pessaries, douches, and condoms, which at the time were made from sheep gut tied at one end.

Initially, these books were bought and read and their advice followed by only a few members of the professional and middle classes, who showed the small reduction in their numbers of children characteristic of Stage 1 of dysgenic fertility. In the second half of the 19th century, knowledge and use of contraception spread widely in the professional and middle classes and, to a lesser extent, to the working classes, bringing about the reduction of fertility in all classes, together with pronounced dysgenic fertility, seen in the 1861–1865 cohort. The increasing use of contraception in the second half of the century was greatly assisted by the development and mass marketing of the modern latex condom in the 1870s and the cervical cap in the 1880s. The third stage of dysgenic fertility took place in the mid-20th century. Knowledge and use of contraception have spread throughout society, with the result that fertility is low. Dysgenic fertility is also small, but it is still present. The major reason for this is probably that less intelligent people use contraception less efficiently so that many have more children by accident. For instance, Vining (1982) found that among both Black and White women, intelligence is negatively associated with fertility, but he found no association between the women's intelligence and their ideal number of children. The explanation for this pattern of correlations is that less intelligent women have more children than they consider ideal. The reason is that the efficient use of contraception is a cognitive task, and more intelligent people perform all cognitive tasks more effectively than do less intelligent people.

INTELLIGENCE AND NUMBER OF SIBLINGS

From 1925 onward, studies began to appear on the relationship between IQ and number of siblings, first in the United States and Britain and subsequently in continental Europe and elsewhere in the economically developed world. In all of these studies, which were mainly but not invariably carried out with schoolchildren, it was found that the correlation between IQ and sibship size was negative: The higher a child's IQ, the fewer the number of his or her siblings. The first report of this association in the United States was published by Chapman and Wiggins (1925) on the basis of 629 12- to 14-year-olds, among whom the correlation was −.33. A year later, Sutherland and Thomson (1926) in Britain obtained a correlation of −.22 for 1,084 10- to 11-year-olds. Numerous subsequent studies over the course of the century have confirmed this inverse association. In the mid-1980s, for instance, Van Court and Bean (1985) obtained a correlation of −.29 for 12,120 adults. In an unpublished study conducted in 1993, my student Lucy Greene found a correlation of −.18 among 517 young adolescents in Britain. Those who collected these data from the 1920s onward believed that children's IQs are similar to those of their parents. Hence, they argued that the fact that children with low IQs typically had a lot of siblings must mean that parents with low IQs had a lot of children. They argued that if this were so, the intelligence of the population must be declining.

A method for estimating the magnitude of the decline in intelligence from the negative association between children's IQs and their sibship size was devised by Theodore Lentz (1927). He began with the assumption that each child has on average the same IQ as his or her parents and siblings. He argued that when data on the IQs of schoolchildren and their numbers of siblings are collected, the parents of one child are less likely to appear in the sample than are the parents of several children. For instance, parents with six children are six times more likely to appear in such a sample of children than are parents of one child. A statistical adjustment needed to be made for this undersampling of parents with few children. This was done by weighting the parental IQs (assumed to be the same as those of their children) by their numbers of children. This adjustment raises the average IQ of the parental generation. Using this technique, Lentz calculated from his data on 4,330 children and adolescents, among whom the correlation between IQ and number of siblings was −.30, that the mean IQ of the parental generation was 4 points higher than that of the children's, indicating a decline of 4 IQ points in one generation.

Lentz's method of calculating the generational decline in intelligence from the negative association of IQ with numbers of siblings was taken up in Britain in the 1930s and 1940s. The first British study on this question was carried out in the mid-1930s by Cattell (1937) using 3,734 11-year-olds. The generational decline was calculated at 3.2 IQ points. However, this was an overestimate because Cattell's test had a standard deviation of 24 rather than the customary 15 or 16 (during this period, the convention of setting the standard deviation of the intelligence at 15 or 16 was not firmly established). When Catell's result is adjusted for the large standard deviation of the test, the rate of decline he obtained is reduced to 2.2 IQ points per generation. In the late 1940s, this problem was addressed again in Britain by Cyril Burt (1952) and Godfrey Thomson (1946), both of whom reached conclusions similar to those of Lentz and Cattell.

It should be noted that the rate of decline of intelligence estimated from the inverse association between intelligence and number of siblings is the notional decline of phenotypic intelligence, that is, the decline that would occur if environmental factors were constant between the two generations. Phenotypic intelligence is of course a product of genotypic intelligence and environmental influences. The decline of genotypic intelligence can be obtained by multiplying the phenotypic decline by the heritability, for which I assume a figure of 0.8. In Britain in the 1930s and 1940s, the notional decline of phenotypic intelligence was approximately 2.0 IQ points per generation. Hence the decline of genotypic intelligence was 1.6 IQ points per generation (2.0 × 0.80). This is an estimate of the actual decline of genotypic intelligence, not of a notional decline.

There have been two criticisms of the use of the "sibling size" method for calculating the decline of intelligence, namely, that the data omit the childless and that the negative association may arise solely from environmental effects. Regarding the first, the objection has been that because the data for making the calculations are derived from children, the sampling procedure does not include individuals in the parental generation who are childless. This problem was recognized by scholars like Cattell, Burt, and Thomson, but they believed that the inclusion of the childless in the calculations would have increased the size of the decline.

The assumption that the childless were of above average intelligence was challenged in the early 1960s in two influential articles, one by Higgins, Reed, and Reed (1962) and the other by Bajema (1963). Both of these studies appeared to show that, contrary to the earlier assumption, childlessness was most prevalent among those with low intelligence and that if these persons were included in the calculation, there was no decline in the intelligence of the population. These two studies involved small and unrepresentative samples, however, and subsequent studies with larger and more representative samples have indicated that the childless are predominantly more intelligent as well as better educated. The first of these, concerning Britain, was published by Kiernan and Diamond (1982), who analyzed the data from a national longitudinal cohort study of 13,687 infants born in 1946. The percentages of people who were childless at age 32 analyzed by low-, average-, and high-IQ categories were 11, 16, and 18% for the female population and 24, 24, and 28% for the male population. Similar results have been obtained for the United States by Van Court and Bean (1985) on the basis of a large representative national sample of about 12,000 adults. Intelligence was measured by a vocabulary test, and the sample was divided into 15 birth cohorts from before 1894 up to 1964. In all 15 cohorts, those without children had higher scores than those with children.

The second objection that has been raised against the method of calculating the decline in intelligence from the negative association between intelligence and number of siblings is that this association is purely an environmental effect. One of those who has advanced this criticism is Blake (1989). The argument is that in larger families parents have less time to devote to each child, which acts as an environmental depressant on the development of the children's intelligence. Conversely, in small families parents can devote more time to each child, which enhances the children's intelligence. This position is based on the confluence model of Zajonc and Markus (1975), in which it is claimed that children's intelligence is a function of the amount of attention they receive from adults, and this is typically less in larger families.

For various reasons the confluence theory is untenable. One objection is that in economically less developed countries there is no relationship between children's IQ and their number of siblings. For instance, it was shown by Ho (1979) that in Hong Kong in the 1970s the IQs of children in 3- to 7-child families were higher than in 1- to 2-child families. This relation has also been found among Vietnamese refugees in the United States in the late 1980s, for whom number of siblings was positively associated with educational attainment in mathematics, a good proxy for intelligence, which was not measured in the study (Caplan, Choy, & Whitmore, 1992). Further reasons for the rejection of the confluence theory have been set out by Galbraith (1982) and by Retherford and Sewell (1991).

I believe that the objections to the sibling size and intelligence argument do not hold up. My conclusion is that the early researchers who studied the problem of dysgenic fertility in the 1920s through the 1940s were right in their argument that the negative association between intelligence and sibling size implies a deterioration in genotypic intelligence.

INTELLIGENCE AND FERTILITY

Even if it was basically sound, the method of using the negative correlations between intelligence and number of siblings to calculate the extent of dysgenic fertility entailed assumptions that can be questioned. A more straightforward method for determining whether genotypic intelligence is declining is to examine the relationship between the intelligence of adults and their numbers of children. If the relationship is negative, genotypic intelligence must be deteriorating. It was not until the second half of the 20th century that studies based on adequate samples began to appear on this issue. I consider first studies in the United States and then studies in Europe.

Studies in the United States

There have been 10 major studies of the relationship between intelligence and fertility in the United States. The results are summarized in Table 3. The relationship is usefully expressed as a correlation, although not all studies present correlational data. The studies are listed in the chronological sequence of their publication. The first four studies obtained eugenic relationships—that is, the association between intelligence and fertility was positive—whereas the remaining six obtained negative, or dysgenic, relationships.

There are two explanations for the conflicting results shown in Table 3. The first, advanced by Vining (1982), is that at some periods fertility has been eugenic, and at others, dysgenic, depending on the state of overall fertility. When overall fertility is low, it tends to be dysgenic, and when it rises, it tends to become eugenic. The data of Van Court and Bean (1985) have provided rather strong evidence against this theory; these authors showed that fertility has been consistently dysgenic among 15 birth cohorts born approximately between 1890 and 1964.

The second explanation for the conflicting results lies in sampling differences. I believe that this explanation is correct and that the results showing eugenic fertility arise from unrepresentative sampling. All four of the studies with eugenic results were based on urban or largely urban White populations. Because rural populations have higher mean fertility rates and lower mean IQ scores than urban ones, the inclusion of rural populations in the samples would have reduced the correlations and possibly turned them negative. McNemar (1942) reported that in the standardization of the Terman-Merrill, there was a rural–urban difference of 12.2 IQ points. The higher fertility of rural populations has been shown in many studies (e.g., Freedman, Whelpton, & Campbell, 1959).

The conclusion that the relationship between intelligence and fertility is positive or neutral was first challenged by Osborne (1975) on the basis of data collected in Georgia. In 1971, Osborne obtained data on the IQs of all children aged 10–14 in Georgia, numbering approximately a quarter of a million. He then calculated the average IQs of

Table 3

Intelligence and Fertility in the United States

Study	Race	Sex	Location	Correlation	Relationship
Higgins et al. (1962)	White	Both	Minnesota	—	Eugenic
Bajema (1963)	White	Both	Kalamazoo, Michigan	.05	Eugenic
Bajema (1968)	White	Both	U.S.	.04	Eugenic
Waller (1971)	White	Both	Minnesota	.11	Eugenic
Osborne (1975)	Mixed	Female	Georgia	−.49	Dysgenic
Van Court & Bean (1985)	White	Both	U.S.	−.16	Dysgenic
Van Court & Bean (1985)	Black	Both	U.S.	−.29	Dysgenic
Retherford & Sewell (1988)	White	Both	U.S.	—	Dysgenic
Vining (1995)	White	Female	U.S.	−.06	Dysgenic
Vining (1995)	Black	Female	U.S.	−.23	Dysgenic

children in each of 159 counties. He also obtained the fertility ratios of these counties: the number of children per 1,000 women aged 15–49 years. The fertility ratio data included children of unmarried women and married women with no children. Six correlations were computed between the average fertility ratios and IQs—for verbal and nonverbal tests for three age groups. The correlations ranged from −.43 to −.54 and averaged −.49, which is highly statistically significant on 159 population units. The population in this study included both Blacks and Whites, who composed 26% and 74% of the population, respectively.

Osborne's (1975) study showed a strong dysgenic trend for intelligence and fertility, in contrast to the four initial studies. Unlike the initial studies, the sample was the entire population of a state and the fertility ratio was based on all women, including the single and the childless. Furthermore, the sample included Blacks as well as Whites. In all these respects, the sample was superior to those of the four initial studies, which were less representative and did not include Blacks. The presence of both Blacks and Whites in a sample examining the relation between intelligence and fertility is likely to make the association negative, because Blacks tend to score lower on IQ tests and to have greater fertility than Whites.

The negative correlation between fertility and intelligence found by Osborne using group data was confirmed 2 years later with individual data by Vining (1982, 1995). The initial study was based on data for 5,172 men and 5,097 women born between 1942 and 1954, for whom intelligence test results and other information were obtained in 1966–1968 and for whom fertility data were obtained in 1976–1978, when the sample was aged 25–34. The data came from the National Longitudinal Study of Labor Market Experience. This was a nationally representative sample including both Blacks and Whites with the exclusion of high school dropouts. For men, fertility was measured on the basis of the number of children living with their own father at the time of the survey. Vining regarded this as an unsatisfactory method for Blacks because of the large number of unstable unions, but as an adequate although imperfect measure for Whites. The results showed statistically significant negative correlations between intelligence and fertility among

White women (−.177), Black women (−.202), and White men (−.140). It was not possible to make a calculation for Black men.

The most satisfactory data on the relationship between intelligence and fertility come from individuals who have completed their childbearing, ideally, those aged 45 and over. Vining's initial sample of 25- to 34-year-olds fell somewhat short of this ideal, but he subsequently published data on the fertility of his female sample at the age of 35–44 (Vining, 1995); for practical purposes, these data represent close to completed fertility. The correlations between intelligence and fertility were significantly negative: −.06 for Whites and −.23 for Blacks (N = 1,839 and 378, respectively). The principal problem with this study concerns the sampling. First, it excluded high school dropouts, who made up about 14% of Whites and 26% of Blacks at this time. Second, only 61% of Whites and 35% of Blacks in the sample provided data for the 26- to 34-year-olds group. Third, the sample was reduced further for the 35- to 44-year-olds. These sampling weaknesses would have distorted the results, but it is difficult to estimate by how much.

The next study to appear on the subject of fertility and intelligence was an attempt to confirm Vining's results. Van Court and Bean (1985) obtained a sample of 12,120 American adults for whom data on intelligence, numbers of siblings, and numbers of children were available. Intelligence was assessed by a 10-item vocabulary test, a short but adequate measure. The data were analyzed for 15 birth cohorts spanning the years of birth from 1890–1894 to 1960–1964. In all the cohorts, the correlations between intelligence and numbers of children were negative; the overall correlations were −.16 for Whites and −.29 for Blacks.

The latest American study of the relationship between intelligence and fertility is that of Retherford and Sewell (1988), which was based on a random sample of 10,317 high school seniors in Wisconsin for whom intelligence test data were collected in 1957. The sample was almost entirely White (98.2%, with 0.9% Black and 0.9% other races). The sample was followed up in 1975, when the participants were approximately 35 years; a response rate of about 90% was secured, and the numbers of children of the members of the sample was recorded. There were statistically significant negative correlations between intelligence and fertility for both male and female participants. The authors calculated the selection differential, which is the difference between the mean IQ of the sample and that of their children. This is calculated by assuming that on average the mean IQ of the children is the same as that of their parents and by weighting the IQ of the parents by their number of children. This process resulted in a selection differential of −1.33 for female and −0.28 for male participants, and an average of −0.81 for both sexes combined. This means that there has been a national decline in the average IQ of the children of 0.81 IQ points as compared with that of their parents.

To calculate the decline in genotypic intelligence, this figure has to be multiplied by the heritability. To do this, Retherford and Sewell adopted a heritability of intelligence of .40, which results in a genotypic decline of −0.32 IQ points per generation ($-0.81 \times .4 = -0.32$). This figure for the heritability of intelligence is much too low for the adults. A more reasonable estimate of the heritability of intelligence among adults is approximately twice as great, namely, .80. The heritability of intelligence can be estimated from the correlation for identical twins reared apart, which Bouchard (1993, p. 58) gave as .72 on the basis of the world literature. This correlation needs to be corrected for test unreliability, which Bouchard estimated at 0.90; the correction increases the correlation to 0.80, the estimated figure for heritability. Estimates derived from identical twin–nonidentical twin differences and from studies of adopted children have given similar results (Bouchard, 1993). The recent Swedish study of 158 identical, and nonidentical twins aged 27–64 yielded a heritability of 81% (Finkel, Pedersen, McGue, & McClearn, 1995).

Studies in Europe

Six studies of the relationship between intelligence and fertility are summarized in Table 4. The study by Young and Gibson (1965) was based on only 100 married couples in Cambridge, England, and found no relationship. The shortcomings of the study included the small sample size and the omission of the unmarried, who are more frequent among more intelligent women (e.g., Kiernan, 1989; Kiernan & Diamond,

Table 4

Intelligence and Fertility in Europe

Study	Sex	Country	Relationship
Young & Gibson (1965)	Both	England	Neutral
Kolvin et al. (1990)	Both	England	Dysgenic
Maxwell (1969)	Both	Scotland	Dysgenic
Nystrom et al. (1991)	Male	Sweden	Eugenic
Nystrom et al. (1991)	Female	Sweden	Dysgenic
Papavassiliou (1954)	Male	Greece	Dysgenic

1982). A second English study, by Kolvin, Miller, Scott, Gatzanis, and Fleeting (1990), was based on 750 adults aged 33 in the city of Newcastle. The participants were divided into three IQ bands. The average numbers of children were 1.9 for the high-IQ group, 2.3 for the intermediate group, and 2.7 for the low group, indicating pronounced dysgenic fertility.

The Scottish study, by Maxwell (1969), was based on a sample of 517 47-year-olds whose IQs were obtained at the age of 11. The data are sketchily reported, but it is possible to calculate that the mean IQ of the sample was higher than that of the children by 0.91 IQ point. Notice the closeness of this figure to the decline of 0.81 IQ point obtained in the leading American study by Retherford and Sewell (1988).

The data for Sweden collected by Nystrom, Bygren, and Vining (1991) consisted of 1,746 adults aged 46–65 living in the county of Stockholm but excluding the city. The results indicated eugenic fertility for men and dysgenic fertility for women. A problem with the study is that the children were confined to those living at home, whether or not they were the biological children of the parents. Many children of this sample would be adult and living away from their parents, and a significant number of them would not have been the biological children of the fathers. For these reasons, the results have to be regarded as inconclusive.

The final European study consisted of a sample of 215 men in Greece (Papavassiliou, 1954). The sample was divided into four IQ bands, and the average numbers of children of the participants, from the highest to the lowest IQ groups, were 1.78, 2.66, 4.00, and 5.56. Clearly, there was strong dysgenic fertility in Greece at about the middle of the century.

Considering the European studies as a whole, there are serious weaknesses in the first English study, based on 100 married couples in Cambridge, and in the Swedish study regarding the men. The remaining four studies are stronger, and all of them have found dysgenic fertility, confirming the results of the better studies in the United States.

EDUCATION AND FERTILITY

The relevance of the relationship between education and fertility lies in the high correlation between educational level and intelligence, which has generally been found to be about .6 (Brody, 1992). Educational level can therefore be regarded as an indirect measure of intelligence.

Studies in the United States

In the United States, data on the relationship between women's level of education and their number of children were collected in the censuses of 1940 and 1960 as well as in a survey carried out by the Bureau of the Census in 1990. The results, analyzed by Osborn (1951), Kiser and Frank (1967), and Bachu (1991), are summarized in Table 5. The first six rows of data are for completed or virtually completed fertility of women aged 35 and older. The last three rows give the anticipated final fertility of 18- to 34-year-old women born between 1956 and 1972. The 1940 census returns were analyzed for White women only; the 1960 data for Whites and Blacks; and the 1990 data for Whites, Blacks, and Hispanics. The earlier studies divided educational attainment into four levels running from less than 4 years of high school to a college degree, whereas the later study added a fifth educational level by subdividing those with tertiary education into basic college graduates and those with

Table 5

Fertility of American Women in Relation to Educational Level

Race or ethnicity	Date of birth	Age (in years)	Educational level					Dysgenic ratio	Study
			1	2	3	4	5		
White	1890–1895	45–49	3.43	1.75	1.71	1.23	—	2.79	Osborn (1951)
White	1915–1919	40–44	2.83	2.44	2.15	1.97	—	1.44	Kiser & Frank (1967)
Black	1915–1919	40–44	3.20	2.69	2.12	1.66	—	1.93	Kiser & Frank (1967)
White	1946–1955	35–44	2.57	2.01	1.72	1.86	1.59	1.49	Bachu (1991)
Black	1946–1955	35–44	2.97	2.28	1.86	1.97	1.70	1.62	Bachu (1991)
Hispanic	1946–1955	35–44	3.07	2.29	2.17	2.47	1.72	2.09	Bachu (1991)
White	1956–1972	18–34	2.38	2.05	2.05	2.02	2.01	1.18	Bachu (1991)
Black	1956–1972	18–34	2.63	2.23	2.00	2.01	1.91	1.34	Bachu (1991)
Hispanic	1956–1972	18–34	2.74	2.21	2.14	2.18	2.01	1.31	Bachu (1991)

Note. Educational levels: 1 = less than 4 years of high school, 2 = 4 years of high school, 3 = some college, 4 = 3 years of college, 5 = 4+ years of college.

4 or more years of university education. The data are for all women rather than for married women only.

Looking first at the general trends, one sees that in all the samples the more poorly educated women had more children than the better educated ones. To provide an approximate measure of the magnitude of the dysgenic fertility, dysgenic ratios were calculated by dividing the fertility of the least educated women by that of the most educated. Because in the first three cohorts all college-educated women were aggregated into one category, whereas in the later cohorts this group was subdivided into 3 years of college versus 4 or more years of college, the last two categories were combined for the purpose of calculating the dysgenic fertility ratios, making these ratios comparable across all the cohorts.

With regard to the secular trends in fertility in relation to educational level in White women, one may observe dysgenic fertility in the first cohort, born between 1890 and 1895, among whom the dysgenic fertility ratio was 2.79, and a slackening of dysgenic fertility among those born 20 years later, for whom there was a dysgenic fertility ratio of 1.42. For the next cohort, born between 1946 and 1955, the dysgenic fertility remained about the same. The final data for White women consisted of the anticipated final fertility of the 1956–1972 cohort; among these, dysgenic fertility slackened to 1.18. The anticipated fertility figures may not be realized because some women may have fewer children than they anticipated, perhaps because of infertility or for other reasons, whereas others may have more, perhaps because of unplanned births. Nevertheless, these data are the best available for recent birth cohorts and will probably turn out to be approximately accurate.

Considering Black women, it may be observed that there was the same initially high dysgenic fertility ratio of 1.93 among the 1915–1919 birth cohort, slackening to 1.62 among those born between 1946 and 1955, and then to 1.34 for anticipated final fertility among the most recent cohort born between 1956 and 1972. The same secular trend was present for the two Hispanic cohorts, among whom dysgenic fertility declined from 2.09 in the 1946–1955 cohort to 1.31 among the 1956–

1972 cohort. The general secular trend for dysgenic fertility to slacken although not to be eliminated was present, therefore, for all three racial groups.

Turning now to the racial differences in fertility, one can observe that the numbers of children of White women were lower than those of Blacks at all educational levels, except for the college-educated cohort, and that the Hispanics had the greatest numbers of children. Numerous studies carried out since World War 1 have shown that Whites score higher on average on IQ tests than Blacks and Hispanics by approximately 15 and 8 IQ points, respectively (Herrnstein & Murray, 1994); therefore, the dysgenic fertility in the U.S. population as a whole has been greater than the dysgenic fertility within the racial and ethnic subpopulations.

Concerning racial differences, the dysgenic fertility ratios were greater for Blacks than for Whites in all three cohorts for which data are available for both groups. This finding is consistent with the greater dysgenic fertility among Blacks for intelligence found by Vining (1995) and by Van Court and Bean (1985), reviewed earlier. The Hispanic dysgenic fertility ratio is the largest among the 1946–1955 cohort, but it is slightly lower than the ratio for Blacks among the 1956–1972 cohort.

Numerous studies have been carried out worldwide concerning the relationship between women's educational level and their fertility. The investigators have all found virtually the same inverse association as found in the United States. These results have been found in Britain (General Household Survey, 1989), Canada (Grindstaff, Balakrishnan, & Dewit, 1991), continental Europe (Nohara-Atoh, 1980), and Japan (Ogawa & Retherford, 1993).

Studies in Developing Countries

The inverse relationship between women's educational level and their fertility is also widely present in the economically less developed world. It has been collated by Ashurst, Balkaran, and Casterline (1984) from studies carried out in the late 1970s in Latin America, the Caribbean,

Table 6

Fertility of Women in Eight Latin American Countries in the Late 1980s in Relation to Years of Education

	Years of education					
Country	0	1–3	4–6	7–9	10+	Dysgenic ratio
Bolivia	6.2	6.4	5.3	4.2	2.8	2.21
Brazil	6.7	5.2	3.4	2.8	2.2	3.05
Colombia	5.6	4.5	3.6	2.5	1.8	3.11
Ecuador	6.4	6.3	4.7	3.5	2.6	2.46
El Salvador	6.0	5.2	3.9	3.5	2.5	2.40
Guatemala	6.9	5.6	4.2	2.8	2.7	2.56
Mexico	6.4	6.3	4.0	2.7	2.4	2.67
Peru	7.4	6.1	4.6	3.7	2.5	2.96

Note. Adapted from data of Martin and Juarez (1995).

Asia, and North Africa. Only in sub-Saharan Africa is the inverse relationship absent or negligible. The reason for this is that natural fertility is generally unchecked by family limitation throughout all educational levels in sub-Saharan Africa.

More recent data (from the late 1980s) for eight Latin American countries on the relationship between women's educational level and their fertility has been published by Martin and Juarez (1995). The data, for all women aged 15–49, were calculated as age-standardized total fertility rates. The results are shown in Table 6. In these countries, the best educated women, with 10 or more years of education, had between 1.8 and 2.8 children, whereas those with no education had 2 or 3 times as many. Dysgenic fertility in these countries is evidently pronounced.

CONCLUSION

Four types of data indicating an inverse relationship between intelligence and fertility have been presented. There is evidence for an inverse relationship between (a) SES and fertility, (b) intelligence and number of siblings, (c) intelligence and fertility, and (d) educational level and fertility. All four lines of evidence point in the same direction and to the same conclusion: Fertility has been dysgenic in the economically developed world since the early decades of the 19th century and in most of the economically developing world during the 20th century.

The data showing an inverse association between SES and fertility go back to the cohorts born in the second quarter of the 19th century. This means that dysgenic fertility has been present for about five generations. Retherford and Sewell's (1988) American study indicated that the genotypic decline was 0.64 IQ points for the generation born in 1940. If this figure is projected back for five generations, it can be concluded that American Whites have suffered a genotypic decline of 3.20 IQ points over the five generations. This is almost certainly an underestimate because dysgenic fertility was considerably greater in the earlier generations than among the 1940 cohort from which the figure of 0.64 IQ points is derived. When this is taken into account, the magnitude of the deterioration of genotypic intelligence in the United States appears to have been about 5 IQ points since the early 19th century. Dysgenic fertility has probably produced a similar deterioration in Europe, considering that dysgenic fertility in relation to SES was present in the early 19th century and that the magnitude of the inverse relationship between intelligence and fertility has been about the same. In the economically developing world, such as the Latin American countries represented in Table 6, dysgenic fertility is very strong, but has probably not been in place for so long.

The proposition that the genotypic intelligence of modern populations is deteriorating is not directly verifiable but is an inference derived from two premises: the inverse relationship between intelligence and fertility, and the heritability of intelligence. Because the two premises are solid, the inference appears to be solid. Recently, however, the inference has been challenged by Preston and Campbell (1993), who claimed to demonstrate that dysgenic fertility is compatible, after some generations, with a stable population IQ. Preston restates this argument in the present volume (chapter 15). In my opinion, the argument is flawed, for the reasons given by Coleman (1993) and Loehlin (chapter 16, this volume). If Preston and Campbell's argument were correct, natural selection by differential reproductive fitness would not work, and the fundamental theorem of biology since the publication of Darwin's *Origin of Species* would be overthrown. I do not believe that biology is ready for such a drastic paradigm shift.

The dysgenic fertility and consequent deterioration of genotypic intelligence that have been in place since the second quarter of the 19th century have been accompanied by environmental improvements that have brought about rises in phenotypic intelligence. It can be predicted that the environmental improvements will in due course show diminishing returns and peter out. If dysgenic fertility is still present when this point is reached, it can be anticipated that phenotypic intelligence will start to decline. Insofar as the maintenance of a high level of civilization depends on the intelligence of its population, the quality of U.S. civilization will also deteriorate. It is a curious fact that the evidence pointing to this conclusion has received no mention in contemporary textbooks of psychology and sociology.

REFERENCES

Ashurst, H., Balkaran S., & Casterline, J. B. (1984). Socio-economic differentials in recent fertility. *Comparative Studies, 42,* 1–61.

Bachu, A. (1991). *Fertility of American women: June 1990.* Washington, DC: U.S. Government Printing Office.

Bajema, C. (1963). Estimation of the direction and intensity of natural selection in relation to human intelligence by means of the intrinsic rate of natural increase. *Eugenics Quarterly, 10,* 175–187.

Bajema, C. J. (1968). Relation of fertility to occupational status, IQ, educational attainment and size of family of origin: A follow-up of the male Kalamazoo public school population. *Eugenics Quarterly, 15,* 198–203.

Blake, J. (1989). *Family size and achievement.* Berkeley: University of California Press.

Bouchard, T. J. (1993). The genetic architecture of human intelligence. In P. A. Vernon (Ed.), *Biological approaches to the study of human intelligence.* Norwood, NJ: Ablex.

Brody, N. (1992). *Intelligence.* San Diego, CA: Academic Press.

Burt, C. (1952). *Intelligence and fertility* (2nd ed.). London: Eugenics Society.

Caplan, N., Choy, M. H., & Whitmore, J. K. (1992, February). Indochinese refugee families and academic achievement. *Scientific American,* pp. 18–24.

Capron, C., & Duyme, L. (1989) Assessment of effects of socio-economic status on IQ in a full cross-fostering design. *Nature, 340,* 552–553.

Carlile, R. (1826). *Every woman's book.* London: Tallis.

Cattell, R. B. (1937). *The fight for our national intelligence.* London: P. S. King.

Chapman, J. C., & Wiggins, D. M. (1925). The relation of family size to intelligence of offspring and socio-economic status of family. *Pedagogical Seminary and Journal of Genetic Psychology, 32,* 414–421.

Coleman, J. S. (1993). Comment on Preston and Campbell's "Differential fertility and the distribution of traits." *American Journal of Sociology, 98,* 1020–1032.

Daly, M., & Wilson, M. (1983). *Sex, evolution and behaviour.* Boston: Willard Grant Press.

Dobzhansky, T. (1962). *Mankind evolving.* New Haven, CT: Yale University Press.

Finkel, D., Pedersen, N. L., McGue, M., & McClearn, G. E. (1995). Heritability of cognitive abilities in adult twins: Comparison of Minnesota and Swedish data. *Behaviour Genetics, 25,* 421–431.

Fisher, R. A. (1929). *The genetical theory of natural selection.* Oxford, England: Clarendon Press.

Freedman, R., Whelpton, P. K., & Campbell, A. A. (1959). *Family planning, sterility and population growth.* New York: McGraw-Hill.

Galbraith, R. C. (1982). Sibling spacing and intellectual development: A closer look at the confluence model. *Developmental Psychology, 18,* 151–173.

Galton, F. (1865). Hereditary talent and character. *MacMillan's Magazine, 12,* 157–166, 318–327.

Galton, F. (1869). *Hereditary genius.* London: Macmillan.

Galton, F. (1874). *English men of science.* London: Methuen.

Galton, F. (1883). *Inquiries into human faculty and its development.* London: Dent.

Galton, F. (1908). *Memories of my life.* London: Methuen.

Galton, F. (1909). *Essays in eugenics.* London: Eugenics Society.

General Household Survey. (1989). *Report.* London: Her Majesty's Stationary Office.

Glass, D. V. (1967). Fertility trends in Europe since the second world war. In *Proceedings of the University of Michigan Conference on Fertility and Family Planning.* Ann Arbor: University of Michigan.

Grindstaff, C. F., Balakrishnan, T. R., & Dewit, D. J. (1991). Educational attainment, age at first birth, and lifetime fertility: An analysis of Canadian fertility survey data. *Canadian Review of Sociology and Anthropology, 28,* 324–339.

Haines, M. R. (1989). Social class differentials during fertility decline: England and Wales revisited. *Population Studies, 43,* 305–323.

Haines, M. R. (1992). Occupation and social class during fertility decline: Historical perspectives. In J. R. Gillis, L. A. Tilly, & D. Levine (Eds.), *The European experience of declining fertility.* Oxford, England: Blackwell Scientific.

Heron, D. (1906). *On the relation of fertility in man to social status.* London: Drapers.

Herrnstein, R. J. (1971). *IQ in the meritocracy.* Boston: Little, Brown.

Herrnstein, R. J., & Murray, C. (1994). *The bell curve: Intelligence and the class structure in American life.* New York: Free Press.

Higgins, J. V., Reed, E. W., & Reed, S. G. (1962). Intelligence and family size: A paradox resolved. *Eugenics Quarterly, 9,* 84–90.

Ho, D. Y. F. (1979). Sibship variables as determinants of intellectual-academic ability in Hong Kong pupils. *Genetic Psychology Monographs, 100,* 21–39.

Huxley, J. S. (1936). *Man stands alone.* London: Harper.

Jencks, C. R. (1972). *Inequality.* New York: Basic Books.

Johnson, D. M. (1948). Applications of the standard score to social statistics. *Journal of Social Psychology, 27,* 217–227.

Jones, D. C., & Carr-Saunders, A. M. (1927). The relation between intelligence and social status among orphan children. *British Journal of Psychology, 17,* 343–364.

Kevles, D. J. (1985). *In the name of eugenics.* New York: Knopf.

Keynes, M. (1993). Sir Francis Galton. In M. Keynes (Ed.), *Sir Francis Galton: The legacy of his ideas.* London: Macmillan.

Kiernan, K. E. (1989). Who remains childless? *Journal of Biosocial Science, 212,* 387–398.

Kiernan, K. E., & Diamond, I. (1982). *Family of origin and educational influences on age at first birth: The experiences of a British birth cohort* (Research paper No. 81). London: Centre for Population Studies.

Kiser, C. V. (1970). Changing patterns of fertility in the United States. *Social Biology, 17,* 302–315.

Kiser, C. V., & Frank, M. E. (1967). Factors associated with the low fertility of non-White women of college attainment. *Millbank Memorial Fund Quarterly, 15,* 427–449.

Knowlton, C. (1877). *The fruits of philosophy.* London: MacMillan.

Kolvin, I., Miller, F. J. W., Scott, D. M., Gatzanis, S. R. M., & Fleeting, M. (1990). *Continuities of deprivation.* Aldershot, England: Avebury.

Lawrence, E. M. (1931). An investigation into the relation between intelligence and inheritance [Monograph supplement]. *British Journal of Psychology, 16,* 1–80.

Leahy, A. (1935). Nature–nurture and intelligence. *Genetic Psychology Monographs, 17,* 4.

Lentz, T. F. (1927). The relation of IQ to size of family. *Journal of Educational Psychology, 18,* 486–496.

Lewontin, R. C. (1993). *The doctrine of DNA.* London: Penguin Books.

Lynn, R. (1996). *Dysgenics: Genetic deterioration in modern populations.* Westport, CT: Praeger.

Martin, T. C., & Juarez, F. (1995). The impact of education on fertility in Latin America. *Family Planning Perspectives, 21,* 52–57.

Maxwell, J. (1969). Intelligence, education and fertility: A comparison between the 1932 and 1947 Scottish surveys. *Journal of Biosocial Science, 1,* 247–271.

McNemar, Q. (1942). *The revision of the Stanford–Binet scale.* Oxford, England: Oxford University Press.

Muller, J. H. (1938–1939). *Out of the night.* New York: Vanguard Press.

Munsinger, H. (1975). Children's resemblance to their biological and adopting parents in two ethnic groups. *Behaviour Genetics, 5,* 239–254.

Nohara-Atoh, M. (1980). *Social determinants of reproductive behaviour in Japan.* Unpublished doctoral dissertation, University of Michigan, Ann Arbor.

Nystrom, S., Bygren, L. O., & Vining, D. R. (1991). Reproduction and level of intelligence. *Scandinavian Journal of Social Medicine, 19,* 187–189.

Ogawa, N., & Retherford, R. D. (1993). The resumption of fertility decline in Japan: 1973–92. *Population and Development Review, 19,* 703–741.

Osborn, F. (1951). *Preface to eugenics* (2nd ed.). New York: Harper.

Osborne, R. T. (1975). Fertility, IQ and school achievement. *Psychological Reports, 37,* 1067–1073.

Papavassiliou, I. T. (1954). Intelligence and family size. *Population Studies, 7,* 222–226.

Pearson, K. (1912). *The groundwork of eugenics.* Cambridge, England. Eugenics Laboratory.

Preston, S. H. (1974). Differential fertility, unwanted fertility, and racial trends in occupational achievement. *American Sociological Review, 39,* 492–506.

Preston, S. H., & Campbell, C. (1993). Differential fertility and the distribution of traits: The case of IQ. *American Journal of Sociology, 98,* 997–1019.

Retherford, R. D., & Sewell, W. H. (1988). Intelligence and family size reconsidered. *Social Biology, 35,* 1–40.

Retherford, R. D., & Sewell, W. H. (1991). Birth order and intelligence: Further tests of the confluence model. *American Sociological Review, 56,* 141–158.

Skipp, V. (1978). *Crisis and development: An ecological case study of the Forest of Arden 1570–1674.* Cambridge, England: Cambridge University Press.

Skodak, M., & Skeels, H. M. (1949). A final follow-up study of one hundred adopted children. *Journal of Genetic Psychology, 75,* 85–125.

Sutherland, H. E. G., & Thomson, G. H. (1926). The correlation between intelligence and size of family. *British Journal of Psychology, 17,* 81–92.

Terman, L. M. (1921). *The intelligence of school children.* London: Harrap.

Thomson, G. (1946). The trend of national intelligence. *Eugenics Review, 38,* 9–18.

Van Court, M., & Bean, F. D. (1985). Intelligence and fertility in the United States: 1912–1982. *Intelligence, 9,* 23–32.

Vining, D. R. (1982). On the possibility of the re-emergence of a dysgenic trend with respect to intelligence in American fertility differentials. *Intelligence, 6,* 241–264.

Vining, D. R. (1995). On the possibility of the re-emergence of a dysgenic trend: An update. *Personality and Individual Differences, 19,* 259–265.

Waller, J. (1971). Differential reproduction: Its relation to IQ test scores, education and occupation. *Social Biology, 18,* 122–136.

Weinberg, R. A., Scarr, S., & Waldman, I. D. (1992). The Minnesota Transracial Adoption Study: A folow-up of IQ test performance at adolescence. *Intelligence, 16,* 117–135.

Weiss, V. (1990). Social and demographic origins of the European proletariate. *Mankind Quarterly, 31,* 126–152.

Wong, D. H. (1980). *Class fertility trends in Western nations.* New York: Arno Press.

Young, M., & Gibson, J. B. (1965). Social mobility and fertility. In J. E. Meade & A. S. Parkes (Eds.), *Biological aspects of social problems.* Edinburgh, Scotland: Oliver & Boyd.

Zajonc, R. B., & Markus, G. B. (1975). Birth order and intelectual development. *Journal of Personality and Social Psychology, 37,* 1325–1341.

14

Problems in Inferring Dysgenic Trends for Intelligence

Irwin D. Waldman

A number of recent publications have highlighted the importance of intelligence to a variety of real-world outcomes, ranging from educational level, occupational status, and income, to divorce, criminality, and welfare dependency (Herrnstein & Murray, 1994; Hunt, 1995). Along with this renewed interest in the predictive power of IQ has come an increased concern for the intellectual level of American society. Indeed, even as secular increases in IQ of about 3 points per decade have been reported (Flynn, 1984), a number of authors (Herrnstein & Murray, 1994; Itzkoff, 1994; Lynn, chapter 13, this volume) have expressed fears of a decline in intelligence and the deleterious impact this will have on U.S. culture. For example, Herrnstein and Murray (1994) stated that

> mounting evidence indicates that demographic trends are exerting downward pressure on the distribution of cognitive ability in the United States and that the pressures are strong enough

I thank Greg Carey, John DeFries, and John Loehlin for their helpful discussions of the relevant issues and literature, and George Vogler for his archaeological efforts at unearthing his 10-year-old Behavior Genetics Association presentation. Much of the background work on this chapter was completed while I was a visiting professor at the Institute for Behavioral Genetics, University of Colorado, Boulder.

> to have social consequences. . . . Putting the pieces together, something worth worrying about is happening to the cognitive capital of the country. (p. 341)

A number of authors have taken these concerns a step further, not only by documenting the social ills that are expected to follow from such a decline but also by making strong assertions regarding its genetic etiology. These authors (e.g., Herrnstein & Murray, 1994; Lynn, chapter 13, this volume) have described this national decline in intelligence as a "dysgenic trend." Even the most casual student of the history of psychology knows that such concerns are not new. As Lynn points out, Francis Galton expressed such concerns toward the end of the 19th century; his belief in eugenic principles and in the desirability of eugenic practices came to be shared by a broad spectrum of English intellectuals that included not only R. A. Fisher and Cyril Burt, but also J. B. S. Haldane and George Bernard Shaw (Kevles, 1986).

It also will come as little surprise to find that eugenic views are held rather strongly by their proponents. Many of these arguments are passionately expressed and readily leave the realm of research and enter that of advocacy. For example, Lynn (chapter 13, this volume) states that "it is my thesis in this chapter that . . . there is an inverse relationship between intelligence and fertility in modern populations and that as a result genotypic intelligence is deteriorating" (p. 336) and that "the inverse relationship between SES and fertility found in the . . . 19th century implies that the genes for high intelligence were replicating less successfully than those for low intelligence and that the genotypic intelligence of the populations was deteriorating" (p. 339). Once the evangelical fervor is cleared away, however, how much evidence is there for a dysgenic trend in intelligence? Furthermore, what do terms like *dysgenic trend* and *genotypic intelligence* really mean? It is my intention in this chapter to take a critical look at these issues and, in particular, to highlight some of the problems in inferring dysgenic trends for intelligence. I focus primarily on two sets of issues: those concerned with differential fertility by intelligence level and those concerned with genetic influences on intelligence. It is my contention that these problems seriously undermine the inflated claims of a dysgenic trend and that the evidence for such a trend is surprisingly weak given the strength of the rhetoric advanced by Lynn and other authors.

WHAT'S IN A "DYSGENIC TREND"?

How do eugenicists define a *dysgenic trend*? Some insight into this question may be gained by considering the conclusion of chapter 13, where Lynn states that "the proposition that the genotypic intelligence of modern populations is deteriorating is not directly verifiable but is an inference derived from two premises: the inverse relationship between intelligence and fertility, and the heritability of intelligence. Because the two premises are solid, the inference appears to be solid" (p. 358).

Authors such as Lynn, therefore, appear to define *dysgenic trend* as the conjunction between (a) increasing fertility with declining IQ and (b) the heritability of IQ. In contrast, I think that it is useful to define the term *dysgenic trend* in a more technical manner, in population genetic terms, as a decrease in the alleles predisposing to higher IQ and an increase in the alleles predisposing to lower IQ in a population from one generation to succeeding generations. I believe this definition to be more accurate; although it requires the two premises described by Lynn, it also depends on additional considerations regarding genetic transmission. Framing a dysgenic trend in population genetic terms is important not only for the sake of increased clarity, but also because the simplified definition offered by Lynn and others glosses over a number of factors that may be critical for determining changes in the genetic influences on intelligence in populations across generations. I return to these population genetic considerations after discussing a number of issues that arise in studies of the relation between differential fertility and intelligence.

ISSUES IN RESEARCH EXAMINING DIFFERENTIAL FERTILITY AND INTELLIGENCE

Sample Size

It is not clear from the tables in chapter 13, this volume, what the sample sizes were in many of the cited studies. Although some of the samples appear to have been large (e.g., that of Vining, 1982, 1995), the samples in other studies that are important to Lynn's argument are quite small. Even in Vining's studies, for example, the sizes of the African American samples (*N*s ranging from 462 to 473 depending on cohort) are quite small for drawing inferences about differential fertility by IQ level. This is especially true when one considers the importance of the data from women with extremely low or high IQs. Given the small numbers of women who have these extreme IQ scores, the fertility–IQ relation in these samples is likely to be highly unstable.

Ecological Correlations

The relation between fertility and intelligence is a research issue at the level of individual participants. Nonetheless, it appears that a number of studies bearing on this issue cited by Lynn used group-level data. The problem is that the relation between two variables can differ considerably when estimated using group-level data versus individual-level data (Gollob, 1991). There is a good example of this in chapter 13, which Lynn in fact mentions: Osborne (1975) had IQ data for a large number of Georgia children and related the average IQs to the average fertility levels by county. It is interesting that the fertility–IQ correlation in this study ($r = -.49$) was by far the largest cited by Lynn, which raises the possibility that this estimate is artifactual owing to a reliance on group-level as opposed to individual-level analyses. It is not clear from Lynn's chapter to what extent this was a problem in other studies.

Corrections for Parental Age at Childbearing and for Differential Infant Mortality

There are two demographic factors—parental age at childbearing and differential infant mortality by IQ level—that could influence the relation between fertility and IQ. Because these issues are given little mention in Lynn's review, it is difficult to know how they were handled in the studies cited. It is likely that children born to parents of lower IQ have higher infant and child mortality rates than those born to parents of higher IQ and that this could offset some of the greater fertility among lower IQ individuals seen in some studies. The issue of parental age at childbearing is a more difficult one. Although individuals with lower IQ tend to have children at younger ages than individuals with higher IQ, a number of investigators examined the fertility–IQ relation in individuals who were early in their potential childbearing years. It is not clear from Lynn's chapter what corrections, if any, were made for parental age at childbearing or what its effect on the observed fertility–IQ relation might be.

The Magnitude of Differential Fertility Within and Across Studies

Perusal of the tables in chapter 13, this volume, reveals a number of facts about the relation between fertility and IQ or education that differ from those emphasized by Lynn. First, the degree of differential fertility appears to be highly dynamic over time and often not in the direction that Lynn suggests. According to Table 5, for example, within each ethnic group the fertility ratio for lower as compared with higher educated women appears to be decreasing in succeeding generations as opposed to increasing, as Lynn suggests. The other tables also suggest that differential fertility is highly dynamic and can change dramatically from generation to generation.

Second, Lynn often uses "dysgenic ratios" to summarize the fertility to IQ/education relation, as opposed to regression coefficients or correlations. Lynn computed the dysgenic ratios by dividing fertility rates in the lowest IQ/education category by those in the highest IQ/education category, thus giving undue weight to the extreme groups, especially considering that these groups are composed of relatively few individuals relative to the middle of the IQ/education distributions. Summarizing the data using regression coefficients or correlations likely would yield fertility–IQ/education relations that are lower in magnitude than those reported by Lynn. Indeed, the studies in which these relations were reported in a correlational metric generally yielded fertility–IQ/education correlations of low magnitude.

Third, the low correlations, as well as examination of the data in Tables 2 and 5, highlight the fact that rather than a steady decrease in fertility from lower to higher IQ/education levels, one sees marked variability in this relation. This implies that even in the face of an overall fertility decline with increasing IQ and education, many individuals of higher IQ and education will have more children than individuals of lower IQ and education. These "residuals" become especially important when one considers the number of individuals at different IQ and educational levels. Because there are so many individuals in the middle of the distribution relative to the extremes, even large differences in fertility at the extremes may be offset by fertility patterns in the middle of the IQ/education distribution. Lynn never mentions the importance of the vastly lower numbers of individuals at the extremes than at the middle of the distribution.

ISSUES CONCERNING GENETIC INFLUENCES ON INTELLIGENCE

As mentioned earlier, a number of important issues regarding genetic transmission, in addition to those concerning differential fertility, are involved in inferring dysgenic trends. Indeed, because Lynn uses terms such as *dysgenic trend* and *genotypic intelligence* throughout his chapter, particular attention needs to be paid to the mechanisms of genetic transmission that bear on generational trends in intelligence levels. These issues center on the inference that such differential fertility patterns, if they do in fact exist, are the result of "dysgenic" rather than "dysenvironmental" trends.

Heritability Estimates for Intelligence

The estimate of heritability that Lynn uses for intelligence, .8, is based on only two studies (Bouchard, Lykken, McGue, Segal, & Tellegen, 1990; Finkel, Pedersen, McGue, & McClearn, 1995), which represent only a small fraction of the available data on familial resemblance for IQ (e.g., Bouchard & McGue, 1981; Loehlin, 1989). Reliance on these studies puts Lynn's heritability estimate at the extreme high end of the range of heritability estimates given for IQ in the literature (e.g., Herrnstein & Murray, 1994). As such, it is likely to be too high; a figure of .6 would be more reasonable on the basis of meta-analyses of the relevant data on familial resemblance for IQ (Chipuer, Rovine, & Plomin, 1990; Loehlin, 1989). A more complicated issue is that heritability for IQ appears to increase dramatically from childhood throughout adulthood, and it is not readily apparent which is the appropriate heritability estimate to use in the equations that Lynn presents.

Additive Versus Nonadditive Genetic Variance for IQ

A bigger problem for Lynn's argument is that the heritability estimate he uses is what is known as a *broad*—as opposed to a *narrow*—heritability estimate (Falconer, 1989). Broad heritability estimates include both additive and nonadditive genetic variance, whereas narrow heritability estimates include only additive genetic variance. Additive genetic variance refers to genetic influences on a trait that are shared by parents and their offspring, whereas nonadditive genetic variance refers to genetic influences produced by the interactive or configural effects of different alleles at the same or at different genetic loci. Nonadditive genetic influences thus can be shared by identical twins, and to a lesser degree by siblings and fraternal twins, but not by parents and their offspring because parents contribute to their offspring only one allele at each genetic locus.

In recent behavior genetic analyses (Chipuer et al., 1990; Loehlin, 1989) in which familial IQ correlations from many studies for a wide variety of familial relationships were used, nonadditive genetic influences have emerged as an important source of IQ variation once assortative mating was considered. Both Chipuer et al. (1990) and Loehlin (1989) found broad heritability estimates of about .5–.6 for IQ, but also estimated that about .15–.20 of this broad heritability was accounted for by nonadditive genetic variance. Thus, the narrow heritability estimate for IQ, the one that is most appropriate for Lynn's equations in that it reflects parent–offspring similarity, would be much closer to .3 than .8.

The Dynamic Nature of Heritability and Assortative Mating for Intelligence

Another assumption implicit in Lynn's thesis is that the factors that affect genetic influences on IQ, primarily narrow heritability and assortative mating, are highly stable across generations. If the genetic influences on IQ vary considerably from generation to generation, drawing accurate inferences regarding dysgenic trends will be difficult if not impossible.

Not much is known about the stability of genetic influences on intelligence from generation to generation. A few studies, however, have shed some light on this issue. Vogler and Rao (1986) examined secular trends in parent–offspring and spousal correlations for IQ using extensive data published by Bouchard and McGue (1981). In Figures 1 and 2, spousal and parent–offspring correlations for IQ are plotted against

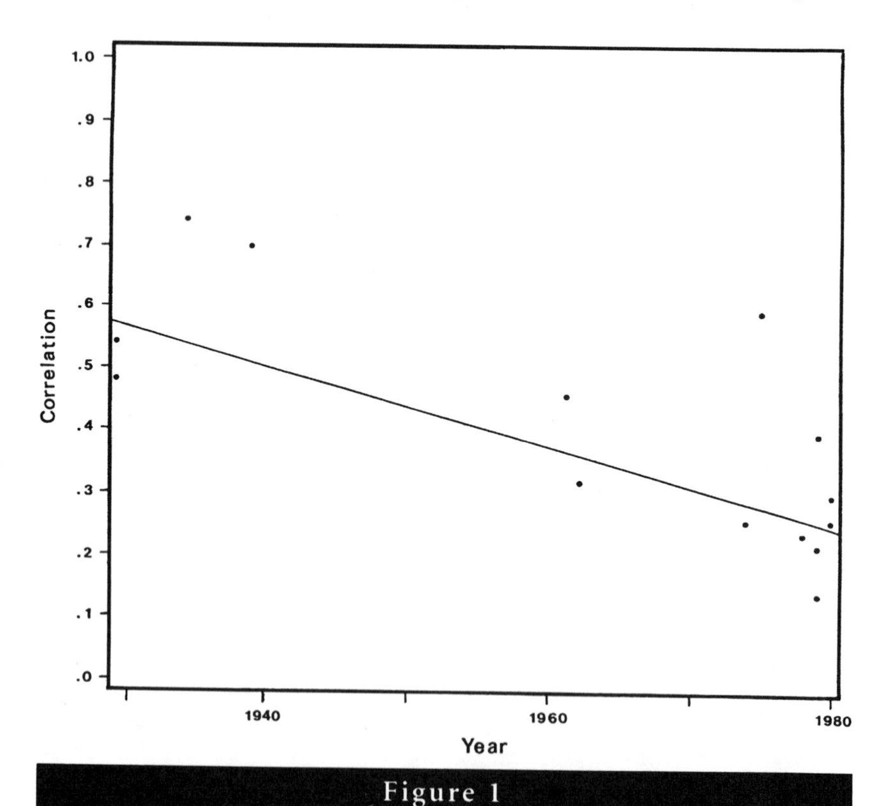

Figure 1

Spousal correlation for IQ by year of publication of study. From *Using Genetic Models to Explore Environmental Sources of Familial Resemblance for Intelligence* by G. P. Vogler and D. C. Rao, paper presented at the 16th annual meeting of the Behavior Genetics Association, Honolulu, HI, June 1986. Reprinted with permission from the authors.

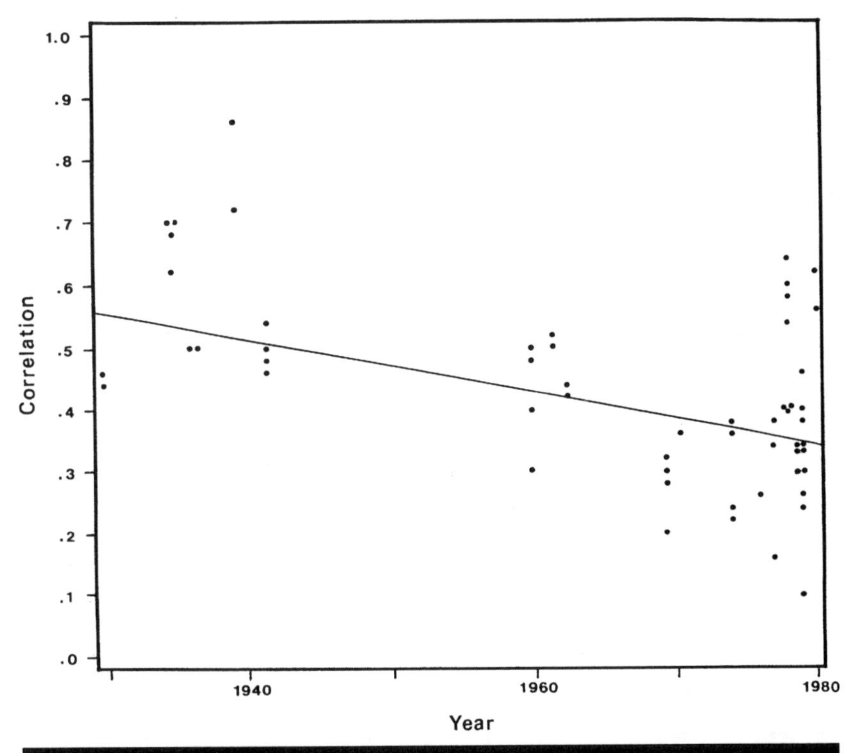

Figure 2

Parent–offspring correlation for IQ by year of publication of study. From *Using Genetic Models to Explore Environmental Sources of Familial Resemblance for Intelligence*, by G. P. Vogler and D. C. Rao, paper presented at the 16th annual meeting of the Behavior Genetics Association, Honolulu, HI, June 1986. Reprinted with permission from the authors.

the date of publication of the study, which is a rough index of participants' age or birth cohort. As shown in the graphs, both the spousal and parent–offspring correlations for IQ appeared to decline over time, suggesting that the magnitude of genetic influences on IQ also is declining. In contrast, there are some data that suggest that assortative mating by educational level has increased in recent generations (Mare, 1991).

A study by Heath et al. (1985) is relevant to the issue of changes in heritability for intelligence across generations. In this study, the effects of secular changes in educational policy on genetic and environmental influences on educational attainment were examined in a large sample of Norwegian twins and their parents. It was found that increases in educational opportunities over time led to a substantial increase in heritability for educational attainment in younger birth cohorts for male but not female twins. (The sex difference in moderation of heritability may have been due to greater accessibility of educational opportunities for males than females in the younger cohorts.) The results of these studies highlight the possibility that genetic influences on IQ may change dramatically from generation to generation, rendering inferences regarding dysgenic trends for intelligence difficult at best.

Further Considerations Regarding Assortative Mating

There are two additional issues concerning assortative mating that bear on this problem. First, researchers have suggested that assortative mating for intelligence may be greater at the higher than at the lower ends of the IQ distribution (Mare, 1991). The consequence for genetic transmission with respect to intelligence would be that children of high-IQ parents receive more of the alleles predisposing to high IQ than children of low-IQ parents receive alleles predisposing to low IQ. It is not known how strongly this pattern would affect dysgenic trends for IQ or whether this heterogeneity in assortative mating is enough to countervail any differential fertility by IQ/educational level, but such compensation is certainly possible.

Second, different forms of assortative mating may occur for intelligence and education, such that some involve only cultural factors and common environmental influences (viz., "social homogamy"), whereas others involve more complex forms of assortative mating that have implications for genetic as well as environmental transmission from parents to children. In understanding the effects of assortative mating on genetic influences on intelligence, it is important to have an accurate model of which forms of assortative mating are operative in a population at a given time. In short, knowledge regarding which form of assortative mating is operative and its effect on genetic influences on intelligence in present-day society is limited; therefore, the difficulties in drawing inferences regarding generational changes in assortative mating and their implications for changes in genetic influences on intelligence are formidable.

CONCLUSION

In this chapter, I have raised a number of issues that challenge inferences regarding dysgenic trends for intelligence. These issues concern both differential fertility by IQ/educational level and genetic influences on intelligence. Problems with inferring differential fertility by IQ/educational level include use of small samples, use of group-level rather than individual-level correlations, the handling of parental age and differential infant mortality, the use of dysgenic ratios to summarize the fertility by IQ relation, and the dynamic nature of this relation over time. Problems with inferring genetic influences on intelligence include choosing an accurate estimate for the heritability of intelligence, distinguishing additive from nonadditive genetic influences, understanding the dynamic nature of heritability and assortative mating for intelligence over time, and understanding the implications of assortative mating. In contrast to authors who have taken an extreme stance on both the existence of dysgenic trends for intelligence and their dire implications for present-day society, I have portrayed a number of the many problems and complexities involved in inferring dysgenic trends rather than take a firm stand on whether dysgenic trends for intelligence are in fact taking place. Nevertheless, I think that the problems and complexities raised in this chapter cast serious doubt on the notion that dysgenic trends are operating and are of sufficient magnitude to cause the levels of concern expressed in the technical and popular literature on intelligence.

REFERENCES

Bouchard, T. J., Jr., Lykken, D. T., McGue, M., Segal, N. L., & Tellegen, A. (1990). Sources of human psychological differences: The Minnesota Study of Twins Reared Apart. *Science, 250,* 223–228.

Bouchard, T. J., Jr., & McGue, M. (1981). Familial studies of intelligence: A review. *Science, 212,* 1055–1059.

Chipuer, H. M., Rovine, M., & Plomin, R. (1990). LISREL modeling: Genetic and environmental influences on IQ revisited. *Intelligence, 14,* 11–29.

Falconer, D. S. (1989). *Introduction to quantitative genetics.* New York: Wiley.

Finkel, D., Pedersen, N. L., McGue, M., & McClearn, G. E. (1995). Heritability of cognitive abilities in adult twins: Comparison of Minnesota and Swedish data. *Behavior Genetics, 25,* 421–431.

Flynn, J. R. (1984). The mean IQ of Americans: Massive gains 1932–1978. *Psychological Bulletin, 95,* 29–51.

Gollob, H. F. (1991). Methods for estimating individual- and group-level correlations. *Journal of Personality and Social Psychology, 60,* 376–381.

Heath, A. C., Berg, K., Eaves, L. J., Solaas, M. H., Corey, L. A., Sundet, J., Magnus, P., & Nance, W. E. (1985). Education policy and the heritability of educational attainment. *Nature, 314,* 734–736.

Herrnstein, R. J., & Murray, C. (1994). *The bell curve: Intelligence and class structure in American life.* New York: Free Press.

Hunt, E. B. (1995). The role of intelligence in modern society. *American Scientist, 83,* 356–368.

Itzkoff, S. (1994). *The decline of intelligence in America.* Westport, CT: Praeger.

Kevles, D. J. (1986). *In the name of eugenics.* Berkeley: University of California Press.

Loehlin, J. C. (1989). Partitioning environmental and genetic contributions to behavioral development. *American Psychologist, 44,* 1285–1292.

Mare, R. D. (1991). Five decades of educational assortative mating. *American Sociological Review, 56,* 15–32.

Osborne, R. T. (1975). Fertility, IQ, and school achievement. *Psychological Reports, 37,* 1067–1073.

Vining, D. R. (1982). On the possibility of the re-emergence of a dysgenic trend with respect to intelligence in American fertility differentials. *Intelligence, 6,* 241–264.

Vining, D. R. (1995). On the possibility of the re-emergence of a dysgenic trend: An update. *Personality and Individual Differences, 19,* 259–265.

Vogler, G. P., & Rao, D. C. (1986, June). *Using genetic models to explore environmental sources of familial resemblance for intelligence.* Paper presented at the 16th annual meeting of the Behavior Genetics Association, Honolulu, HI.

15

Differential Fertility by IQ and the IQ Distribution of a Population

Samuel H. Preston

Among the factors that affect the IQ distribution of a population are the rates of childbearing among persons of different IQ categories. IQ scores among their offspring are, on average, lower for persons who themselves have low scores on IQ tests. The tendency for IQ scores to be correlated across generations reflects a combination of genetic and environmental influences (Neisser et al., 1996).

The fact that IQ scores are correlated across generations, combined with the fact that lower IQ individuals have had higher fertility rates throughout most of the 20th century in the United States and elsewhere (chapter 13, this volume; Van Court & Bean, 1985; Vining, 1982), has given rise to fears that the IQ distribution is, or soon will be, deteriorating as a result of fertility differentials. This fear was vividly expressed by Richard Herrnstein in a widely circulated article in *Atlantic Monthly.* He was referring to U.S. fertility differentials by IQ in his final words: "We ought to bear in mind that in not too many generations differential fertility could swamp the effects of anything else we may do about our economic standing in the world" (Herrnstein, 1989, p. 79).

The expectation that higher fertility for persons of lower IQ will produce a declining IQ distribution in the population is perfectly reasonable and commonsensical. However, it is not always correct, and in fact it may not be correct in most populations. Nature is full of examples in which members of a species gain a reproductive advantage by virtue of some trait, perhaps unusually large antlers or colorful plumage, without any perceptible change in the distribution of the trait in the population. Evolutionary biologists have shown that differential fertility according to some trait is often consistent with an equilibrium, or constant, distribution of that trait in the population (e.g., Roughgarden, 1979). Such examples do not suspend or contradict processes of natural selection; if conditions were to change in such a manner that the reproductive advantage of the trait increased, then a new equilibrium would be established in which that trait was (almost certainly) more prevalent.

In this chapter, I extend this insight to the case of IQ. I present a simple example of a situation in which differential fertility by IQ is accompanied by a constant IQ distribution in the population. I then extend this example to somewhat more complex and realistic circumstances. For a more detailed analysis, see Preston and Campbell (1993a).

THE EXAMPLE

The simplest illustration of how differential fertility by IQ can be consistent with a constant IQ distribution in the population assumes that there are only two IQ classes, high (H) and low (L). The numbers of high and low individuals in Generation 1 are designated as $H(i)$ and $L(i)$. One needs to show how $H(i)$ and $L(i)$ in one generation relate to $H(i)$ and $L(i)$ in the previous generation. Two sets of values are needed to do this: the relative reproduction rates in the two classes and the probabilities that the offspring of H and L individuals will themselves be H or L. In particular,

$$H(2) = H(1)F(H)P(H/H) + L(1)F(L)P(H/L) \quad (1)$$

$$L(2) = H(1)F(H)P(L/H) + L(1)F(L)P(L/L), \quad (2)$$

where $F(H)$ and $F(L)$ are the mean numbers of surviving offspring per high- and low-IQ person, respectively; $P(H/H)$ and $P(L/H)$ are the probabilities that an offspring of an H will be H and L, respectively; and $P(H/L)$ and $P(L/L)$ are the probabilities that an offspring of an L will be H and L, respectively.

Assume the following values in Generation 1: Sixty individuals in Generation 1 are in the low-IQ class, and 40 are in the high-IQ class: $L(1) = 60$ and $H(1) = 40$. The mean number of surviving offspring for an individual in the low class is 1.2, and the mean number for a person in the high class is 0.7: $F(L) = 1.2$, and $F(L) = 0.7$. Thus, fertility is nearly twice as high among low-IQ persons as among high-IQ persons. Finally, assume that 75% of children born to low-IQ persons remain in the low class and that 79% of children born to high-IQ persons remain in the high class: $P(L/L) = .75$, $P(H/L) = .25$, $P(H/H) = .79$, and $P(L/H) = .21$.

Inserting these values in the equation that determines the IQ distribution of Generation 2 results in the following equations:

$$H(2) = 40(0.7)(.79) + 60(1.2)(.25) = 40 \qquad (3)$$

$$L(2) = 40(0.7)(.21) + 60(1.2)(.75) = 60. \qquad (4)$$

Thus, the distribution of IQ scores in Generation 2 is identical to the distribution in Generation 1, despite the fact that a low-IQ parent has nearly twice the number of children of a high-IQ parent, *and* IQ scores are highly correlated across the generations. The IQ distribution is in equilibrium despite fertility differentials. Generations 3, 4, and so on will have the same distribution if the *F*s and *P*s remain constant.

This equilibrium is maintained by intergenerational mobility in IQ scores. Because there are more offspring from the low class (both because the class is larger to begin with and because fertility is higher), there are many more individuals who “migrate” from L to H (18) than individuals who migrate from H to L (6). This net migration of 12 individuals exactly offsets the change in the distribution that would occur if all people remained in their own IQ class; $L(2)$ would then be 72 (60×1.2) rather than 60, and $H(2)$ would be 28 (40×0.7) rather than 40.

Intergenerational mobility vitiates the increasing concentration of population in the lower IQ class that would occur in its absence. I believe that the expectation that differential fertility will lead to a deterioration in the IQ distribution is based on the (implicit) assumption that all persons remain in the IQ class of their parents. The IQ classes are, in effect, assumed to be isolated subpopulations with their own unique demography. If that were true, then the expected deterioration would come to pass. In the preceding example, the whole population would soon be in the lower IQ category; however, the assumption is not true empirically.

Is the example just presented a curiosity based on an unusual set of values that enabled the IQ distribution to be replicated from one generation to the next? The answer is *no.* A remarkable set of theorems in the field of population dynamics demonstrate that an equilibrium distribution will *always* result from the continuation of constant fertility differentials and constant mobility rates. Even if I had chosen a set of numbers in the example that produced a change in the distribution from Generation 1 to Generation 2, those changes would eventually cease if the same set of fertility values (*F*s) and outcome–origin probabilities (*P*s) were maintained through time. Eventually, the distribution would become constant. Therefore, the population will accommodate itself to whatever pattern of fertility differentials exists. Once it has done so, fertility differentials do not produce changes in the IQ distribution from generation to generation; only changes in fertility differentials could do so. More generally, what matters for trends in the IQ distribution is not the existence or size of differentials but how recent patterns of fertility differentials relate to those of the past.

ELABORATIONS AND EMPIRICAL EXAMPLES

The example presented above is extremely simplified. Most obvious, there are many more than two IQ classes, and it takes two parents to produce a child, not one. The first problem is not serious; the formal mathematics of the process can be used to show that the same sort of result will apply regardless of the number of classes. The second problem is far more difficult both conceptually and empirically. Because reproduction is a function of both sexes, it becomes necessary to specify how men and women combine to produce offspring. In particular, it is necessary to specify a rule of "assortative mating" that determines how men and women of different IQ classes pair up to reproduce. Once paired, they must then be assigned a fertility level.

Preston and Campbell (1993a) used two different assumptions about assortative mating that represent polar extremes. One is complete endogamy: Individuals in a particular IQ class marry only persons in the same IQ class. The other is random mating: Individuals marry randomly with respect to IQ, so that the distribution of pairings for women of a particular IQ class is simply proportional to the number of available men by IQ class. For implementing both types of mating rules, an equal number of men and women is assumed to be produced each generation; everyone is assumed to pair up (though not necessarily to reproduce); and male and female offspring have the same IQ distribution (given the IQs of their mother and father).

The process by which the IQ distribution of Generation $T + 1$ is produced from the IQ distribution of Generation T can be represented in the following way:

$$\mathbf{Q}(T + 1) = f(\mathbf{Q}[T], \quad R, \quad H, \quad M), \tag{5}$$

where $\mathbf{Q}(T)$ is a vector representing the number of persons by IQ class in Generation T; R is a matrix that assigns fertility levels to pairs of men and women by their joint IQ scores; H is a three-dimensional matrix that provides the probability distribution of offspring's IQ scores, given the IQ class of their mother and father; and M is the mating rule in effect. This equation would, if properly implemented with accurate data, provide an exact representation of IQ transmission from Generation T to Generation $T + 1$. All the information required to derive the IQ distribution of Generation $T + 1$ from that of Generation T is contained in this formulation. That is to say, it represents an identity and not a model.

Modeling features arise when assumptions are made about how R, H, and M vary through time. To illustrate the impact of differential fertility on IQ distributions, Preston and Campbell assumed that *R* and *M* are fixed through time. For the most part, they also assumed that *H* is fixed, although in one formulation they allowed it to vary randomly with different amounts of variability introduced. The assumption of fixity in *H* is undoubtedly the most controversial of the assumptions. Coleman (1993) argued that *H* will not be fixed if the population consists of isolated subgroups (e.g., racial subgroups) among which *H* may differ and among which rates of reproduction vary. Preston and Campbell (1993b) noted that in such circumstances it is necessary analytically to disaggregate the population. Once it is disaggregated, the analytic results obtained in the nondisaggregated model would still apply, although the empirical details may differ.

The major objection to assuming fixity of *H*, I believe, is that there may be grandparent effects on IQ. That is to say, knowing the IQ of grandparents may add information to the prediction of IQ scores in one generation over and above the information provided by parents' scores. This would violate the Markov assumption used by Preston and Campbell (1993a)—the assumption that to predict outcomes in Generation $T + 1$, it is necessary to know only the distribution in Generation T. Introducing grandparent effects into the model would be a useful elaboration, but it has not been attempted.

John Loehlin's comment on my chapter (chapter 16, this volume) introduces a model that also violates the Markov assumption: The expected IQ of offspring is a function not only of parents' IQ but also of the mean of the IQ distribution in the parents' generation. In other words, the IQ of children of 120-IQ parents depends on how many 80-IQ persons there are in the parents' generation. I am not aware of a plausible mechanism that would create such a dependency. Regression toward the mean is a frequently observed statistical property, but it does not deserve elevation to an inviolable scientific principle. In the empirical *H* matrix drawn from Reed and Reed (1965) and used in the next paragraph, "regression" occurs not to the mean but to the 106- to 115-IQ class.

Preston and Campbell recognized seven IQ classes, so that their *H* matrix contained $7 \times 7 \times 7 = 343$ cells. (Introducing grandparent effects would require $7 \times 7 \times 7 \times 7 \times 7 \times 7 \times 7 = 1{,}176{,}649$ cells.) They used an empirical *H* matrix based on Reed and Reed's database of Minnesota families. It contains IQ data on 1,986 individuals and their parents. Their results can be summarized as follows:

1. With endogamous mating, the population's IQ distribution will always reach an equilibrium under the assumptions identified earlier. That is, a constant pattern of fertility differentials always produces an unchanging IQ distribution, given constant *H*. This result is not only empirical but analytic: It must occur under the assumptions made. The length of time before an equilibrium distribution is established depends on the disparity between the initial distribution and the equilibrium distribution. For most purposes, an equilibrium distribution to two decimal places is established quickly, within three or four generations. The population "forgets its past" rather quickly. Fertility differentials before, say, 1870 are essentially irrelevant to today's distribution.

   Figure 1 illustrates the evolution of the IQ distribution from a bizarre initial distribution in which 70% of the population is assumed to reside in the lowest IQ class and 5% in each of the remaining six classes. The empirical Reed and Reed *H* matrix is used; mating is endogamous; and fertility rates are additive across the IQ-class rank. Fertility levels are 3.5 for the lowest IQ class and 0 for the highest. Note that for the first five generations—until approximate equilibrium is established—the IQ distribution is improving despite the fact that lower IQ groups have much higher fertility than higher IQ groups.
2. With random mating, the population's IQ distribution always reaches an equilibrium under the assumptions identified earlier. However, this result is empirical, not analytic; Preston and Campbell were unable to prove that it will always occur. Nevertheless, they experimented with a wide variety of *R* and *H* matrices and used very different initial distributions, so that the result appears robust.
3. The equilibrium distribution of IQ is more sensitive to fertility

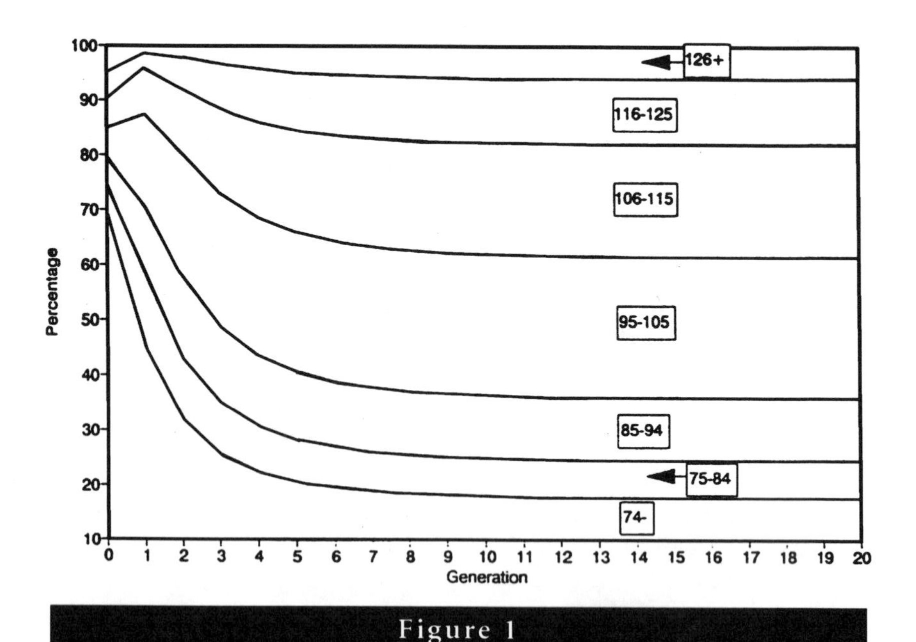

Figure 1

Percentage of populations in various IQ classes in each generation when fertility rates are highest in the lowest IQ groups and mating is endogamous. From "Differential Fertility and the Distribution of Traits: The Case of IQ," by S. H. Preston and C. Campbell, 1993a, *American Journal of Sociology, 98*. Copyright 1993 by the University of Chicago Press. Reprinted with permission.

differentials in the case of endogamous mating than in the case of random mating. Figure 2 shows the mean value of IQ at equilibrium for the two different mating rules when they are combined with the same fertility differentials. Fertility differentials assumed in the figure vary from the extremes of four children per couple in the highest IQ combination and none in the lowest to none in the highest and four in the lowest. (In all cases shown, fertility is a linear function of the IQ-class rank of both father and mother.) When there are no fertility differentials, the mean IQ in equilibrium is nearly the same for the two mating regimes. As fertility differentials are introduced, endogamous mating produces a fairly rapid change in mean IQ at equilibrium, but the mean changes very little when mating is random. This result supports an intuition of Herrn-

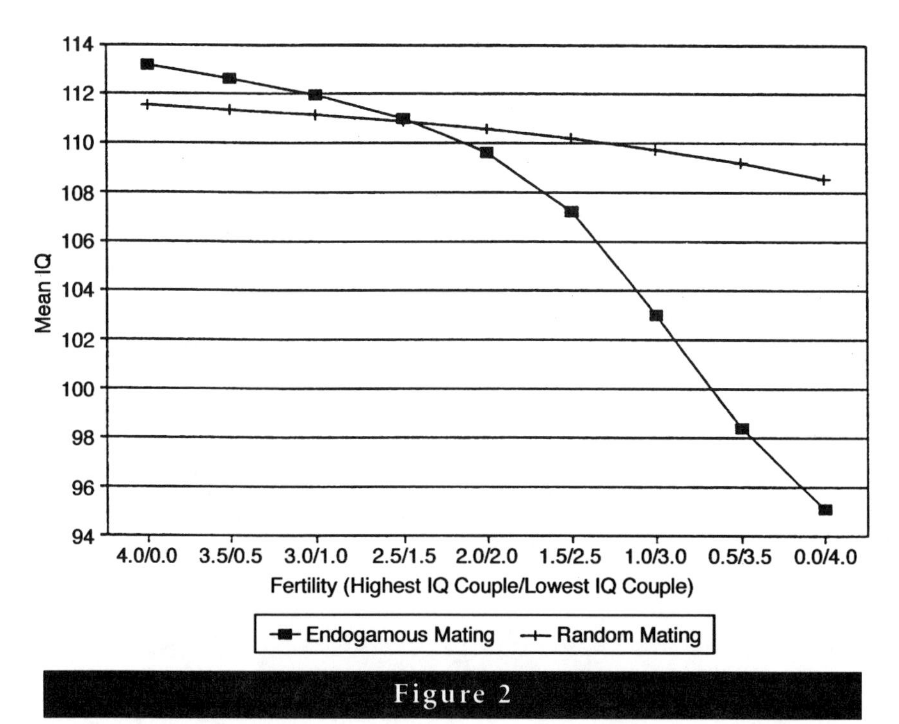

Figure 2

Comparison of mean IQ levels at equilibrium for various patterns of fertility differentials by IQ: Endogamous and random mating. From "Differential Fertility and the Distribution of Traits: The Case of IQ," by S. H. Preston and C. Campbell, 1993a, *American Journal of Sociology, 98*. Copyright 1993 by the University of Chicago Press. Reprinted with permission.

stein and Murray (1994) that the IQ distribution is more sensitive to fertility differentials in situations of greater endogamy. Presumably, the greater reproductive mixing of IQ classes when mating is random is responsible for this robustness to fertility differential.

4. With one exception, the equilibrium associated with many different combinations of *R, M,* and *H* is unique. That is, the IQ distribution will evolve to one and only one steady state, a state that can be identified uniquely when *R, M,* and *H* are known. In the case of complete endogamy, the uniqueness of the equilibrium can be established analytically; it will always be observed. However, one example of multiple equilibriums involving random mating was identified. This occurred

when fertility levels were highest at the extremes of IQ and an *H* matrix was assumed that manifested high retention of IQ. In this case, two equilibriums with very different IQ distributions were produced. The significance of multiple equilibriums is that one generation's fertility regime could push the system toward one of two very different outcomes. The example is likely to be of only theoretical import. It could not be reproduced when the empirical *H* matrix from Reed and Reed (1965) was used. Furthermore, fertility levels that are highest at extremes of IQ are rarely observed.

5. Introduction of time-varying random error into the empirical *H* matrix does not alter the basic result. Although the system does not reach an equilibrium when *H* varies from generation to generation, there is no trend in the outcome; the distribution wanders about randomly in a trendless pattern even in the presence of differential fertility.

CONCLUSION

Intuition suggests that higher fertility among persons of lower IQ will lead to a deterioration in the IQ distribution over time. This intuition appears to be based on an assumption that IQ groups are isolated from one another and that children always inherit the IQ of their parents. However, there is substantial mobility from parent to child in IQ scores, and this mobility is sufficient to offset the effect of fertility differentials on IQ trends. Any pattern of fertility differentials, when combined with a constant probability distribution of children's IQ given that of their parents, produces an equilibrium in which the IQ distribution is unchanging. This result applies in the case of complete endogamy and in the case of random mating.

There is no reason to believe that fertility differentials by IQ will inevitably or even typically produce a deterioration in the IQ distribution over time. What matters for trends in the IQ distribution is how the current pattern of fertility differentials relates to those of the past —those that have given rise to the current distribution. It is necessary to understand IQ transmission as a macro-level population process to appreciate how fertility differentials affect the IQ distribution.

REFERENCES

Coleman, J. (1993). Comment on Preston and Campbell's "Differential fertility and the distribution of traits." *American Journal of Sociology, 98,* 1020–1032.

Herrnstein, R. J. (1989, May). IQ and falling birth rates. *Atlantic Monthly,* pp. 73–79.

Herrnstein, R., & Murray, C. (1994). *The bell curve: Intelligence and class structure in American life.* New York: Free Press.

Neisser, U., Boodoo, G., Bouchard, T. J., Jr., Boykin, A. W., Brody, N., Ceci, S. J., Halpern, D. F., Loehlin, J. C., Perloff, R., Sternberg, R. J., & Urbina, S. (1996). Intelligence: Knowns and unknowns. *American Psychologist, 51,* 77–101.

Preston, S. H., & Campbell, C. (1993a). Differential fertility and the distribution of traits: The case of IQ. *American Journal of Sociology, 98,* 997–1019.

Preston, S. H., & Campbell, C. (1993b). Reply to Coleman and Lam. *American Journal of Sociology, 98,* 1039–1043.

Reed, E. W., & Reed, S. C. (1965). *Mental retardation: A family study.* Philadelphia: Saunders.

Roughgarden, J. (1979). *Theory of population genetics and evolutionary ecology: An introduction.* New York: Macmillan.

Van Court, M., & Bean, F. D. (1985). Intelligence and fertility in the United States: 1912–1982. *Intelligence, 9,* 23–32.

Vining, D. R. (1982). On the possibility of the reemergence of dysgenic trends with respect to intelligence in American fertility differentials. *Intelligence, 6,* 241–264.

16

Whither Dysgenics? Comments on Lynn and Preston

John C. Loehlin

The possible effect on the human gene pool of different patterns of reproduction has been a controversial topic in the Western world for well over a century. Professors Lynn and Preston (chapters 13 and 15, this volume, respectively) make provocative contributions to this dialogue. I first comment briefly on Lynn's chapter and then in somewhat more detail on Preston's.

LYNN'S DISCUSSION OF DYSGENICS

Professor Lynn (chapter 13, this volume) has done a service by bringing together a variety of data relevant to a familiar argument: that in the economically developed world during the last hundred years or so, in most times and most places, persons of lower IQ have been contributing proportionately more offspring to the next generation than have persons of higher IQ. To the extent that IQ differences reflect genetic differences—and I see no reason to doubt that, in part, they do—this implies at least some dysgenic trend. What is more, there is also good reason to believe, although Lynn does not especially emphasize this fact, that persons of less education and lower IQ also tend to start having their children at earlier ages, which would result in a dysgenic trend even in the absence of differences in completed family size.

The other thing that struck me in looking through Lynn's tables is how fast things can change. In a mere 300 years, a brief blip on the time scale of human evolution, the pattern has gone from eugenic to dysgenic to partway back again. What has happened, apparently, is that effective birth control started at the higher IQ end of the scale and worked its way down. During such a process, dysgenic trends will be present; as it nears completion, they will attenuate. Consider the trends of family size by mother's education in the United States and in Mexico. Mexico, which is presumably at an earlier stage in this transition, shows a decidedly more dysgenic trend. Kenya, still earlier, shows no dysgenic trend because reproduction still remains high at all educational levels.

The important point is that reproductive patterns in the modern world can change rapidly in response to changes in knowledge and attitudes. Even if things can and do change, however, it surely is important to try to understand what has happened in the past, what is happening now, and what might happen in the future if things continue as at present. For this, readers are in Professor Lynn's debt.

One or two more comments might be made on the issue of dysgenics. First, I think it was proper—and thoroughly civic-minded—of Galton and his successors to be concerned about this matter, even though, in retrospect, their alarm may have turned out to have been excessive. Second, I do not believe that for the U.S. population as a whole there is anything especially urgent to worry about at the moment. According to the data Lynn summarizes, whatever dysgenic IQ trend exists currently is quite weak. Its effects could easily be reversed in a generation or two by a shift to mild positive eugenics if we as members of the society desire to bring this about by any of the variety of methods available, ranging from discouraging reproduction by low-IQ teenagers to the provision of child-care facilities in graduate schools. So, if we wish, we can afford to wait and see. One of the things we can afford to wait and see about is the effect of *within-family* eugenics. This eugenic effect would follow from a choice by parents among the many potential combinations of sperm and ova that they can produce, assuming that they would give some preference to combinations likely to lead to higher intelligence. This is only possible to a very limited extent now, in the selective abortion of fetuses bearing genes for certain mental defects, but it seems likely that with increased knowledge will come a wider range of opportunities for positive choice. It seems likely that some parents, at least, will elect to make use of such opportunities in the hope of enhancing the chances of their offsprings' success in an increasingly complex and changing world. The net effect of such choice will be eugenic for the trait concerned.

For the U.S. population as a whole, there may be no immediate cause for alarm about dysgenic trends, but does this conclusion hold for subpopulations, such as African Americans or Hispanic Americans? Maybe, maybe not. Lynn's tables suggest that dysgenic trends might be more severe within these groups than for the U.S. population as a whole. If so, this could in principle *introduce* a genetic component into the average IQ difference between Blacks and Whites or between Whites and Hispanics, even if there were none to begin with. Because the trends are weak, changes would be slow, and as everyone knows there is often a considerable gap between *in principle* and *in fact*. Further investigation is essential before proposing policy recommendations. For starters, it is important to know how much of a change is involved (rather small, on the face of it) and how genetic and environmental changes contribute to it. One should remember that, at least for ability and personality traits, individual variation within groups dwarfs the average differences between groups, so that the question of whether there is or is not some genetic component to a group difference is of little or no practical importance in dealing with individuals as individuals. However, because many people seem to get extremely excited about the question of whether group differences involve *any* genetic component, perhaps this is not a matter to be taken altogether lightly.

PRESTON'S MODEL OF IQ TRANSMISSION

I discuss Professor Preston's (chapter 15, this volume) intriguing contribution in a little more detail. First, I underline some of the merits of using models like his rather than vague intuitions in thinking about intergenerational transmission of IQ. Second, I explain how I think his particular model goes wrong.

An Inheritance Matrix

I start with a simple artificial matrix (Figure 1) that tells how a population gets from one generation to the next—an inheritance matrix like those in Preston's models, except that to make it easier to follow I have reduced everything to simple small integers. (Artificial examples such as these are not, of course, meant to be taken seriously in their details but are intended to clarify ideas about how important real-world processes may operate. The reader might find it instructive to go step-by-step through the example with pencil and paper in hand; nothing more formidable than ordinary arithmetic is involved.)

Across the bottom of the matrix in Figure 1 are parent IQ cate-

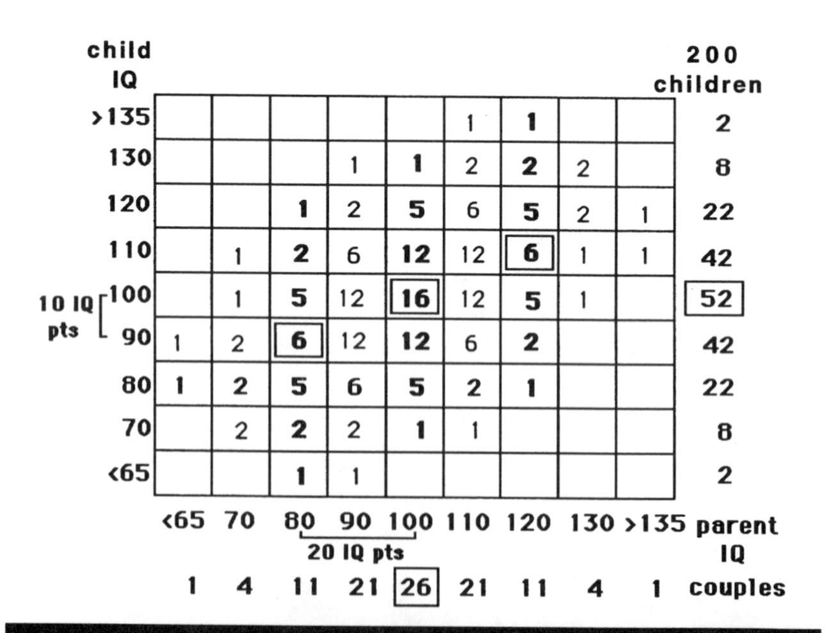

Figure 1

An inheritance matrix illustrating intergenerational stability and regression to the mean. Squares indicate array means. Arrays referred to in the text are in bold type.

gories. Each (except the end ones) is 10 IQ points wide, centered on the IQ given. Below these are the number of couples out of a hypothetical 100 who fall in each category. I am assuming here Preston's simplest condition, in which both parents have the same IQ. Things become a little more complicated when spouses' IQs are merely correlated to some lesser degree, as is the case in the real world, but these additional complexities do not affect the essential points of Preston's arguments or mine. The underlying IQ distribution is the traditional one: approximately normal with $M = 100$ and $SD = 15$.

Each column of the figure shows how the children of parents in the IQ class at the bottom of the column will tend to be distributed among the children's IQ categories labeled at the left. I am assuming that the couples in each category average two children apiece, just enough to keep the population size the same in the next generation. The column above parental IQ of 80 says that of the 22 children of the 11 couples in that IQ category, on average, 1 would have an IQ less than 65, that 2 would be between 65 and 75 in IQ, that 5 would be between 75 and 85, and so on, up to 1 in the 115- to 125-IQ category. Note that many individual children are not much like their parents. The parents' IQs all lie within a 10-point range from 75 to 85, but the children range from below 65 to the neighborhood of 120, and the most frequent category (85–95 IQ) is not even the category of the parents.

The rows of the table tell where the children in a given IQ category originate. Five of the children in the 80-IQ category come from parents in that category, 6 from 90-IQ parents, 2 from 70-IQ parents, and so on, to yield a total of 22 children in the 80-IQ category (the totals are shown to the right of the table). Note that when the offspring are paired up into couples in the next generation, they will constitute 1, 4, 11, 21 couples, and so on, exactly reproducing the parental generation. If the inheritance matrix stays the same, the grandchildren that these children eventually produce will be distributed in exactly the same way as the children and their parents were. This is a population at equilibrium. Of course, with small numbers like these, chance could mess things up, but in a large population with an equivalent inheritance matrix, the proportions should remain stable from generation to generation—forever, if nothing happens to change matters. One of the things that *could* happen is for some parent IQ classes to have more children than others do; this is an eventuality that is central to Preston's modeling, and I discuss it later. First, it is important to point out another feature of the matrix in Figure 1. Even though particular children may be quite different from their parents, there is a general correlation for IQ between parents and children, in that the children of below-average-IQ parents are themselves more often below than above average in IQ, although not exclusively so.

Regression Toward the Mean

As everyone since Francis Galton (who first drew up tables like these) has noticed, when parents imperfectly transmit characteristics to their children, the mean of the children of a given category of parents is less deviant from the population mean than is that of the parents. In this example, which has a regression of .5 built in, parents with an IQ of 80, 20 points below the population mean, have children whose average IQ is 90, one half as far below the mean. (The means of the columns are indicted by the small squares.) Parents whose IQ is at the population mean (100) will have children whose mean is also at the population mean. Parents with IQs of 120, 20 points above the mean, will have children with IQs averaging above the mean, but less so—in this case, the mean is 110, as shown in Figure 1.

It is important to realize that this regression has nothing to do with whether IQ is transmitted between the generations by way of the genes or the environment. There is reason to believe that the genes do play an important role, but the arguments I am making do not presume this; the mechanism could be entirely environmental. All that is required is that parents' advantages (or disadvantages) be imperfectly transmitted to their offspring.

Why does it work this way? When one selects parents who have extreme scores, one is selecting outcomes, not causes. The reasons for the parents' extreme scores include some factors that they will pass on to their children and some that they will not. The factors that parents do pass on through genes or the environment will ensure that their children (on average) lie in the same direction from the mean as their parents. The factors that contributed to a parent's extreme score but are not passed on will ensure that the children are (on average) less extreme than the parent. These nontransmitted factors might include such things as the effects of unique genetic configurations that break apart each generation because of the random draw of half one's genes from each parent, errors in measuring the parent's IQ, and sheer accidents that happened mostly to go in one direction in the particular parent's development but would be expected to balance out next time around. Why is it that regression toward the mean does not make the whole distribution shrink in each successive generation? The reason is that there are individual cases that move out from the center as well as those that move in. Consider the 100-IQ category of parents. Most of their children will have *more* extreme scores than their parents, although equally so in both directions so that the mean will not change. The inheritance matrix merely summarizes these ongoing individual stories.

Unequal Reproduction

Next, I examine the consequences of unequal reproduction among parents at different IQ levels, the main point of Preston's modeling. I am going to be even more extreme about it than Professor Preston was: I am only going to let the couples in the 80-IQ category have children (see Figure 2). Of course, these 11 couples will have to have a lot more than 2 children apiece if the total numbers in the next generation are to stay constant—in fact, about 18 per couple—but assuming the parents face up to their responsibilities, the distribution of IQs in the next generation will be as shown to the right of the mean. Notice that the mean of this distribution is 90. For this group of parents who are 20 IQ points below the mean, the children will average 10 IQ points below.

These children are entered below the inheritance matrix as the parents of a second generation. The question arises of how to locate them. Professor Preston made one assumption about this, but I make a different one. Which is the category in this generation that is at the mean, and whose children will therefore show no regression? It is the 90-IQ

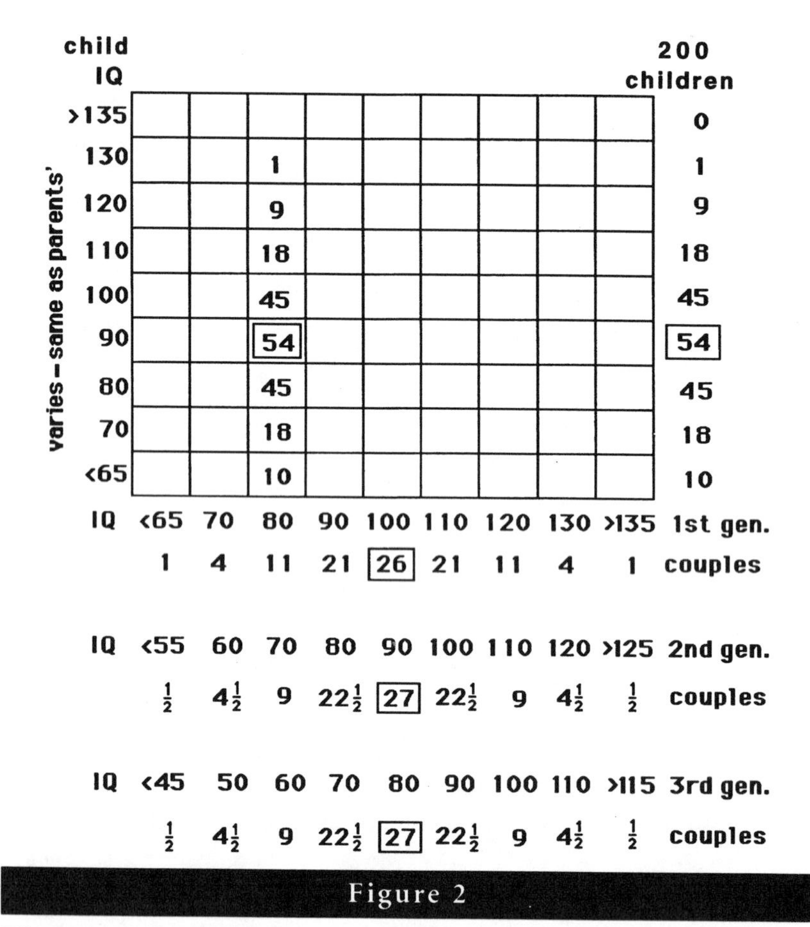

Figure 2

Effect of biased reproduction; only parents in one below-average-IQ category reproduce. Squares indicate array means. gen. = generation.

category; therefore, that must go in the middle column, where 100 went before, and the others must shift appropriately. Now it is the 70-IQ category that is 20 points below the mean and for which the inheritance matrix should indicate an offspring distribution averaging 10 points below the mean, and it is 110 that is as favorably selected, up 20 points, as 120 was for the parents.

Comparing the second to the first generation, it is seen that the spread of the distribution has shrunk a little bit. Theoretically, the standard deviation should be about 87% of its original value, given a regression of .5. (This is obtained as $\sqrt{[1 - .5^2]}$.) If the process is repeated for another generation, again with reproduction restricted to the category 20 IQ points below the mean, the nine couples will have to have about 22 children apiece to maintain the population size. The distribution of their children will have a mean IQ 10 points below that of their parents (i.e., it will be 80), but the size and shape of the distribution will be otherwise identical, as shown to the right of the matrix and in the third generation of parents below. The process will continue, down 10 IQ points per generation but otherwise stable, until someone notices that this way of arranging reproduction is probably not a good idea. If people then switch to 2 kids per couple across the board, the distribution will stay at whatever mean IQ level it has reached. If they should decide to switch to having the parents who are 20 points above the mean do all the reproducing, the mean will go back up at the rate of 10 points per generation.

In short, if one takes the view that where parents stand relative to the mean of their generation is what determines in which direction and how much regression occurs, one predicts that if parents from categories below the mean tend to have more children, the population mean will move downward from generation to generation. With no bias in reproduction, it will stay steady. With greater above-mean reproduction, it will rise. I am assuming, as Preston does, a constant inheritance matrix, but it is one expressed in terms of relative, rather than absolute, scores along the marginals.

My example also illustrates another interesting and somewhat nonintuitive point that Preston emphasizes: the rapidity with which a population distribution can stabilize after violent perturbation of its reproductive pattern. In just one generation, according to Figure 2, the size and shape of the population distribution takes on its equilibrium form, with only the mean shifting steadily downward.

I hope that Preston and I have persuaded you that inheritance matrices are worth looking at and that they provide a useful way of tracing out implications of different assumptions about the processes of population stability and change. Although my examples differ from Preston's in a number of details, the quite different conclusions to which they lead depend largely on one critical difference in assumptions: Preston's matrices remain fixed with respect to the original distribution at the time the unequal reproduction was imposed; mine are relative to the current population mean. I believe that mine reflect more realistically the phenomena that lead to regression to the mean.

CONCLUSION

If I am correct about Preston—and if Lynn is right about the typical pattern of reproduction with respect to IQ in this century—this does not help to explain the Flynn effect. In fact, it exacerbates matters. Not only must the environment be boosting scores on IQ-type tests, but it must be doing so sufficiently to offset a (modest) tendency for IQs to decline because of below-average-IQ parents transmitting their characteristics to a larger number of children—by genetic means, by environmental means, or both.

Thus, there remains a paradox here: a rising tide raising leaking boats, if you will. I do not think that the analogy holds in all respects: that there is some point at which the boats sink to the bottom and the rising tide becomes irrelevant. Nevertheless, I do not believe that it is sensible to ignore the possibility altogether, and I think that it is a notion like this one that underlies Lynn's concern.

Author Index

Numbers in italics refer to listings in reference sections.

Subject Index

About the Editor

Ulric Neisser has recently returned to Cornell University (where he taught from 1967 to 1983) after 13 years as Woodruff Professor of Psychology at Emory University in Atlanta, Georgia. Widely regarded as one of the founders of cognitive psychology, Neisser received his PhD from Harvard University in 1956. He is best known for his books *Cognitive Psychology* (1967), *Cognition and Reality* (1976), *Memory Observed* (1982), and *The School Achievement of Minority Children* (1986) as well as for his studies of memory in natural settings. In 1995–1996, he headed an American Psychological Association task force that reviewed controversial issues in the study of intelligence. Neisser is a member of the National Academy of Sciences as well as a former Guggenheim and Sloan Fellow.